D1267460

Time, Temporality, Now

Springer

Berlin
Heidelberg
New York
Barcelona
Budapest
Hong Kong
London
Milan
Paris
Santa Clara
Singapore
Tokyo

Harald Atmanspacher
Eva Ruhnau (Eds.)

Time, Temporality, Now

Experiencing Time
and Concepts of Time
in an Interdisciplinary Perspective

With 40 Figures

Springer

Dr. Harald Atmanspacher
Max-Planck-Institut für extraterrestrische Physik
D-85740 Garching, Germany

Dr. Eva Ruhnau
Institut für med. Psychologie
Ludwig-Maximilians-Universität
Goethestrasse 31
D-80336 München, Germany

ISBN 3-540-62486-4 Springer-Verlag Berlin Heidelberg New York

Die Deutsche Bibliothek – CIP-Einheitsaufnahme

Time, temporality, now: experiencing time and concepts of time in an interdisciplinary perspective / Harald
Atmanspacher; Eva Ruhnau (ed.). – Berlin; Heidelberg; New York; Barcelona; Budapest; Hong Kong; London;
Milan; Paris; Santa Clara; Singapore; Tokyo: Springer, 1997
ISBN 3-540-62486-4
NE: Atmanspacher, Harald [Hrsg.]

© Springer-Verlag Berlin Heidelberg 1997
Printed in Germany

Typesetting: Data Conversion by Kurt Mattes, Heidelberg
Cover design: Erich Kirchner, Heidelberg
Production editor: P. Treiber, Heidelberg

SPIN 10567127   55/3144 – 5 4 3 2 1 0 – Printed on acid-free paper

Table of Contents

List of Contributors

Amann, Anton   anton.amann@uibk.ac.at
ETH Hönggerberg, Postfach 164, CH–8093 Zürich, Switzerland

Antoniou, Ioannis   antoniou@solvayins.ulb.ac.be
International Solvay Institute for Physics and Chemistry, CP 231, ULB,
Campus Plaine, Boulevard du Triomphe, B–1050 Brussels, Belgium

Arecchi, F. Tito   arecchi@firefox.ino.it
Istituto Nazionale di Ottica, Largo E. Fermi 6, I–50125 Firenze, Italy

Atmanspacher, Harald   haa@mpe-garching.mpg.de
Max-Planck-Institut für extraterrestrische Physik, D–85740 Garching,
Germany

Barbour, Julian B.
College Farm, South Newington, Banbury, Oxon, OX15 4JG, England

Dalenoort, Gerhard J.   g.j.dalenoort@ppsw.rug.nl
Department of Psychology, University of Groningen, P.O. Box 72,
NL–9700 AB Groningen, The Netherlands

Ehlers, Jürgen   office@aei-potsdam.mpg.de
Max-Planck-Institut für Gravitationsphysik (Albert-Einstein-Institut),
Schlaatzweg 1, D–14473 Potsdam, Germany

Engel, Andreas   engel@mpih-frankfurt.mpg.de
Max-Planck-Institut für Hirnforschung, Deutschordenstrasse 46,
D–60528 Frankfurt, Germany

Euler, Manfred   eu-eu@physik.uni-paderborn.de
Fachbereich Physik, Universität GH Paderborn, Warburger Strasse 100,
D–33098 Paderborn, Germany

Fernandes, Marco
Physics Department, Birkbeck College, University of London,
Malet Street, London WC1E 7HX, England

Graudenz, Dirk   dirk.graudenz@cern.ch
Theoretical Physics Division, CERN, CH–1211 Geneva 23, Switzerland

Hiley, Basil J.   b.hiley@physics.bbk.ac.uk
Physics Department, Birkbeck College, University of London,
Malet Street, London WC1E 7HX, England

Kiefer, Claus   kiefer@phyq1.physik.uni-freiburg.de
Fakultät für Physik, Universität Freiburg, Hermann-Herder-Strasse 3,
D–79104 Freiburg, Germany

Klose, Joachim   klose@ikts.fhg.de
Department of Philosophy, Ludwig-Maximilians-Universität,
Geschwister-Scholl-Platz 1, D–80539 München, Germany

König, Peter   peterk@nsi.edu
The Neurosciences Institute, 10640 John Jay Hopkins Drive,
San Diego, CA 92121, USA

Kronz. Frederick M.   kronz@mail.utexas.edu
Department of Philosophy, University of Texas at Austin,
Austin, TX 78712, USA

Kull, Andreas   kul@mpe-garching.mpg.de
Max-Planck-Institut für extraterrestrische Physik, D–85740 Garching,
Germany

Lyre, Holger   holger.lyre@rz.ruhr-uni-bochum.de
Institute of Philosophy, Ruhr-Universität Bochum, D–44780 Bochum,
Germany

Mahler, Günter   mu@theo.physik.uni-stuttgart.de
Institut für Theoretische Physik, Universität Stuttgart, D–70550 Stuttgart, Germany

Malin, Shimon
Department of Physics and Astronomy, Colgate University,
Hamilton, NY 13346, USA

Pöppel, Ernst   secret@imp.med.uni-muenchen.de
Research Center Jülich, D–52425 Jülich, Germany

Primas, Hans   primas@phys.chem.ethz.ch
Laboratorium für physikalische Chemie, ETH Zentrum, CH–8092 Zürich,
Switzerland

Roelfsema, Pieter R.
Department of Medical Physics, University of Amsterdam, P.O.Box 12141,
NL–1100 AC Amsterdam, The Netherlands

Ruhnau, Eva   eva@imp.med.uni-muenchen.de
Institut für medizinische Psychologie, Ludwig-Maximilians-Universität,
Goethestrasse 31, D–80336 München, Germany

Singer, Wolf   singer@mpih-frankfurt.mpg.de
Max-Planck-Institut für Hirnforschung, Deutschordenstrasse 46,
D–60528 Frankfurt, Germany

Suchanecki, Zdzislaw   zsuchane@ulb.ac.be
International Solvay Institute for Physics and Chemistry, CP 231, ULB,
Campus Plaine, Boulevard du Triomphe, B–1050 Brussels, Belgium

Szelag, Elzbieta
Department of Neurobiology, Nencki Institute of Experimental Biology,
3 Pasteur Street, PL–02-093 Warsaw, Poland

Weizsäcker, Carl Friedrich von
Maximilianstrasse 14c, D–82319 Starnberg, Germany

Introductory Remarks

Harald Atmanspacher and Eva Ruhnau

The topic of "time" has always been a topic of interest to human beings, though certainly under changing perspectives and with changing motivations. It is not only an issue in the sciences and humanities, but also in the arts, in religion, and – of course – in everyday life. In all areas of human culture, time bears riddles that address most basic questions and problems, sometimes even with a strong paradoxical flavor. "What is time? If no one asks me, I know; but if I want to tell it to someone who asks me, I cannot. One thing, however, I can tell with confidence: I know that there would be no past if nothing passed by, no future if nothing existed. But what are those two tenses, past and future, if the past is no longer, and the future is not yet? The present, however, if it were always present and did not pass into past, would be eternity rather than time." – thus goes the famous quote from the *Confessions* by St. Augustine (1982), one of the first voices clearly expressing the tension between time as an abstract, explicit concept and its concrete experience at a (more or less) implicit level.

Although it is fair to say that "problems of time have been examined much less than the problems of space" (Reichenbach 1957), uncountably many treatises, articles, monographs, and proceedings volumes have been devoted to time and its many aspects such as temporality, the now, irreversibility, eternity, and others. "In any attempt to bridge the domains of experience belonging to the spiritual and physical sides of our nature, time occupies a key position" (Eddington 1968). Accordingly, time is one of the most outstanding paradigms for the significance of interdisciplinary work that are conceivable. Its realms of relevance range from religion, arts, philosophy, and ethics over the social, cognitive, and life sciences to chemistry, physics, and (back to) metaphysics. It is impossible to find this enormous scope comprehensively covered in a single reference. For general introductory reading, *The Natural Philosophy of Time* by Whitrow (1961) or *The Discovery of Time* by Toulmin and Goodfield (1965) are recommended. As a special highlight, let us also emphasize a remarkable and well illustrated volume on time in the arts, edited by Baudson (1985), unfortunately only published in German so far.

This broad significance notwithstanding, time as a topic of the natural sciences and its neighboring areas is of particular significance, especially in regard of the progress that has provided a deeper understanding of many of its features in our century. Physics may be considered as most fundamental among those sciences insofar as its distance from direct imagination and experience is extremely large, although many of its technological applications pervade our daily lives. Physics focuses on various different concepts and notions of time rather than dealing with the perception of time and its experience as such. As far as the latter is concerned, major areas in the cognitive and neural sciences as well as in the philosophy of mind are basic fields of current research. Until not long ago, mainstream attitudes in physics and cognitive science maintained that there is almost no overlap between their different approaches. Today there is a clearly visible tendency to acknowledge intimate relationships between the study of time as an abstract concept and the study of its concrete experience.

In modern cognitive science, the perception and experience of time is currently considered as a paradigmatic example for the concept of qualia, and although this is a controversial issue itself, there is a clearly visible and steadily growing interest in this field of research. In a certain sense, this discussion represents a bifurcation in the historical development of the cognitive sciences, which started with early cognitivism à la Simon and Minsky. In addition to the various kinds of computer metaphors for the human mind/brain, there is an increasing tendency to take non-algorithmic, non-functionalist, non-reductive approaches (like those concerning qualia) seriously.

Traditionally, the main emphasis in the cognitive sciences has been put on processes operating on mental representations. This attempt to avoid the restrictions imposed by behaviorism led to the consequence that the environmental aspects of cognitive systems have largely been disregarded. An essential implication of this neglect was a bias toward perception and the consideration of motor components as peripheral and secondary. At present, the situation is changing. The situatedness of cognitive systems is now frequently referred to as the crucial concept for understanding perception and action, each separately and both together as a unity. In this respect, not only the computation of information, but also the emergence of meaning from contextual experiences is important.

The historical development of physics has led to two basically different concepts of time. The distinction between them can be drawn in different ways, for instance according to groups of notions like universality, reversibility, continuity, and infinity versus contextuality, irreversibility, discreteness, and finiteness. The first of the two groups of notions refers to time as an external parameter, used to describe the evolution of a (closed) system. The second has to do with time as a property intrinsic to a system, regardless of its evolution as considered from an external viewpoint. Those fields of physics for which the second viewpoint has gained increasing significance are thermodynamics, nonlinear dynamics, and certain modern developments in quantum theory. With respect to our everyday experience of a flow of time that is irreversibly directed from past to future, this appears to be more adequate than reversible theories like classical mechanics, electrodynamics, and the (special and general) theory of relativity.

It is a strange fact that all reversible theories of physics are empirically confirmed with overwhelming precision, but this confirmation exclusively refers to processes that are observed as being directed forward in time. This is even more perplexing since reversible and irreversible dynamical laws in physics exist next to each other without any ultimate explanation (i.e., more than mere selection rules) of how the broken symmetry of an irreversible process emerges from a time-reversal invariant, symmetric description. In addition to these features, it is obvious that the abstract, conceptual character of any notion of time, reversible or irreversible, is in principle not suitable to cover any concrete, qualia-like features which some consider to be even more fundamental.

The essays collected in this volume address crucial questions that arise at the borders between all the dichotomous concepts mentioned so far. They are organized in four sections. The first two sections reflect perspectives of a modern natural philosophy of time and of certain achievements in the cognitive sciences and neurosciences. The remaining two sections address questions as to the physical nature of time in a relativistic framework and in non-relativistic quantum theory. It is clear that a number of interesting aspects within these ranges had to be left aside, for instance the entire field of computational and connectivist approaches to cognitive science, various new developments in chronobiology (circadian rhythms, aging), or psychopathological impairments of the perception of time. Instead, there is a distinct focus on progress in neuropsychology and neurophysiology toward a better understanding of our experience of time. Within this scope, some of the articles reflect a more or less phenomenological perspective (cf. Petitot et al. 1997).

Among the physics-oriented approaches in this collection, origins of irreversibility in thermodynamcis, statistical physics, cosmology, and complexity are underrepresented in this volume. For information about these topics the reader may consult other sources, e.g., Zeh's classic *The Physical Basis for the Direction of Time* (Zeh 1989) or the more recent collection of articles *Physical Origins of Time Asymmetry*, edited by Halliwell, Pérez-Mercader, and Zurek (1994). A certain specific flavor of the present volume is due to some emphasis on algebraic approaches to nonlinear dynamics and quantum theory, mainly in the last section. In this respect, the reader may think of Hamilton's dictum that "the mathematical science of time ... will ultimately be found to be co-extensive and identical with algebra" in his essay on *Algebra as the Science of Pure Time* (Hamilton 1837). Needless to add that such an algebraic understanding of time alone can be only a first step toward the synthesis of algebraic and geometric elements required for an understanding of the dynamics and structure of spacetime as a whole.

This volume is based on elaborated and refereed manuscripts of lectures presented at an international workshop on "Time, Temporality, Now", held at Ringberg Castle near Tegernsee, south of Munich, from February 26 until March 1, 1996. Ringberg castle is operated as a conference center of the Max Planck Society, whose hospitality is gratefully acknowledged. In particular, it is our pleasure to thank Mrs. Remberger, Mr. Hörmann, and the personnel of the center for their invaluable help in all little and big matters upon which the success of such an event decisively depends. Further thanks go to the Max-Planck-Institut für extraterrestrische Physik (Garching) and to the Institut für Grenzgebiete der Psychologie und Psychohygiene (Freiburg) for financial support. Finally, this volume owes its timely appearance to the encouragement from Professor Wolf Beiglböck and to the technical help from Urda Beiglböck, both at Springer-Verlag.

References

Aurelius Augustinus (1982): *Bekenntnisse* (Artemis, Zürich), 11. Buch, p. 312. English translation: *Confessions* (Harvard University Press, Cambridge, MA, 1912).

Baudson M., ed. (1985): *Zeit – die vierte Dimension in der Kunst* (VCH, Weinheim).

Eddington A. (1968): *The Nature of the Physical World* (University of Michigan Press, Ann Arbor), p. 91.

Halliwell J.J., Pérez-Mercader, and Zurek W.H., eds. (1994): *Physical Origins of Time Asymmetry* (Cambridge University Press, Cambridge).

Hamilton W.R. (1837): Theory of conjugate functions, or algebraic couples; with a preliminary and elementary essay on algebra as the science of pure time. *Trans. Roy. Irish Acad.* Vol. XVII, 293–422. Reprinted in *The Mathematical Papers of Sir William Rowan Hamilton, Vol. III: Algebra*, ed. by H. Halberstam and R.E. Ingram (Cambridge University Press, Cambridge 1967), 3–96, here: p. 7.

Petitot J., Roy J.M., Pachoud B., and Varela F., eds. (1997): *Naturalizing Phenomenology: Contemporary Issues in Phenomenology and Cognitive Science* (Stanford University Press, Stanford).

Reichenbach H. (1957): *The Philosophy of Space and Time* (Dover, New York), p. 109.

Toulmin S. and Goodfield J. (1965): *The Discovery of Time* (Hutchinson, London; second edition by Penguin, Harmondsworth 1967).

Whitrow G.J. (1961): *The Natural Philosophy of Time* (Nelson, London; second edition by Clarendon, Oxford 1980). See also *Time in History* by the same author (Oxford University Press, Oxford 1988).

Zeh H.D. (1989): *The Physical Basis for the Direction of Time* (Springer, Berlin).

Part I

Natural Philosophy of Time

Theory and Experience of Time: Philosophical Aspects

Frederick M. Kronz

Department of Philosophy, The University of Texas at Austin,
Austin, TX 78712, USA

1   Introductory Remarks

The concept of time is among the most fundamental elements of the set of philosophical concepts. It has a very rich history and is closely connected with a large, diverse set of ongoing philosophical debates that involve other fundamental concepts such as space, motion, substance, causation, the infinite, and consciousness. It is notorious for leading to philosophical perplexity, despite its familiarity and ubiquitous use in thought, conversation and action. In addition, it has been examined from a wide variety of perspectives including the logical, mathematical, physical, metaphysical, epistemological, psychological, linguistic, sociological, and biological. Those perspectives are by no means mutually disjoint or jointly exhaustive, nor is any one of them pre-eminently philosophical. On the contrary, research on the concept of time from one perspective is enhanced by a familiarity with insights gained from other perspectives.

This essay provides a brief introduction to some key discussions of the concept of time in the history of philosophy from the time of Aristotle to the beginning of the twentieth century. It will focus on the views of six major figures: Aristotle, Newton, Leibniz, Hume, Kant, and Mach. Certainly there are other major figures whose views on time are worth considering such as Plato, Plotinus, Augustine, Locke, Bradley, McTaggart, Bergson, and Husserl. Six figures were chosen in order to strike a proper balance between breadth and depth, and those figures in particular in order to facilitate the development of two specific themes. One asserts that a substantial part of the confusion surrounding the problem of characterizing the nature of time was due to a failure to appreciate and to satisfactorily resolve the tension existing between metaphysical, mathematical, and empirical aspects of time. The other asserts that as the awareness of this tension and the desire to resolve it grew, so did the relative importance of the empirical aspect, which became the predominating one of the three by the beginning of the twentieth century. The essay concludes by indicating some of the key problems of empiricism and their relevance for furthering our understanding of the nature of time.

2   Aristotle's Broad-Ranging Views on the Nature of Time

Aristotle's writings on time serve as a natural starting place for the following reasons. Prior to the emergence of the philosophical thinkers in Ancient Greece, time was often personified in quasi-mythical cosmogonies. A fairly common character was Chronos, a substitution for Kronos, the father of Zeus, and there are many examples in which Time is a figure of might who is generator and ruler of the cosmos (Peters 1967, pp. 30–33) and (Kirk et al. 1983, p. 22f). The representation of time in ancient thought was gradually transformed from personification to mathematization beginning with the writings of the pre-Socratic philosophers (such as Anaximander, Empedocles, Pythagoras, Parmenides, Zeno, Heraclitus, and Democritus) and culminating in the writings of Plato and Aristotle, the pre-eminent philosophers of that period. Unfortunately, all that remains of their writings are fragments (the words of the philosophers themselves) and somewhat unreliable reports of their teachings in later writers. Those fragments and reports have been collected together, and around that collection has arisen a formidable body of highly specialized and technical literature involving much speculation (Robinson 1968, pp. vi-viii; and Kirk et al. 1983, pp. 1-6). Aristotle's view on time was the predominating view for centuries. Those of Plato lacked the breadth and depth of Aristotle's, and were considerably less influential.

Aristotle discusses time throughout the *Physics*. The most important discussions occur in Books III–IV, with the most concentrated discussion occurring in Chapters 10–14 of Book IV.[1] The following list summarizes some of the major themes concerning time that emerge in the *Physics*:

1. Time is a measure of motion, and this means that
   (a) time actually exists only if motion exists, and
   (b) time actually exists only if a measurer exists, where
   (c) a measurer is a conscious, rational agent that is capable of counting.
2. Time is not identical with or reducible to motion, since
   (a) regular motions can be used to measure time, but time is not a measure of itself;
   (b) motion can actually exist without a measurer and, consequently, without time; and
   (c) motions can vary from one thing to another at different places and times, but the passage of time is the same everywhere for all things.
3. The present moment actually exists, the past and future only potentially exist; moreover,

---

[1] Hope's English translation of Aristotle's *Physics* (Aristotle 1961) is very readable and uses an effective method (a numbering system with an analytical index) for handling the translation of technical terms. Hussey's translation of Books III–IV of the *Physics* (Aristotle 1983) contains an insightful commentary.

    (a) a moment is a point-like temporal entity.

    (b) Time is not constituted of moments, since a part must measure the whole.

    (c) The past and future are infinite because there is eternal motion.

    (d) The present moment separates the past from the future.

4. Time is a continuous flux, since

    (a) motion is a continuous flux; moreover,

    (b) space is infinitely divisible but not infinitely divided, and

    (c) time is infinitely divisible but not infinitely divided.

5. There is only one time, and

    (a) the elements of time are ordered with respect to before and after; because

    (b) there is only one universe.

The above list is not intended to be exhaustive, but it does represent the most important themes concerning the nature of time that were put forth by Aristotle.

The first of the five major themes posits a crucial link between time and the perception of motion, and that is indicative of a tendency towards empiricism in Aristotle's thinking. The second thesis is primarily ontological. It says that motion and consciousness are ontologically prior to time; that is to say, the former are elements, the latter a construct. The rest are primarily mathematical, and are indicative of a tendency towards rationalism in Aristotle's thinking. In each of the three types of cases (empirical, ontological, and mathematical), there are other elements present; for example, in the empirical case, there is a reference to counting which is, arguably, a mathematical operation. In addition, the themes are rather eclectic, ranging from common-sense to abstract idealization.

There is substantial tension between the empirical and mathematical themes. Aristotle's analysis of the concept of time is primarily empirical, since it is founded upon the perception of motion. But, the punctual nature of the present, the infinite divisibility of time, and the infinite extension of time into the past and the future seem to be beyond the bounds of experience. One might try to analyze those notions of potential infinity in terms of processes of dividing or traversing, but it seems that such an analysis must involve other notions such as perpetual, eternal, endless, everlasting, and permanent that seem to be just as irreducible to experience as the potentially infinite. Clearly, finite but indefinite is too weak to capture Aristotle's intent. In any case, given Aristotle's rejection of Platonism, and specifically his rejection of a separately existing realm of forms, it is difficult to see these mathematical notions as idealizations. What, then, is his basis for countenancing them? It is difficult to say, even in light of his detailed discussion of motion in the latter half of the *Physics*. Nevertheless, it does seem clear that it would not be appropriate to regard them merely as convenient fictions with no ontological consequences given his rejection of atomism, which is based on arguments for the infinite divisibility of matter, space, and motion.

3   Newton on Absolute Time

The Aristotelian system was the predominating cosmology for centuries. Among the first to offer important and influential alternatives to it were Galileo and Descartes. A new concept of time was not elaborated by either of them. But, it must be emphasized that although Galileo's leanings were predominantly empirical, the conception of time that he presupposed in his empirical studies was predominately mathematical (Galileo 1914, p. 265). The irony is compounded by a brief remark by Descartes, who was very much a rationalist, in which he maintains the core of the Aristotelian notion of time, that it is the measure of a motion (Descartes 1967, p. 242).

The mathematical motif that was implicit in Galileo's writings is elaborated in Newton's. Like Aristotle, Newton was primarily an empiricist. He also was concerned with the empirical, ontological, and mathematical aspects of time. But, in the famous Scholium following Definition VIII at the beginning of Book I of the *Principia*, Newton explicitly characterizes two different notions of time, absolute time and relational time (Newton 1966, pp. 6–12). Newton regards the former as mathematical, the latter as empirical. He explicitly attributes ontological status to the mathematical notion, which makes his view as puzzling as Aristotle's, given his strong leanings towards empiricism. Later, he seems reluctant to reify some of his mathematical representations of time; but it is difficult to explain what underlies that reluctance, ultimately.

The Scholium does not make clear why Newton regards absolute time as mathematical, since he merely says there that everything exists in it, that it flows equably without relation to any motion, and that the order of its parts is unchanging. No mention is made as to what the parts of time are, whether there are moments, what moments are, or whether the past and future are infinite. This is peculiar because it is those issues that have a mathematical character; whereas, the features that he mentions appear to be based primarily in common sense.

Even more peculiar are his empirical arguments for the existence of absolute motion in the Scholium, which are supposed to serve as indirect support for the existence of absolute space and time. He asserts that two globes bound by a rope are in absolute rotation if and only if there is tension in the rope, and this is so whether or not there are other bodies present in the universe. If there are no other bodies, the rotational motion or lack thereof can only be with respect to absolute space. Presumably, the same is supposed to go for time. But, even granting that there would be motion with respect to absolute space, it is not clear that the motion would have to be with respect to absolute time, since it seems that the motion could serve to measure itself, if a spring scale is used to measure the tension in the rope. In any case, this argument is crucial, and will be returned to several times below.

For some insights concerning the mathematical properties that Newton ascribed to absolute time, it is best to have a brief look at earlier works of Newton and other parts of the first edition of the *Principia*. Later, it will be necessary to examine Newton's change in attitude towards some of those properties in his later writings, including some important differences between the first edition of the *Principia* (which was published in 1687) and subsequent editions that appeared during Newton's lifetime (the second edition in 1713 and the third in 1726).[2]

In "De Gravitatione", an unfinished work that very likely preceded his work on the *Principia*, and may have been written during his student days at Cambridge (from 1661 to 1664), Newton was more explicit about the nature of time. In that work, he says that time is eternal because time is an emanent effect of God and God is eternal (Newton 1962, pp. 136–137). He indicates that each moment of duration is the same at each part of space, thereby indicating that time involves the idea of absolute simultaneity. Finally, he says that the parts of time derive their individuality from their order alone, but he does so without indicating what a moment is or whether moments are parts of time. Since these are metaphysical and mathematical aspects of time rather than empirical aspects, it is rather likely that these are regarded by Newton as features of absolute time.

In "The October 1666 Tract on Fluxions" (Newton 1967), Newton presented his results of his research on the calculus, which he worked on during the previous two years. The use of o as an infinitesimal (i.e., infinitely small but non-zero) unit of time explicitly plays a crucial conceptual role in that essay. He regarded time as passing with a steady flow (following his teacher Barrow) and that enabled him to regard the fluxion of a quantity as the rate at which the quantity changes with respect to the uniform flow of time. Thus, if p and q represent fluxions of the flowing quantities x and y, respectively (in later works, Newton introduces the familiar dot notation in which p and q are replaced by $\dot{x}$ and $\dot{y}$), then the expressions po and qo represent moments of x and y. The moment of a quantity corresponds to the infinitely small increment of the quantity that is generated in a moment of time o. Thus, infinitesimal quantities are eliminated by reduction using fluxions and an infinitesimal unit of time.

De Morgan reports, in his essay "On the Early History of Infinitesimals in England" (De Morgan 1852), that in the first edition of the *Principia*, moments were understood as infinitesimally small quantities and were used throughout. But, in the second and third editions, most arguments involve re-

---

[2] For a brief discussion of the changes introduced in subsequent editions, see notes 4–10 (pp. 628–638) of Florian Cajori's historical and explanatory appendix to Newton's *Principia* (Newton 1966). For a detailed discussion of those changes, see I.B. Cohen's *Introduction to Newton's Principia* (Cohen 1971).

lations that are only true in the appropriate limit. The explanation of this difference is Newton's growing skepticism about infinitesimal quantities, which became manifest in his "De Quadratura Curvarum", a work on the foundations of the calculus which was begun around 1691 and published in 1704.[3] In that work, he attempts to eliminate infinitesimal quantities in favor of a limit notion. It is curious that he almost made the work a supplement to the second edition, but abandoned that idea at the very last minute.[4] It is even more curious that even in the third edition of *Principia*, there are passages, such as in Example I of Corollary II of Proposition X of Book II, where Newton continues to make use of infinitesimal quantities (Newton 1966, pp. 260–262). Both occurrences are suggestive of some ambivalence as to whether he saw the limit approach as the true one.

If it is assumed that Newton held a view of time that was relatively constant from his student days until he began to rethink the foundations of the calculus, then the views expressed by Newton concerning time in various works prior to 1691 can be summarized in the following table:

1. Time is an emanent effect of God, and
   (a) time is eternal because God is eternal, and
   (b) time exists whether or not motion exists, and
   (c) everything exists in time.
2. Moments are parts of time, and
   (a) a moment is an infinitesimal unit of time,
   (b) the order of the parts of time are unchanging, and
   (c) the parts derive their individuality from their order alone.
3. Time flows equably without relation to any motion, and
   (a) motion is a continuous flux, so
   (b) quantities have moments (infinitely small increments).
4. Time is linear, since
   (a) time is ordered with respect to before and after, and
   (b) each moment of duration is the same at each part of space, meaning that
   (c) there is only one time.

Theses 2 through 4 are primarily mathematical in character. The first thesis is primarily ontological, but clearly has a substantial mathematical consequence, the infinity of time. The second leaves some important issues unaddressed. For example, Aristotle asserted that a part must constitute a measure of the whole. Is there a sense in which infinitesimals measure a finite length? If so, that needs to be explained. If not, then it is not clear why

---

[3] See (Struik 1969, pp. 303–312) for a representative selection from this essay with a brief commentary, or the entire essay in (Newton 1981, pp. 92–167).

[4] David Gregory, an enthusiastic supporter of Newton, indicates this in a memorandum written following a meeting with Newton. See (Cohen 1971, pp. 191–194).

temporal points of zero length should not be regarded as parts, in Newton's sense, or why Newton's notion should be preferred to Aristotle's.

Although Newton is careful to distinguish between an empirical and a mathematical notion of time, it is puzzling (to elaborate on a point mentioned earlier) that he should give ontological preference to the mathematical notion given his strong empirical leanings, which are reflected in his four methodological precepts in Book III of the *Principia*, and given his repeated pronouncements against the use of hypotheses (i.e., propositions not obtained by induction from experience). Except for his argument to show that absolute acceleration has observable effects, which was characterized above, it seems that a relational theory of time such as the one advocated by Leibniz, one of Newton's chief antagonists, would be a more suitable alternative.

4   Leibniz's Relational Theory of Time and His Criticisms of Newton

Leibniz's views on time, which are abstracted from the famous Leibniz-Clarke correspondence (Alexander 1956), are very close to those of Aristotle. Indeed, it would not be misleading to say that Leibniz's account is an elaboration of Aristotle's since they share so many theses concerning time in common. As with Aristotle, there is a close tie between time and motion, but they are clearly distinct notions (Alexander 1956, p. 77). Leibniz says that time is merely a relation of objects with regards to succession, and that succession requires relative motion between two things (Alexander 1956, pp. 25–26, 69–70); i.e., it requires a continual change in the distance relation between two things.

There is also a tie between time and perception, though Leibniz's view does differ somewhat in so far as he holds that motion must have observable, as opposed to observed, effects (Alexander 1956, pp. 73–74). This difference is questionable since what Aristotle said could be interpreted that way. Concerning the present moment and the infinity of the past and future, Leibniz agrees with Aristotle that only the present moment, an instant (which he says is not a part of time, presumably, because it does not measure the whole, and is therefore likely to be a temporal point), exists (Alexander 1956, pp. 72–73); unlike Aristotle, he implies that time is not eternal, since the world has both a beginning and an end (Alexander 1956, pp. 75–76). Nevertheless, there are some subtle and important differences.

One key difference between them is that Leibniz uses two metaphysical principles, the principle of sufficient reason (PSR) and the principle of the identity of indiscernibles (PII), to ground his system of philosophy. He regarded them as great principles that would change the state of metaphysics by making it a real and demonstrative science (Alexander 1956, p. 37), and used them in a variety of different contexts, including his discussion of the nature of time. The principle of the identity of indiscernibles asserts that "there are not in nature two indiscernible real absolute beings" or that "two completely similar substances are identical" (Alexander 1956, pp. 36–37). The principle of sufficient reason asserts that "all truths have a reason why they are rather than are not" (Alexander 1956, pp. 26–27). Leibniz indicates that a reason for a truth must show why that truth contributes to making this world the best, suggesting that final causality plays as crucial a role for him as it does for Aristotle.

Samuel Clarke, an advocate of the Newtonian ideas of absolute space and time, suggested in the famous correspondence with Leibniz that this same world could have been created earlier in absolute time than it was (Alexander 1956, p. 49). Leibniz counters using PSR, asserting that there would be no reason for God to choose one time rather than another to create the universe, and PII, asserting that there would be no choice since all possible creation times are identical (Alexander 1956, pp. 75–78). Clarke responds by accepting PSR, asserting that God would have a reason,[5] and by questioning the status of PII, asserting that the points of time are distinct (Alexander 1956, pp. 99–100).

Clarke also points to the type of inertial effects discussed by Newton in his thought experiment involving the two globes to make a case for absolute time or space (Alexander 1956, p. 48). Leibniz responds cryptically and, as a result, ineffectually that the source of inertial effects are in the bodies themselves and, consequently, that those effects need not be referred to absolute space and time (Alexander 1956, pp. 74–75). Clarke responds by referring Leibniz back to Newton's famous Scholium (Alexander 1956, pp. 104–105).

Thus, the metaphysical stratagems of Leibniz are unconvincing. It took the empiricist stratagems of Hume and their effective implementation by Mach to develop what came to be regarded as a devastating reply to Newton's thought experiment.

5 Hume's Empirical Analysis of Time

Hume, like Newton, was an empiricist. His empiricism was strongly motivated by the success of Newtonian mechanics. In section 1 of the *Enquiry Concerning Human Understanding* (Hume 1993), he asserts his intent: to apply the Newtonian method to discover the rules governing the operation of the mind. It is striking that Hume's views on time, which are fully expressed

---

[5] It is rather puzzling that Leibniz refuses Clarke's response after appealing to the depth and abyss of divine wisdom in responding to an objection to his claim that this is the best of all possible worlds (Sect. 30, *Discourse on Metaphysics* (Leibniz 1991)). The objection is based on the claim that evil exists; Leibniz replies that we can know that God will derive a greater good from the evil, but that the underlying explanation is beyond our understanding.

in Part 2 of Book I of his *Treatise Concerning Human Nature* (Hume 1975), are consonant with his empirical outlook, unlike Newton's.[6]

In the second section of the *Enquiry* (Hume 1993), Hume explains that all perceptions ("perception" is broadly construed to include whatever is presented to the mind including sensations, emotions, feelings, thoughts, judgments, etc.) have their source in experience. They are divided into two classes, impressions and ideas, which are distinguished by their degree of force and vivacity; e.g., the pain of a pin prick is an impression, the perception invoked when recalling the pin prick three weeks later is an idea. Each idea was either preceded by a corresponding impression, or can be analyzed into parts each of which was so preceded. Hume asserts that the latter type of idea is formed by the imagination through its abilities to compound, transpose, augment, or diminish ideas of the former. This analysis is applied to space and time, and leads to the following views.

Since the capacity of the mind is limited, it is not possible for the mind to form an adequate conception of the infinite. There is a lower limit on division into parts and an upper division on additions to extension. One way to produce a sensation of an indivisible length is to put a dot of ink on a wall and then recede from the wall. At some distance from the wall (the distance will depend on variety of factors), the dot will not be visible, but moving one step closer will bring the dot back into the visual field. The resulting sensation is an indivisible length. Such lengths have a finite measure in the sense that placing two end to end results in a length that is double the original length of either one of them. Hume suggests that similar considerations hold for time; that is to say, he held that time must be composed of indivisible moments having a finite measure. This is the first of two major themes that Hume asserted concerning the nature of time. The second makes change a necessary condition for time. An interval of time, or duration, is a succession of moments. It is derived from a perceivable succession of perceptions, meaning that each element in the sequence differs from its predecessor in some perceivable manner. The time interval is nothing more than the perceived order of the succession of differing perceptions.

Contrary to Hume's expressed views, it does not appear that there are perceived moments of time as there are perceived indivisible lengths. He does not present a temporal counterpart to the example involving the ink dot, and it is difficult to imagine one. The empiricist Newton seems to have had it right in so far as he emphasized the flowing aspect of time. Another difficulty with Hume's point of view is that it is entirely subjective. But, it seems that experienced time has an objective aspect that Hume does not try to explain or even acknowledge. The flowing aspect of experienced time is emphasized

---

[6] Hume wrote the *Treatise Concerning Human Nature* between 1734 and 1737, and it was first published in three volumes between 1738 and 1740. The *Enquiry Concerning Human Understanding* is a revised version of the first part of the *Treatise*, and it was first published under a different title in 1748.

in the writings of Bergson and Husserl which are not discussed in this essay. The objective aspect is discussed by Kant, and it is tied to the notion of causation.

6 Kant on the Empirical Reality and Transcendental Ideality of Time

In Sect. 2 of the "Transcendental Aesthetic" of the *Critique of Pure Reason* (Kant 1965, pp. 74-82), Kant explains that time is an empirically real and transcendentally ideal a priori intuition. Time is an intuition and not a concept because concepts are multiply instantiable, but there is only one time; it is infinitely extendible but not infinitely extended, and infinitely divisible but not infinitely divided. It is empirically real in that all experiences are in time. It is a priori because it is a subjective condition (i.e., a condition of the subject) that is a prior condition for all experiences. That is to say, time is not merely the observed order of a sequence of observationally distinct perceptions; rather, it is the subjective condition that makes such an ordering possible. The a priori status of time is also crucial for Kant because it serves to make counting possible and thereby provides a suitable basis for arithmetic.[7] Finally, time is transcendentally ideal in that it is nothing whatsoever in reference to what is outside of the bounds of experience.

Kant regards Newton's views on time as metaphysically extravagant. That objection is weak, given his superficial treatment of Newton's two-globe argument in the fourth chapter of his *Metaphysical Foundations of Natural Science* (Kant 1985). He substitutes the two globes in Newton's universe with a single rotating sphere, and claims that the rotation would have observable inertial effects because there is relative motion between the center of the sphere and its other parts. Perhaps that is what Leibniz had in mind. In any case, the response does not mesh well with his view that matter is infinitely divisible but not infinitely divided, and does not work for a rotating ring.

Kant rejects Hume's view on the grounds that there is an objective succession of appearances that cannot have its source in perception, and because it does not provide a proper foundation for arithmetic. His account of the objective succession of appearances occurs in the famous "Second Analogy" in the *Critique*, and is based on the view that causality is an a priori concept of the understanding whose function is to order objectively the sequence of perceptions given by the sensibility according to a rule. That ordering makes possible the perception of motion; so, time is more fundamental than motion for Kant. Finally, he rejects Leibniz's views because they are based on the assumption that there is an isomorphism between things in themselves and appearances.

---

[7] Kant develops this view in Sect. 1 of Chap. 1 of the "Transcendental Doctrine of Method" of the *Critique*; and in Par. 3 and Pars. 7–11 of the *Prolegomena to Any Future Metaphysics* (Kant 1977).

7   Mach's Critique of Newton and the Influence of Machian Empiricism

Kantian rationalism and Humean empiricism were the two predominating philosophies of science in the nineteenth century. In the writings of William Whewell, a predecessor of Mach and the first philosopher of science to base his philosophy of science on a thorough study of the history of science and regarded by some as being the father of the historiography of science, the Kantian influence is hard to miss.[8] So is his originality as a philosopher. His notion of consilience of inductions had a profound impact on Darwin, and it has much insight to offer with regards to how to think about the relationship between physical theory and physical reality. That notion and related themes are discussed briefly later in assessing Mach's view.

Empiricism became the predominating influence on the development of philosophy of science and physics in the first half of the twentieth century. The most influential proponent of empiricism in the nineteenth century was Ernst Mach, whose philosophical writings had a major impact on Einstein in connection with the development of relativity theory. Einstein says in his "Autobiographical Notes" (Einstein 1949), which were written around 1947 and published in 1949 in Schilpp's *Albert Einstein: Philosopher–Scientist*, that the type of critical thinking required for discovering the central point needed to clarify and then to resolve a paradox was decisively furthered by reading the philosophical writings of Hume and Mach (Einstein 1949, p. 53). The paradox turned out to involve absolute time in a crucial manner, and its resolution ultimately led to Einstein's discovery of the relativity of simultaneity, a consequence of the theory of relativity, which was first developed in his article "On the Electrodynamics of Moving Bodies" (Einstein 1905). The Machian influence is clearly present in the introductory paragraphs of that essay, where he affirms the adequacy of appealing only to relative motions, and in the kinematical part of the essay, where he affirms both the conventionality of the travel time for a one way trip of a light beam from one place to another and the relativity of simultaneity for distant events.

The success of relativity was a major factor in the institutionalization of empiricism, and led to the formation of the *Vienna Circle*, which was formed in the early twenties around Moritz Schlick, and consisted primarily of formal, natural, and social scientists with strong empiricist leanings. The institutionalization of empiricism also influenced Heisenberg in his student years and later in his development of quantum mechanics during the mid-twenties, which no doubt served to bolster the empiricist movement. By the late twenties, the influence of Carnap and Wittgenstein led to the formation of

---

[8] An excellent collection of selections from Whewell's works on the philosophy of science is *William Whewell: Theory of Scientific Method*, which was edited by Robert Butts (Butts 1989). Whewell's works went under-appreciated for quite some time, but that is changing.

another group with an official name, philosophy, and set of goals. That group was the *Verein Ernst Mach*, its philosophy logical positivism, and its goal to become the driving force behind an international movement. By the early thirties, it was organizing important international congresses and developing influential publications in which the philosophical program was elaborated and applied. It was extremely influential throughout the forties and fifties, and began to wane in the early sixties.[9]

Mach was the first to construct what was seen as a decisive criticism of Newton's argument for the existence of absolute motion from the inertial effects of rotational motion in Sect. 6 of Chap. 2 of his *The Science of Mechanics* (Mach 1960, pp. 271–297). His criticism involves two principles, a verification principle and an inertial principle. The verification principle asserts that a judgment has scientific significance if and only if there is some observable phenomena that is relevant to determining whether that judgment is true or false. Since there is no way to set up an experiment involving a universe with only two globes joined by a rope, the claim concerning absolute rotation is not scientifically significant (Mach 1960, pp. 280–283). A Newtonian could respond by asserting that Newton's assertion about what would happen in the hypothetical universe is a straightforward extrapolation from what happens in the actual one. Mach forestalls that response by introducing an alternative inertial principle that is observationally equivalent to Newton's as long as there is a sufficient number of bodies that are practically fixed with respect to one another, sufficiently massive and distant, and appropriately distributed (Mach 1960, pp. 286–288). Observational equivalence is relative to the available data, meaning in part, relative to the existing technology. That is to say, it may eventually be possible to construct a crucial experiment that decides in favor of one of the two inertial principles. Until then, pragmatic considerations such as simplicity give the edge to Newton's inertial principle over Mach's; that is to say, the issue is a matter of convention, not confirmation. What is most important, however, is that it will always be possible to formulate an observationally equivalent alternative to the inertial principle that wins favor at any stage (Mach 1960, pp. 289–290; the italicized sentence is most telling). Thus, it will never be possible to establish what will happen in the hypothetical world; i.e., absolute motion and absolute time cannot be established as scientifically significant in the manner suggested by Newton.

Aside from the critical component, Mach's views on time had a affirmative component. In several publications, he contrasts physiological time with physical time.[10] Physiological time is the time-sensation experience,

---

[9] A good, brief history of the movement is Joergen Joergensen's *The Development of Logical Empiricism* (Joergensen 1951).
[10] See Chap. 12 and Sect. 26 of Chap. 14 of his *Analysis of Sensations* (Mach 1959), and his two essays "Physiological Time in Contrast with Metrical Time" and "Space and Time Physically Considered" in *Knowledge and Error* (Mach 1976).

of which he offers brief but suggestive remarks that fall under three categories: phenomenological characterization, evolutionary-biological accounts, and physico-chemical conjectures. Those remarks are, for the most part, summary statements of the work of other writers of that period. Physical time is what is obtained when mathematical and metrical concepts are applied to physical processes, and the result is a refinement of the time sensation that serves as a useful tool or an economically efficient means to organize sensations and regulate behavior. From an ontological point of view, those benefits mean nothing. The world consists only of our sensations. The representation of time as a continuum, for example, is scientifically significant to the extent that it facilitates the development and use of physical theories that are in agreement with experience, but that could never suffice to give it ontological significance.[11]

Although Mach's stance may be quite reasonable with regards to inertial principles and the temporal continuum, there are other cases where it is too extreme. To the extent that sensations are inescapably theory laden is the extent to which Mach's position cannot be metaphysics free. If all sensation has that character as Whewell (and others before him, such as Francis Bacon (*Novum Organon*, Part I, Sects. 41ff. (Bacon 1857–1874)), and after him, such as Popper (Popper 1965, pp. 93–111) and Hanson (Hanson 1958, pp. 4–30)) have claimed, then there cannot be a fundamental distinction between theoretical constructs and sensations.[12] Also, there are theoretical constructs that the majority of scientist do not and will not fail to reify such as amoebas, DNA molecules, hydrogen atoms, electrons, and neutrinos. Thus, Mach's empiricism (or that attributed to him) must be mitigated.

Whewell suggested that when several seemingly unrelated phenomenological laws are subsumed under a theoretical law, that has ontological significance. When several seemingly unrelated lower level theoretical laws are subsumed under another theoretical law, that has much greater ontological significance. Both situations are cases of what Whewell calls a consilience of inductions.[13] At what level of generality are we justified in assuming that the theory reflects the underlying nature of physical reality? It is difficult to say; but, it is clear that the empiricist solution of giving ontological significance only to sensations is too simple minded. It is best to approach that question (as with the more specific question concerning the nature of time) from a

---

[11] Feyerabend asserts of Mach, in his essay "Mach's Theory of Research and its Relation to Einstein" (Feyerabend 1987), that his philosophy of science was really much more sophisticated than the positivism that is generally attributed to him. Feyerabend's version of Mach's philosophy is closer to Whewell's point of view, which is characterized below, than it is to positivism.

[12] See Whewell's essay "On the Fundamental Antithesis in Philosophy", reprinted in (Butts 1989, pp. 54–75).

[13] See Sect. 3 of Chap. 5 of Book II of Whewell's *Novum Organon Renovatum*, which is reprinted in (Butts 1989, pp. 149–160).

variety of perspectives, including that of the experimentalist as well as that of the theorist, and hope to gain some deeper insights.[14]

Perhaps Einstein's very deep insights concerning the nature of time, such as the relativity of simultaneity and the resulting need to replace the independent realms of space and time by space-time, are or soon will be generally regarded as being beyond question. Other matters, such as whether Heisenberg's uncertainty relations are fundamental in the same way, whether there must be a time-energy uncertainty relation to complement the position-momentum uncertainty relation, whether time must be represented as an operator in addition to or even rather than a parameter, are far from being regarded as characterizing other aspects of the true nature of time.[15] Yet, it does not seem that such matters are forever outside the bounds of theory-laden experience. That is to say, it is reasonable to suppose that there are some features of time that can be settled within a more sophisticated version of empiricism, one that is suitably mitigated by theory and practice.

References

Alexander H.G., ed. (1956): *The Leibniz-Clarke Correspondence* (Manchester University Press, Manchester).

Aristotle (1961): *Physics*, translated by Richard Hope (University of Nebraska Press, Lincoln).

Aristotle (1983): *Physics, Books III and IV*, translated with a commentary by Edward Hussey (Oxford University Press, Oxford).

Bacon F. (1857–1874): Novum Organon. In *The Works of Francis Bacon, Vol. VIII*, ed. by J. Spedding, R. Ellis, and D. Heath (Longman and Company, London), 57–350.

Butts R.E., ed. (1989): *William Whewell: Theory of Scientific Method* (Hackett, Indianapolis).

Cohen I.B. (1971): *Introduction to Newton's Principia* (Harvard University Press, Cambridge).

De Morgan A. (1852): On the early history of infinitesimals in England. *Philosophical Magazine* **4**, 321–330.

Descartes R. (1967): Principles of Philosophy. In *The Philosophical Works of Descartes*, ed. by E.S. Haldane and G.R.T. Ross (Cambridge University Press, Cambridge), 201–302.

Einstein A. (1905): Zur Elektrodynamik bewegter Körper. *Ann. Phys.* **17**, 891–921. English translation: On the electrodynamics of moving bodies. In *The Principle of Relativity*, ed. by W. Perrett and G.B. Jeffery (Dover, New York 1952), 37–65.

---

[14] Hacking's *Representing and Intervening* (Hacking 1983) does an admirable job of discussing this question and related questions from the prespective of the experimentalist as well as that of the theorist.

[15] For a brief overview of some of the current quandaries regarding time in physical theory, see Sect. 2.3 of Isham's "Prima Facie Questions in Quantum Gravity" (Isham 1994).

Einstein A. (1949): Autobiographical Notes. In *Albert Einstein: Philosopher-Scientist*, ed. by P.A. Schilpp (Open Court, La Salle, Ill.), 2–94.

Feyerabend P. (1987): Mach's theory of research and its relation to Einstein. In P. Feyerabend, *Farewell to Reason* (Verso, New York), 192–218.

Galileo Galilei (1914): *Dialogues Concerning Two New Sciences* (Dover, New York).

Hacking I. (1983): *Representing and Intervening* (Cambridge University Press, Cambridge).

Hanson N.R. (1958): *Patterns of Discovery* (Cambridge University Press, Cambridge).

Hume D. (1975): *A Treatise Concerning Human Nature* (Oxford University Press, Oxford).

Hume D. (1993): *An Enquiry Concerning Human Understanding* (Hackett, Indianapolis).

Isham C. (1994): Prima facie questions in quantum gravity. In *Canonical Gravity: From Classical to Quantum* (Springer, Heidelberg), 1–21.

Joergensen J. (1951): *The Development of Logical Empiricism* (Chicago University Press, Chicago).

Kant I. (1965): *Critique of Pure Reason*, translated by N.K. Smith (St. Martin's Press, New York).

Kant I. (1977): *Prolegomena to Any Future Metaphysics*, translated by J.W. Ellington (Hackett, Indianapolis).

Kant I. (1985): *Metaphysical Foundations of Natural Science*, published with the *Prolegomena in Philosophy of Material Nature*, edited and translated by J.W. Ellington (Hackett, Indianapolis).

Kirk G.S., Raven J.E., and Schofield M. (1983): *The Presocratic Philosophers* (Cambridge University Press, Cambridge, 2nd edition).

Leibniz G.W. (1991): *Discourse on Metaphysics* (Open Court, La Salle, Ill.).

Mach E. (1959): *The Analysis of Sensations*, first translated by C.M. Williams and revised by S. Waterlow (Dover, New York).

Mach E. (1960): *The Science of Mechanics*, translated by T.J. McCormack (Open Court, La Salle, Ill.).

Mach E. (1976): *Knowledge and Error*, translated by T.J. McCormack and P. Foulkes, and edited by B. McGuiness (Reidel, Boston).

Newton I. (1962): De gravitatione et aequipondio fluidorum. In *Unpublished Scientific Papers of Isaac Newton*, edited and translated by A.R. Hall and M.B. Hall (Cambridge University Press, Cambridge), 89–156.

Newton I. (1966): *Mathematical Principles of Natural Philosophy, Volumes I and II*, translated by A. Motte in 1729 from the third edition, which was published in 1726, and revised by F. Cajori and published in 1934 (University of California Press, Los Angeles).

Newton I. (1967): The October 1666 tract on fluxions. In *The Mathematical Papers of Isaac Newton, Volume I*, ed. by D.T. Whiteside (Cambridge University Press, Cambridge), 400–448.

Newton I. (1981): De quadratura curvarum. Reprinted in *The Mathematical Papers of Isaac Newton, Volume VIII*, ed. by D.T. Whiteside (Cambridge University Press, Cambridge), 92–167.

Peters F.E. (1967): *Greek Philosophical Terms: A Historical Lexicon* (New York University Press, New York).

Popper K. (1965): *The Logic of Scientific Discovery* (Harper, New York).

Robinson J.M. (1968): *An Introduction to Early Greek Philosophy* (Houghton Mifflin, Boston).

Struik D.J., ed. (1969): *A Source Book in Mathematics, 1200-1800* (Harvard University Press, Cambridge).

Whitehead's Theory of Perception

Joachim Klose

Department of Philosophy, Ludwig-Maximilians-Universität,
Geschwister-Scholl-Platz 1, D–80539 Munich, Germany

Abstract. Like Locke and Hume, Whitehead accepts that the stream of thought
has temporal extension. Perception, however, does not take place in a nontemporal
moment; it rather takes time. Following William James, Whitehead calls this duration "specious present". It is a prerequisite of perception in the specious present that
not only the reality of the percipient event but also of the perceived event is temporally extended. The problem of how distinctions in the stream of events appear
and why the apparent world does not collapse in one moment disappears when we
observe that the present itself is duration, and therefore includes directly perceived
time-relations between events contained within reality. The temporality of nature is
one of the major insights which shapes Whitehead's whole metaphysics. Classical
physics has reasons to disregard temporality and to use a continuous concept of
time. But "both science and metaphysics start from the same given groundwork
of immediate experience, and in the main proceed in opposite directions on their
diverse tasks" (MT, p. 108). In his metaphysics Whitehead tries to avoid any bifurcation between a perceiving consciousness and an independent reality. He develops
a theory of perception more general than those of the British empiricists. This paper shows how changes in the understanding of perception provoked Whitehead's
epochal theory of time. As a result, the ultimate constituents of reality are no longer
non-temporal events but actual occasions which possess spatio-temporal extension
and creativity. Thus causality becomes a real feature of the world rather than a
judgement of consciousness.

1   Antibifurcationism

In his publications on natural philosophy, *An Enquiry Concerning the Principles of Natural Knowledge* (PNK), *The Concept of Nature* (CN), and *The
Principle of Relativity* (R), Whitehead was "concerned only with nature,
that is, with the object of perceptual knowledge, and not with the synthesis
of the knower with the known. This distinction is exactly that which separates natural philosophy from metaphysics" (PNK, p. vii). Nature is in a
sense independent of thought and therefore independently describable: "...
we can think about nature without thinking about thought" (CN, p. 3). The
sciences are not concerned with the reasons of knowledge, but with a coherent
explanation of nature. They try to understand how the events of nature are
related to each other.

The development of physical theories, especially the introduction of the
transmission theory of Newton's mechanism, resulted in a bifurcation of na-

ture. Whitehead delivers two similiar formulations of such a bifurcation (CN, p. 26ff), which he both rejects categorically:

1. the distinction between the events of nature and the events as they are formulated in scientific theories;[1]
2. the distinction between the events of nature as they exist by themselves and as they *appear* to us.[2]

The first concept of bifurcation maintains a pure conceptual existence of physical entities. On one hand, there are phenomena and, on the other, logical terms of scientific formulae (cf. CN, p. 45). Whitehead firmly rejects the theoretical postulate of physical objects as atoms or molecules. Each system of notations represents the relationship between its terms and makes statements about the truth or falsehood of the phenomena. An atom or a molecule as a term of a chemical formula represents the sum of all true statements of a phenomenon which has the expected properties: "...scientific laws, if they are true, are statements about entities which we obtain knowledge of as being in nature; ... if the entities to which the statements refer are not to be found in nature, the statements about them have no relevance to any purely natural occurrence" (CN, p. 45-46).

Here Whitehead supports a simple realism. Concepts – insofar as they are true – refer directly to the phenomena; otherwise they are nonsense. Well-established concepts of science refer to the factors which will be found in nature. Scientific theories represent the relationships between these factors. They do not represent the phenomena completely, but rather under certain considerations.[3] Reality is the unit of fact which includes many factors. The factors are the aspects of reality which we perceive. Natural philosophy is concerned with the totality of fact. "For natural philosophy everything perceived is in nature. ... For us the red glow of the sunset should be as much part of nature as are the molecules and electric waves by which men of science would explain the phenomenon" (CN, p. 29).

Scientific concepts and theories are attempts to structure this totality. They do not form a reality in opposition to the reality of nature. They are derived by logical abstraction from nature. Whitehead argues against the bifurcation of reality into a mathematical logical world and the apparent world: "Nature is thus a totality including individual experiences, so that we

---

[1] "One reality would be the entities such as electrons which are the study of speculative physics. This would be the reality which is there for knowledge; although on this theory it is never known. For what is known is the other sort of reality, which is the byplay of the mind." (CN, p. 30)
[2] "Another way of phrasing this theory which I am arguing against is to bifurcate nature into two divisions, namely into the nature apprehended in awareness and the nature which is the cause of awareness." (CN, p. 30-31)
[3] "For example, the wave-theory of light is an excellent well-established theory; but unfortunately it leaves out colour as perceived." (CN, p. 45)

must reject the distinction between nature as it really is and experiences of it ... Our experiences of the apparent world are nature itself" (R, p. 61/62).

The fundamental concepts of physics are not free inventions of the mind, but logical abstractions from sense experiences. One of the main pillars in Whitehead's natural philosophy is the method of extensive abstraction. The method is a logical model which links the content of perception as a whole with the conceptual framework of science. By the method of extensive abstraction, sense objects are arranged in converging series wherein each object contains a smaller one. For example the sense object "house" contains the sense object "cabinet", the sense object "cabinet" contains the sense object "bottle," and so on. Finally this series is perceived (or conceived) to terminate in a point or some other basic element (cf. Griffin 1985, p. 95). In this way Whitehead is able to connect scientific objects like electrons or atoms with sense objects. "He distinguishes between them, not as does Newtonian and Einsteinian physics on the old bifurcationary basis, according to which the colours and sounds are sensed data and the electrons and protons are unobserved postulated entities too small for the senses to detect, but upon the basis that both atoms and sense data are immediately sensed adjectives of immediately sensed events, the difference between them being that the adjectives which are colours and sounds depend upon the relation between the event which is their sensed locus and the percipient event which are scientific objects such as electrons are more persistent and are a function only of the events which are their sensed locus" (Schilpp 1991, p. 190).

The second formulation of bifurcation is a direct consequence of the first. After separating the realm of apparent nature from the realm of physical description of this nature, Locke asked how both realms could be connected. The question arises as to how the unperceivable atoms, which, according to Newton's kinematic theory, move in an absolute space and constantly perish time, are connected with our space-time experiences. Locke realised that a moving particle can only set other particles in motion and does not produce redness, for example. Therefore he thought that there are secondary qualities which are not in the things by themselves but rather psychic additions of the mental substance of an observer. The entities have the power to produce ideas in our mind. Locke wrote: "...the Power to produce any Idea in our mind, I call Quality of the Subject wherein that power is" (Locke 1975, p. 134). The primary qualities like form, strength, and extension are in the things alone and cannot be separated from them. The secondary qualities such as colours, tones, and tastes are not in the entities, but the entities have the power to produce these impressions.[4]

---

[4] "...the *Ideas of primary Qualities* of Bodies, *are Resemblances* of them, and their Patterns do really exist in the Bodies themselves; but the *Ideas, produced* in us *by* these *Secondary Qualities, have no resemblance* of them at all. There is nothing like our *Ideas*, existing in the Bodies themselves. They are in the Bodies, we denominate from them, only a Power to produce those Sensations in us: And

The secondary qualities, which do not exist in nature, are supplied by the mental substance. Whitehead calls this theory the theory of psychic addition. To make this form of bifurcationism understandable, one is forced to use a causal theory. The British empiricists did not require causal explanation for touch-perceptions. "These touch-perceptions are perceptions of the real inertia, whereas the other perceptions are psychic additions which must be explained on the causal theory" (CN, p. 43). The preference of touch-perceptions over visual perceptions is a result of the materialism of the 18th and 19th century; it is not acceptable to Whitehead. One could think that touch-perceptions in combination with muscular effort should give reliable information about reality. But for Whitehead, information derived from touch is just as free from any psychic addition as information derived from vision.[5] "This means a refusal to countenance any theory of psychic additions to the object known in perception" (CN, p. 29). In the notion of an "apparent nature" one should erase the attribute, "for there is but one nature, namely the nature which is before us in perceptual knowledge" (CN, p. 40). The senses provide the mind with information which does not need any additions. Sense data are the ultimate things which we know about nature.

The splitting up of reality into primary and secondary qualities has transformed the question of how consciousness and nature are related into the question of how consciousness and the human body are related (cf. CN, p. 27). In addition to the first bifurcation between scientific concept and sense perception, a second bifurcation appears between sense perception and reality itself. This results in banishing the observer (mental substance) out of nature. Berkeley concluded correctly that the observer can have knowledge only of his sense impressions but not of the objects which produced them. The knowledge of reality now requires a theory: Newton's theory of transmission.

To avoid the contradictions in Locke's theory, Leibniz developed his monads. "Leibniz noted that the material substances, by definition, are located in space and hence their action upon anything else can occur only in space. The mental substances, on the other hand, are not in space; instead, for them, space is merely a relational item within the notion of consciousness" (Schilpp 1991, p. 179). Whitehead calls the assumption, that something has a fixed position in time and space, the *fallacy of simple location*. If the material substances are only in space then a material substance can only act upon a material substance but not upon a mental substance. It follows that "either the material substances must be brought within the space in the field

---

what is Sweet, Blue, or Warm in *Idea*, is but the certain Bulk, Figure, and Motion of the insensible Parts in the Bodies themselves, which we call so." (Locke 1975, p. 137)

[5] "I am not denying that the feeling of muscular effort historically led to the formulation of the concept of force. But this historical fact does not warrant us in assigning a superior reality in nature to material inertia over colour or sound." (CN, p. 44)

of awareness of the mental substance ... or the mental substance must be defined in terms of the material substances" (Schilpp 1991, p. 179). The latter is impossible for Leibniz because he would be confronted with Locke's and Hobbes's problem. "Accordingly he concluded that space is a relational item within the consciousness of the observer conceived as a windowless monad" (Schilpp 1991, p. 179/180).

The natural philosophers of the 17th century were mistaken in taking the abstract for the concrete. This is an example of the *fallacy of misplaced concreteness*" (SMW, p. 52). "There are the dualists, who accept matter and mind as on an equal basis, and the two varieties of monists, those who put mind inside matter, and those who put matter inside mind. But this juggling with abstractions can never overcome the inherent confusion introduced by the ascription of *misplaced concreteness* to the scientific scheme of the seventeenth century" (SMW, p. 56). With his antibifurcationism, Whitehead shows complete commitment in his efforts to avoid any separation between the entities of nature. Reality as a fact is a unit of factors, and we are one of them.

Whitehead, like Einstein, is a realist in the sense that reality, for him, is given independently of consciousness. But there is only one nature. There is no such thing as an "independent reality"; there is no reality behind our perceived reality which we try to reach by improving our "perception systems". What we perceive is reality. The only thing we can do is analyze our perception. The only material which we can work with is sense data and reflections on those. We have an experience of reality that is wider than our sense perceptions, which are representations of aspects of this reality. Therefore, we can perceive a difference between our model of reality and the event from which we have taken the different aspects to form this model; but sometimes the difference seems to pale into insignificance. At this point philosophers come into danger of confusing their model of reality with reality itself.

The whole discussion so far leaves the question open whether Whitehead can indeed avoid bifurcations in such a radically empirical scheme. Northrop argues he cannot, because he cannot account for causality: "... the concept of causality in the science of physics," Northrop concludes, "is not given empirically as a deliverance of sense awareness. This means that Whitehead has not succeeded in avoiding bifurcation" (Schilpp 1991, p. 187). I hold that Northrop is wrong, because he does not take into account Whitehead's theory of perception. This will be discussed in detail in the following sections.

2   The Specious Present

If we want to understand what time is, we must first answer the question, "What is the origin of our experience of the past?" From where do we get the meaning of the term "time"? (Cf. James 1890, p. 605). William James suggests that simple sensations are abstractions of a complex reality, and "all our concrete states of mind are representations of objects with some amount of complexity. Part of the complexity is the echo of the objects just past, and, to a lesser degree, perhaps, the foretaste of those just to arrive" (James 1890, p. 606). He also states that we have a constant feeling *sui generis* (James 1890, p. 605) of the past. James notices, like Locke, that a constant stream of ideas passes through our mind. "If the present thought is of A B C D E F G, the next one will be of B C D E F G H, and the one after that of C D E F G H I – the lingerings of the past dropping successively away, and the incomings of the future making up the loss" (James 1890, p. 606). But if we look at our idea of the time-stream, what are we talking about when we talk about the present? What does it mean when we say "now", as if to capture the present moment? If we have spoken the word, it is already past. When can we speak of a present moment? The present, only a point in the stream of time, appears to be an abstraction. We know that it must exist, but the fact that it *does* exist can never be a fact of our immediate experience (cf. James 1890, p. 609). We can only say that the sensible present has duration.[6]

James calls the sensible present, which has duration, the *specious present*. Different authors had studied the duration of such a specious present at James' time. Using different experimental techniques, they had come to the conclusion that the minimum amount of duration we can distinctly feel lasts 1/500 of a second; the maximum amount is 6 to 12 seconds (cf. James 1890, p. 612). These units of duration are *the maximal extent of our immediate distinct consciousness of successive impressions*. Minutes, hours, and days are constructed from these units of duration by mental addition.[7] But perception of the actual world in the specious present alone does not guarantee perception of causality. McTaggart has shown this in his essay about the unreality of time (see chapter *Time* in McTaggart 1921). In addition, we perceive retentions which continuously fade away. At this point, I must mention Husserl's theory of perception, which leads to an inner-time consciousness (Husserl 1985), but also, of course, Whitehead's own theory of perception (SYM, PR), which I shall present in the following section.

---

[6] "The practically cognized present is no knife edge, but a saddle-back, with a certain breadth of its own on which we sit perched, and from which we look in two directions into time. The unit of composition of our perception of time is a *duration*, with a bow and a stern, as it were – a rearward- and a forward-looking end." (James 1890, p. 609)

[7] "We are constantly conscious of a certain duration – the specious present – varying in length from a few seconds to probably not more than a minute, and this duration (with its content perceived as having one part earlier and the other part later) is the original intuition of time. Longer times are conceived by adding, shorter ones by dividing, portions of this vaguely bounded unit, and are habitually thought by us symbolically." (James 1890, p. 642)

3   Whitehead's Theory of Perception

To find the key to Whitehead's philosophy and to show how he can avoid the bifurcations discussed above, one has to consider how he conceives of the origin of every possible knowledge. Whitehead regards this origin to lie in everyone's daily experiences. "The ... set of metaphysical notions rests itself upon the ordinary, average experience of mankind, properly interpreted" (AOI, p. 194). Whitehead directly takes up the starting point of the British empiricists:

1. Every experience has its origin in perceptions;
2. The primary ideas of perception join secondary ideas which are deduced by reflection to put the sense data into an order;
3. In addition to these two starting points of the British empiricists, White-head integrates psychic impressions like emotions, beauty, love, satisfaction, and others.

Whitehead's theory of perception is of greater universality than the theory of perception of the British empiricists because he integrates all moments of experience. Whitehead is, like Kant, an empirical realist, but in contrast to Kant he rejects the position of transcendental idealism. He avoids Kant's idealism-realism doubling through enlarging the theory of perception by a mode which he calls "causal efficacy", in addition to the traditional mode of "presentational immediacy". If Whitehead were to restrict himself to the theory of perception of the British empiricists he would be subjected to the false conclusion that reality is constituted out of static, isolated substances. But it is one of Whitehead's goals to break up the nominalistic concept of reality and to oppose bifurcation by simple location. One can only avoid the fallacy of simple location if one takes reality as a network of relations. Relations do not fit into a substance philosophy which takes the final constituents of reality as static substrata. Relations have a certain independence with regard to substances because they are not tied to the presence of their relata. Whitehead can do justice to the relations in sense perception due to the extension of the theory of perception by the mode of causal efficacy. However, it is important to note that *actual* sense perceptions do neither occur in presentational immediacy nor in causal efficacy. They occur in the mode of "symbolic reference", which renders presentational immediacy and causal efficacy as abstractions and combines the two into a concept refering to what perception means concretely.

3.1   Presentational Immediacy

In presentational immediacy we perceive our surrounding directly. Presentational immediacy is a two-termed relation between the observer and the sense datum. "Presentational immediacy ... expresses how contemporary events are relevant to each other, and yet preserve a mutual independence" (SYM,

p. 16). Sense perceptions have their origin in perceiving external sense data by the sense organs. In contrast to the empirical understanding of given sense data, which are transmitted to consciousness by sense organs, the transmission is an integral part of the situation and is as original as the sense data. In addition, sense perception depends on the present state of the bodily organisation. It takes place against a background of preconceptions. In the mode of presentational immediacy the percipient is not aware of these preconceptions of the body as an amplifier (PR, p. 119) of sense data. What appears to us directly as a sense datum is already an abstraction of a temporally extended process. It is an abstraction of an extended reality.

Although presentational immediacy in fact has the duration of the specious present, this temporal extension is short enough to provide apparently static representations of reality. "The shorter the stretch of time, the simpler are the aspects of the sense-presentation contained within it. The perplexing effects of change are diminished and in many cases can be neglected. Nature has restricted the acts of thought which endeavour to realise the content of the present, to stretches of time sufficiently short to secure this static simplicity over the greater part of the sense-stream" (AOE, p. 128). Sense perceptions perceived in presentational immediacy are like mental photographs which abstract from the camera (sense organs of the body), and the components of the exposure process, especially the exposure time (temporality). At each instant of time, the camera delivers sharp representations. Sense perception in the mode of presentational immediacy abstracts from temporal experiences. It does not deliver any information about future or past, but only about the sense data and their spatial positions (cf. PR, p. 61).

One important element in judging a photograph is the choice of the motive. This points to a degree of freedom in presentational immediacy. "It is ... to a large extent controllable at will" (SYM, p. 22). This degree of freedom is attention. "In our own lives, and at any one moment, there is a focus of attention, a few items in clarity of awareness, but interconnected vaguely and yet insistently with other items in dim apprehension, and this dimness shading off imperceptibly into undiscriminated feeling" (FR, p. 78). By means of attention, the body becomes an integral part of sense perception. Attention directs the perception of the perceiver and determines the selection of the sense data. In this process "the penetration of intuition follows upon the expectation of thought. This is the secret of attention" (FR, p. 79). The direction of perception depends on the interest of the perceiver. In this sense, empiricism is confined within immediate interests (cf. FR, p. 11).

Interest is connected to an expectation of sense data. Attention comprises a teleological and temporal aspect. The analysis of past data directs the attention to future data. This results in the selection of relevant information for the integration process of perception. But the analysis of past data is not the subject of presentational immediacy any longer – it belongs to the mode of causal efficacy. Attention is the relation between presentational immediacy and causal efficacy.

Directed attention in Whitehead's philosophy must not be understood according to the normal usage of the word "attention". When a hungry cat watches a mouse, she does not have the freedom to turn away. Normally animals do not perceive in the mode of presentational immediacy, but only in causal efficacy. If the mouse did not act on the cat, the cat would not see her. This attention is a kind of reaction by reflex action. Animals react to their environment by adapting it. Conformity is only perceivable by means of causal efficacy. Humans perceive in the mode of presentational immediacy in addition to causal efficacy: "...all scientific observations, such as measurements, determinations of relative spatial position, determinations of sense data such as colours, sounds, tastes, smells, temperature feelings, touch feelings, etc., are made in the perceptive mode of presentational immediacy" (PR, p. 169). In contrast, physical theories exclusively refer to causal efficacy.

Summary: Perceptions in presentational immediacy provide the spatial aspects of the perception process. The main topics concerning presentational immediacy are (SYM, p. 23):

- Sense perceptions depend on the spatial relationships between the perceiver and the sense data;
- Temporal aspects are ignored;
- Perception in presentational immediacy contributes to the experience of only a few high developed organisms.

3.2   Causal Efficacy

Whitehead's deliberations about perception in the mode of causal efficacy arise from his criticism of the perception theory of the British empiricists, as well as the transcendental idealists. The result is an extension of their theory of perception (cf. SYM, p. 31). Hume's fundamental philosophy is "that (i) presentational immediacy, and relations between presentationally immediate entities, constitute the only type of perceptive experience, and that (ii) presentational immediacy includes no demonstrative factors disclosing a contemporary world of extended actual things" (SYM, p. 34). The consequence is that only sense perceptions of pure, private nature will be uncovered in visual perception. It is impossible to found space, time, and identity with regard to reality upon these perceptions. Hume is restricted to the present moment. Whitehead calls this restriction, following Santayana, the "solipsism of the present moment" (SYM, p. 33).

The starting point of Whitehead's philosophy is his observation that sense perception in presentational immediacy does not do justice to all moments of experience. Therefore, he introduces the mode of causal efficacy in addition to presentational immediacy. The ontological basis of his realism is to be found in his theory of the extensive continuum. Generally all items of the world are temporally and spatially extended and connected with each other. For this reason one has to consider the universe as process. If reality is empirically perceivable then relations have to be perceivable, too. This is impossible if all knowledge is founded in presentational immediacy alone. In this case, there were no empirical knowledge of an extensive continuum, and it were impossible to perceive the phenomena of the world properly. In contrast to Hume and Kant, Whitehead finds sufficient evidence for causal connections and temporal continuity in direct sense perception. The tacit presupposition for the demonstration of this evidence is the experience of temporal and spatial extension. Temporal extension in particular underlines the significance of perception in the mode of causal efficacy. Temporally adjoining events are present contemporaneously in the specious present, and they are perceived together. Later events confirm the content of earlier ones. This is a basic datum of our experience. Spatio-temporal relations and their perception are the reason for our knowledge of an extensive continuum.

Actually, Kant accepts that causal efficacy is a fact of the phenomeno-logical world, but in Whitehead's opinion he insists erroneously that it is not a part of sense perception. For Kant, causal efficacy is only a way of our thinking because he accepts uncritically Hume's presupposition that the phe-nomenological world is constituted of a series of momentary events (cf. SYM, p. 37). Hume's presupposition is attributed to a reduced concept of time. For Whitehead, momentary events are already abstractions from actual occasions (cf. Sect. 4.2). To accept this reduced concept of time as ultimate is an exam-ple of the "fallacy of misplaced concreteness". Kant justifies his position by the experience that in each instance one perceives only details of reality, and never reality as a whole. Partial perceptions present only partial data of real-ity. But one makes valid statements about the whole of reality. Such general knowledge cannot be deduced from finite partial aspects of reality. Any uni-versal validity of the description of nature can never be legitimated by sense data alone. For this reason Kant founds the phenomenological world upon an effort of coherent judgements (SYM, p. 38). For Whitehead, Kant's phi-losophy is an attempt to show how subjective experiences are processed into objectivity. Whitehead tries the opposite: He wants to show how objectively experienced entities produce subjective capacity as they are experienced in a temporal act, which is not reducible to intellectual faculties: the specious present.[8]

---

[8] "The philosophy of organism is the inversion of Kant's philosophy. ... For Kant, the world emerges from the subject; for the philosophy of organism, the subject emerges from the world." (PR, p. 88)

"... for Kant, the process, whereby there is experience, is a process from subjec-tivity to apparent objectivity. The philosophy of organism inverts this analysis, and explains the process as proceeding from objectivity to subjectivity, namely, from the objectivity, whereby the external world is a datum, to the subjectivity, whereby there is one individual experience." (PR, p. 156)

The starting point for any consciousness consists of the data of the external world. In the specious present, the events with their spatial and temporal extension are directly perceived. The track of a spark emitted by a fire, as an example, can be perceived as one form in the specious present. But perception of forms does not answer the question as to whether we can really perceive causality. One cannot identify temporally extended experience with the experience of causality. However, the specious present does not contain unique perceived events alone; it also includes the immediate past (James 1890, p. 606). "Causal efficacy is the hand of the settled past in the formation of the present" (SYM, p. 50). The presence of immediately past events shows that present and future events have to fit with earlier events in the same way that immediately past events had to fit with more distantly past events. "Causal efficacy is the confirmation of the present to the immediate past, and both the immediate past and its relevance to the present are experienced" (Lee 1961, p. 61). Whitehead calls the presence of past events in causal efficacy "non-sensuous perception" (AOI, p. 180). Causal efficacy contains a concept of causality which is not intended to predict future events but to describe past events. Causality in Whitehead's philosophy means that we never perceive a series of events alone, but that later events arise from earlier events in the specious present. Perceptions in causal efficacy are vague, persistent, urgent, and uncontrollable, whereas perceptions in presentational immediacy are exactly defined, easy to reproduce, and to manage.

Causality is perceivable in the comparison of present occasions with immediate past occasions: "...the present event issues subject to the limitations laid upon it by the actual nature of the immediate past. ...the complete analysis of the past must disclose in it those factors which provide the conditions for the present" (SYM, p. 46). Conformity finds its expression in the behaviour of lower developed organisms. Plants and animals come to terms with facts as they find them in their surroundings. "In its lowest form, mental experience is canalized into slavish conformity. It is merely the appetition towards, or from, whatever in fact already is. The slavish thirst in a desert is mere urge from intolerable dryness. This lowest form of slavish conformity pervades all nature. ... It is degraded to being merely one of the actors in the efficient causation" (FR, p. 33-34).

Besides the conformity of temporally neighbouring events, the transmission process of sense data into the mind is an additional argument for perception in causal efficacy. The percipient experiences the sense perceptions as depending on his body with its physiological properties. "For example, in touch there is a reference to the stone in contact with the hand, and a reference to the hand; but in normal, healthy, bodily operations the chain of occasions along the arm sinks into the background, almost into complete oblivion. Thus M, which has some analytic consciousness of its datum, is conscious of the feeling in its hand as the hand touches the stone. According to this account, perception in its primary form is consciousness of the causal efficacy of the external world by reason of which the percipient is a concrescence from a definitely constituted datum. The vector character of the datum is this causal efficacy" (PR, p. 120). Immediate sense perceptions are the result of a transmission process which is already past. "Our bodily experience is primarily an experience of the dependence of presentational immediacy upon causal efficacy" (PR, p. 176). Perception in causal efficacy avoids the solipsism of the present moment because it points to the fact that sense perception comprises more than one perceives in presentational immediacy, and that presentational immediacy depends on this "more".

Summary: Perceptions in causal efficacy contain the temporal aspects of the process of reality. These aspects are directly perceived in sense perceptions. Whitehead gives three arguments for perception in causal efficacy:

- perception in the specious present;
- perception that the present comes out of the past;
- perception of stimulus-reaction chains.

3.3   Symbolic Reference

Sense perception never takes place in one of the pure modes of presentational immediacy or causal efficacy, but only in the complex mode of symbolic reference which connects the two. Sense perceptions in symbolic reference possess only a symbolic content. As a result of the fusion of the two pure and abstract modes of perception, real perception in symbolic reference is the reason of errors and misinterpretations. Symbolic reference is an active synthetic element of the perceiver which produces emotions, convictions, and beliefs concerning other elements of reality. It delivers a secure ground of experience only if it fulfills certain criteria which are demanded by the pure modes of perception. One of these criteria is the consistency of perceptions. Indistinctness and ambiguity in symbolic reference lead to mistakes which are not to be found in the reference of concepts but in the process of perception.

Sense perceptions are not such basic elements as the British empiricists believed; they are rather located at a higher level of abstraction. Sense perceptions are more than sense impressions of sensed data. "They also represent the conditions arising out of the active perceptive functioning as conditioned by our own natures" (SYM, p. 58). In sense perception, the data of reality are transformed by the subjective form of perception in which the emotional state of the perceiver and the specific state of the sense organs are included.

An event of reality emits an emotional form, which produces a stimulus in the sense organ and results in a corresponding sense perception. "The more primitive types of experience are concerned with sense-reception, and not with sense-perception" (PR, p. 113-114). Emotional forms contain the rough data as they are emitted by the events: they are not spatialized and not distinguished in different objects. "Symbols" are those components of the sense process which produce emotions, convictions, and beliefs. Elements of reality that produce sense stimuli are denoted as "meanings" by Whitehead. The relation between "symbols" and "meanings" indicates symbolic reference (SYM, p. 8). Symbolic reference describes the transition from the more simple elements, the meanings, to the simple elements, the symbols. Knowledge of this relation alone does not designate which relatum is the symbol and which the meaning. There is no element in our experience which is only symbol or only meaning. In symbolic reference, the more complex elements point to the simple ones; thereby sense perception proceeds from the simpler elements to the more complex (cf. SYM, p. 10): "...the transition from without to within the body marks the passage from lower to higher grades of actual occasions. ... Thus the transmitted datum acquires sense enhanced in relevance or even changed in character by the passage from the low-grade external world into the intimacy of the human body" (PR, p. 119-120).

Analyzing sense perceptions in the mode of symbolic reference and searching for the meanings, one has to concentrate on those elements which both pure modes of perception have in common. One finds these elements in the sense datum: "...'presentational immediacy' deals with the same datum as does 'causal efficacy' ..." (PR, p. 173). The elements themselves are the "presented loci"[9] and the "eternal objects"[10]. "The partial community of structure, whereby the two perceptive modes yield immediate demonstration of a common world, arises from their reference of sense-data, common to both, to localizations, diverse or identical, in a spatio-temporal system common to both" (SYM, p. 53).

Presentational immediacy abstracts totally from sense objects. It marks the immediate phenomena, whereas causal efficacy refers to their spatio-temporal relationships.[11] Both modes of perception are interwoven with each other, and it is difficult to separate them: "For example, colour is referred to an external space and to the eyes as organs of vision. In so far as we are dealing with one or other of these pure perceptive modes, such reference is direct demonstration; and, as isolated in conscious analysis, is ultimate fact against which there is no appeal. Such isolation, or at least some approach to it, is fairly easy in the case of presentational immediacy, but is very difficult in the case of causal efficacy. Complete ideal purity of perceptive experience, devoid

---

[9] "...all that is perceived is that the object has extension and is implicated in a complex of extensive relatedness with the animal body of the percipient." (PR, p. 122)

[10] "In sense perception we discern the external world with its various parts characterized by form of quality, and interrelated by forms which express both separation and connection. These forms of quality are the sensa, such as shades of blue, and tones of sound. The forms expressing distinction and connection are the spatial and temporal forms." (MT, p. 73)

[11] "We shall find that generally – though not always – the adjectival words express information derived from the mode of immediacy, while the substantives convey our dim percepts in the mode of causal efficacy." (PR, p. 179)

of any symbolic reference, is in practice unobtainable for either perceptive mode" (SYM, p. 54). "Our natures must conform to the causal efficacy. Thus the causal efficacy *from* the past is at least one factor giving our presentational immediacy *in* the present. The *how* of our present experience must conform to the *what* of the past in us" (SYM, p. 58).

Critics of Whitehead's philosophy very often fail to see that he considers actual perception as a synthesis of two different modes of perception. Empirical scientific research takes place in presentational immediacy exclusively, but its representation takes place in causal efficacy: "...all scientific theory is stated in terms referring exclusively to the scheme of relatedness, which, so far as it is observed, involves the percepta in the pure mode of causal efficacy" (PR, p. 169). Whitehead does not introduce a new kind of bifurcation here because the basic unity of sense perception is presented in symbolic reference. Causal efficacy and presentational immediacy are abstractions. Causality is an integral component of reality and an aspect which is uncovered in immediate perception.

4   Time

4.1   Time in Whitehead's Natural Philosophy (1914–1925)

In each stage of the development of his philosophy, Whitehead expresses that space and time do not exist by their own: "... space-time cannot in reality be considered as a self-subsistent entity. It is an abstraction, and its explanation requires reference to that from which it has been abstracted" (SMW, p. 64). Reality is extended. Space and time are abstractions from extended events, and they are experienced empirically. In his natural philosophy, nature exists only as a whole, as a continuously flowing process. "Thus an entity is an abstraction from the concrete, which in its fullest sense means totality. ... any factor, by virtue of its status as a limitation within totality, necessarily refers to factors of totality other than itself. It is therefore impossible to find anything finite..." (R, p. 17). Every part receives meaning only in relation to the whole. What the scientist accepts as the elements or parts of a whole, are in reality abstractions. In actuality the parts exist and have meaning by virtue of the whole, and vice versa. Temporally extended events do not exist by their own. They are parts of an extended nature and their duration is the specious present of a perceiver. For this reason, time does not have any reality in nature but is only a property of a perceiver. Therefore, Whitehead does not have any problems in considering momentary events as ultimately given data for philosophical analysis. Reality is an extensive continuum and events are arbitrary parts which are simplified by abstraction.

> "The continuity of nature arises from extension. Every event extends over other events, and every event is extended over by other events.
> ... every duration is part of other durations; and every duration has

other durations which are parts of it. Accordingly there are no maximum durations and no minimum durations. Thus there is no atomic structure of durations, and the perfect definition of a duration, so as to mark out its individuality and distinguish it from highly analogous durations over which it is passing, or which are passing over it, is an arbitrary postulate of thought. Sense-awareness posits durations as factors in nature but does not clearly enable thought to use it as distinguishing the separate individualities of the entities of an allied group of slightly differing durations." (CN, p. 59)

Events do not have any reality independent of a consciousness, and they do not have definite temporal extensions. Time relations are an expression of an ordering relation imposed by a perceiver. "... space-time is nothing else than a system of pulling together of assemblages into unities. But the word *event* just means one of these spatio-temporal unities" (SMW, p. 70-71). Physical time only deals with certain formal relational aspects of our changing human experience (cf. Mays 1970, p. 509). Relative to other abstractions, space and time offer a comparatively simple structure, which is suitable as a basis for objective distinctions in reality.[12]

In the specious present, one perceives a unity which is already separated into its parts by the activity of the perceiver. The parts entertain certain characteristics of which time and space are an example. The common structure of space-time conforms to the uniform experiences of sense perception. "Time and space are necessary to experience in the sense that they are characteristics of our experience..." (AOE, p. 163). But the unity of sense perceptions is not free from problems. Whitehead's early concept of time has the same problems as Kant had. It is not clear how one can proceed from individual experiences to a uniform space-time-structure. In the article *Space, Time, and Relativity* (1917), Whitehead wrote:

"But I admit that what I have termed the "uniformity of the texture of experience" is a most curious and arresting fact. I am quite ready to believe that it is a mere illusion; and ... I suggest that this uniformity does not belong to the immediate relations of the crude data of experience, but is the result of substituting for them more refined logical entities, such as relations between relations, or classes of relations, or classes of classes of relations. By these means it can be demonstrated – I think – that the uniformity which must be ascribed to experience is of a much more abstract attenuated character than is usually allowed. This process of lifting the uniform time and space of the physical world into the status of logical abstractions has also the advantage of recognising another fact, namely, the extremely fragmentary nature of all direct individual conscious experience.

_____

[12] "This simplicity of time and space is perhaps the reason why thought chooses them as the permanent ground for objectival distinction..." (SI, p. 134)

My point in this respect is that fragmentary individual experiences
are all that we know, and that all speculation must start from these
disjecta membra as its sole datum. It is not true that we are directly
aware of a smooth running world..." (AOE, p. 163f)

As a consequence, Whitehead's natural philosophy of the period before
1925, accepting that reality is a continuous process, does not represent considerable progress with respect to Kant's concept of time. Temporality of
perception is not sufficient to account for causality and time as properties of
nature.

4.2   The Epochal Theory of Time (After 1925)

The transition from momentary events to extended actual occasion in Whitehead's philosophy is not only initiated by the knowledge of perception in the
specious present, but also by logical difficulties within physical theories and
metaphysical outlines. The difficulties with the traditional concept of continuous time are:

1. Perception takes place in a duration (specious present).
2. Physical descriptions of dynamical processes presuppose the existence of
   temporal events (momentum, velocity, etc.).
3. The description of simple physical structures (like atoms) or biological
   organisms needs temporal extension.
4. Becoming and continuity are incompatible. Becoming is only possible if
   reality is constituted of atomic temporal events (Zenon's paradox).
5. Causal interactions are directly perceivable in the specious present.
6. Momentary events are abstractions. They can be deduced from exten-
   sional events by means of the method of extensive abstraction.

All these points forced Whitehead to conclude that reality is not based
on momentary events but on temporally extended events. Yet he needed
additional justification to transfer the notion of duration to the events of
reality. He did not find such justification until the early development of the
quantum theory, whose first results motivated him to apply basic elements
of philosophy and psychology to all events of reality. Although Whitehead
probably did not study the theoretical foundations of quantum theory in
detail,[13] some developments of the early quantum theory led him to his speculative philosophy. Especially Bohr's atom model (1913) and de Broglie's
wave theory (1924) inspired him to critically reexamine his natural philosophy and gained explicit entry into the chapter "The Quantum Theory" in

[13] "The internal evidence of Whitehead's writings suggests, in fact, that he never
became acquainted with the post-1924 development of quantum theory or perhaps even with de Broglie's 'matter waves,' since there is not a single allusion
to any of these developments or to their authors in all of Whitehead's published
work." (Palter 1970, p. 215-216)

Science and the Modern World (SMW, p. 119ff). The "particles" of reality are no longer material static forms, but spatio-temporally extended events.[14] Whitehead explains that "the change from materialism to 'organic realism' ... is the displacement of the notion of static stuff by the notion of fluent energy. Such energy has its structure of action and flow, and is inconceivable apart from such structure. It is also conditioned by 'quantum' requirements" (PR, p. 309). This shows that he got his inspiration from scientific discoveries without going into specific details of the formalism. His doctrine of the epochal character of time depends on the analysis of the intrinsic character of an event, considered as the most concrete finite entity (cf. SMW, p. 118), which he calls "actual occasion".

In his epochal theory of time, Whitehead unifies four different aspects of time, which one finds in the experience of an actual occasion.

- Aspects of internal time:
  - passage of thought (becoming and perishing, retentions),
  - experience of extension (unlimited act, inner time consciousness, retentions and protentions);
- Aspects of external time:
  - actual physical time (perception of past actual occasions, passage of nature, becoming and perishing),
  - potential physical time (extensive continuum).

To avoid any bifurcations, every actual occasion has to be a factor equal to other factors in the fact of nature. A perceiving consciousness is no longer an exception, which is outside of nature, but an inherent part of it. Therefore, every internal aspect of time has an external aspect as an equivalent. Experience of extension corresponds to potential physical time; and passage of thought corresponds to passage of nature.

The physical concept of time unifies two experiences: the experience of an extensive continuum and the perception of concrete actual occasions. Thereby it unifies discontinuity and continuity of the external world in one concept. The actual separation of the extensive continuum depends on the prehensions[15] of the occasion of becoming actual. "It is by means of 'extension' that the bonds between prehensions take on the dual aspect of internal relations, which are yet in a sense external relations" (PR, p. 309). An actual occasion presents a physical *and* a mental pole. This means that an actual occasion is both a physical occasion as well as a mental occasion. Only the physical pole has potential extension over the whole extensive continuum and can be divided coordinately. The mental (conceptual) pole does not share the

---

[14] "There is a spatial element in the quantum as well as a temporal element. Thus the quantum is an extensive region." (PR, p. 283)

[15] "Prehension" is a shortened form of "apprehension" and unifies conscious and unconscious perceptions in one concept.

coordinate divisibility of the physical pole, from which the extensive continuum is derived.

Similarly, passage of thought in the internal concept of time is confronted with the experience of an unlimited temporal act. Consciousness as a whole is an actual occasion which is constituted by a society of subordinated actual occasions which are the single states of mind. Every state of mind, every thought, is a mental pole of an actual occasion whose passage determines the experiences of mind. A mentale pole of an actual occasion has its equivalent in a thought of mind and is an act of attention which has the duration of the specious present.

"The actual entity is the product of the interplay of physical pole with mental pole. In this way, potentiality passes into actuality, and extensive relations mould qualitative content and objectifications of other particulars into a coherent finite experience." (PR, p. 308)
"Each actuality is essentially bipolar, physical and mental ... So, though mentality is non-spatial, mentality is always a reaction from, and integration with, physical experience which is spatial." (PR, p. 108)

Every actual occasion is a spatio-temporal unit which possesses an indivisible volume and time quantum. Coordinate divisibility provides that each actual occasion is composed of a number of subordinated actual occasions. But "just as, for some purposes, one atomic actuality can be treated as though it were many coordinate actualities, in the same way, ... a nexus of many actualities can be treated as though it were one actuality" (PR, p. 287). Actual occasions express the uniform space-time structure of the universe because their external relations, which fit the actual occasion into a superordinate actual occasion, and their internal relations, which represent the coordinate divisibility into subordinate actual occasions, merge into an extensive continuum.[16]

"Blind prehensions, physical and mental, are the ultimate bricks of the physical universe. They are bound together within each actuality by the subjective unity of aim which governs their allied genesis and their final concrescence. They are also bound together beyond the limits of their peculiar subjects by the way in which the prehension in one subject becomes the objective datum for the prehension in a later subject, thus objectifying the earlier subject for the later subject. The two types of interconnection of prehensions are themselves bound

---

[16] "There is, in this way, one basic scheme of extensive connection which expresses on one uniform plan (i) the general conditions to which the bonds, uniting the atomic actualities into a nexus, conform, and (ii) the general conditions to which the bonds, uniting the infinite number of coordinate subdivisions of the satisfaction of any actual entity, conform." (PR, p. 286)

together in one common scheme, the relationship of extension." (PR, p. 308f)

Physical time, physical space, and creative advance[17] are abstractions which presuppose the more general relationship of extension. "The extensiveness of space is really the spatialization of extension; and the extensiveness of time is really the temporalization of extension. Physical time expresses the reflection of generic divisibility into coordinate divisibility" (PR, p. 289). The spatio-temporal extensive continuum is the general structure to which all actual occasions must conform.[18]

The assignment of the different aspects of time to the ultimate units of reality becomes possible by transforming the natural philosophical concept of momentary events into actual occasions. In Whitehead's metaphysics, actual occasions are no longer momentary cuts through reality, but forms which have the properties of spatio-temporal extension and creativity. In Whitehead's natural philosophy, the events were dependent on the activity of a perceiver. They had no independent reality. Only reality as a whole was real. In Whitehead's metaphysics, the actual occasions are the ultimate units of reality. They are the real things in the world which have their own being, temporality, and creative activity. Whitehead's metaphysics is a consequence of the internal experiences of man with respect to all entities of nature. If one takes into account that the actual world consists of actual occasions which have different temporal extensions overlapping each other, and that one perceives this world in the specious present, one is able to perceive causal connections directly. One can recognize the development of actual occasions along their historical routes and realize how an actual occasion passes and another one becomes.

References

Eisendrath C.R. (1971): *The Unifying Moment – The Psychological Philosophy of William James and Alfred North Whitehead* (Harvard University Press, Cambridge).

Griffin D.R., ed. (1985): *Physics and the Ultimate Significance of Time* (SUNY Press, Albany).

Husserl E. (1985): *Texte zur Phänomenologie des inneren Zeitbewußtseins (1893–1917)* (Meiner, Hamburg).

James W. (1890): *The Principles of Psychology* (Henry Holt, New York).

---

[17] "... the self-enjoyment of an occasion of experience is initiated by an enjoyment of the past as alive in itself and is terminated by an enjoyment of itself as alive in the future. This is the account of the creative urge of the universe as it functions in each single individual occasion." (AOI, p. 193)

[18] "To be an actual occasion in the physical world means that the entity in question is a relatum in this scheme of extensive connection." (PR, p. 288)

Lee H.N. (1961): Causal efficacy and continuity in Whitehead's philosophy. *Tulane Studies in Philosophy* **10**, 59–70.

Locke J. (1975): *An Essay Concerning Human Understanding* (Oxford University Press, Oxford).

Mays W. (1970): Whitehead and the philosophy of time. *Studium Generale* **23**, 509–524.

McTaggart J.E. (1921): *The Nature of Existence, Vol. 1* (Cambridge University Press, Cambridge).

Palter R. (1970): *Whitehead's Philosophy of Science.* University of Chicago Press, Chicago).

Schilpp P.A., ed. (1991): *The Philosophy of Alfred North Whitehead* (Open Court, LaSalle, Ill.)

Whitehead's publications, here arranged in the sequence of their first, original publication, are referred to as follows:

SI   Whitehead A.N. (1967): The anatomy of some scientific ideas. In *The Aims of Education and Other Essays* (Free Press, New York), Chap. IX, 121–153. (First edition 1916.)

PNK   Whitehead A.N. (1955): *An Enquiry Concerning the Principles of Natural Knowledge* (Cambridge University Press, Cambridge). (First edition 1919.)

CN   Whitehead A.N. (1990): *The Concept of Nature* (Cambridge University Press, Cambridge). (First edition 1920.)

R   Whitehead A.N. (1922): *The Principle of Relativity With Applications to Physical Science* (Cambridge University Press, Cambridge).

SMW   Whitehead A.N. (1964): *Science and the Modern World* (New American Library, New York). (First edition 1925.)

SYM   Whitehead A.N. (1927): *Symbolism – Its Meaning and Effect* (Fordham University Press, Fordham).

AOE   Whitehead A.N. (1967): *The Aims of Education and Other Essays* (Free Press, New York). (First edition 1929.)

FR   Whitehead A.N. (1958): *The Function of Reason* (Beacon Press, Boston). (First edition 1929.)

PR   Whitehead A.N. (1979): *Process and Reality* (Free Press, New York), ed. by D.R. Griffin and D.W. Sherburne. (First edition 1929.)

AOI   Whitehead A.N. (1967): *Adventures of Ideas* (Free Press, New York). (First edition 1933.)

MT   Whitehead A.N. (1968): *Modes of Thought* (Free Press, New York). (First edition 1938.)

Delayed-Choice Experiments and the Concept of Time in Quantum Mechanics

Shimon Malin

Department of Physics and Astronomy, Colgate University,
Hamilton, New York 13346, USA

Abstract. Delayed-choice experiments seem to have a puzzling feature: the present seems to change the past. This is shown to be a result of the paradigm in which such experiments are described. In particular, it is shown that in a description which is based on the philosophy of A.N. Whitehead, this puzzling feature does not occur.

1 The Puzzle

To introduce the puzzle presented by delayed-choice experiments we need to invoke the quantum mechanical "principle of superposition". This principle states that if a quantum system, such as an electron or a photon, is presented with two (or more) alternatives, it will not choose between them; it will enter into a state of "superposition" of the two alternatives. For example, if a photon entering an apparatus is presented with two possible paths to follow, path (a) and path (b), it will enter into a superposition of the two paths, a state in which it can be said to be following partly path (a) and partly path (b). This will occur provided no observation is made as to which path was taken. If such an observation is made, a state of superposition is no longer possible, and the photon will be found either along path (a) or along path (b).

A delightful explanation of the principle of superposition is contained in Robert Gilmore's book *Alice in Quantumland* (Gilmore 1995):

> Alice entered the wood and made her way along a path which wound among the trees, until she came to a place where it forked. There was a signpost at the junction, but it did not appear very helpful. The arm pointing to the right bore the letter "A", that to the left the letter "B", nothing more. "Well, I declare," exclaimed Alice in exasperation, "that is the most unhelpful signpost I have ever seen." She looked around to see if there were any clues as to where the paths might lead, when she was a little startled to see that Schrödinger's Cat was sitting on a bough of a tree a few yards off.
> "Oh, Cat," she began rather timidly. "Would you tell me please which way I ought to go from here?"
> "That depends a great deal on where you want to get to," said the Cat.

"I am not really sure where..." began Alice.

"Then it doesn't matter which way you go," interrupted the Cat.

"But I have to decide between these two paths," said Alice.

"Now that is where you are wrong," mused the Cat. "You do not have to decide, you can take all the paths. Surely you have learned that by now. Speaking for myself, I often do about nine different things at the same time. Cats can prowl around all over the place when they are not observed. Talking of observations," he said hurriedly, "I think I am about to be obs..." At that point the Cat vanished abruptly.

Let us proceed now to describe a paradigmatic delayed-choice experiment. The apparatus is drawn in Fig. 1. It can be used as a classical or quantum experiment; the delayed-choice feature occurs only at the quantum level.

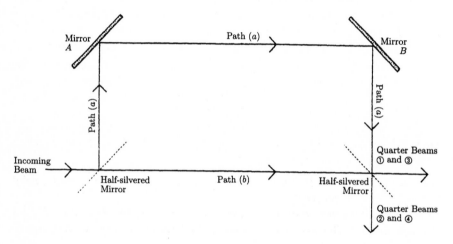

Fig. 1. Schematic representation of the set-up of a delayed-choice experiment.

A classical experiment can be performed as follows. A beam of light, approaching from the left, hits a half-silvered mirror, a mirror that is partly reflective and partly transparent. Half of the beam is reflected, following path (a), and the other half follows path (b). Along path (a) the half beam meets two ordinary mirrors, mirror A and mirror B, and then meets the other half beam at the half-silvered mirror on the right. Here each half beam is split in half again, and two of the quarter beams combine and leave the apparatus, moving towards the right.

The experiment becomes a quantum one when the intensity of the incoming beam is lowered to the point where one photon at a time enters the apparatus. Let us consider such a photon. Having entered the apparatus, will it follow path (a) or path (b)? According to the principle of superposition, it follows both paths, since nothing in the apparatus can be used to observe which path was taken. Consider, however, the following scenario: While the photon is inside the apparatus, approaching mirror B, we change the mounting of mirror B, replacing the rigid mounting by a spring mounting. The spring that mirror B is now mounted on is so light that the impingement of one photon is enough to make it vibrate. Now mirror B can serve as a means of observation of the photon having taken path (a), and a superposition is no longer possible.

What will happen? The choice between a state of superposition and a state of taking one of the paths was presumably made when the photon entered the apparatus! And yet, once the change in the mounting of mirror B is performed, the photon will be found either along path (a), as the vibration of the mirror would show, or along path (b), as the absence of such a vibration would indicate. It seems as if the present changes the past! This is the puzzle (for a more detailed discussion see Wheeler 1978).

Let me hasten to emphasize that this is a thought experiment, not an experiment that can be performed. The change in the mounting of mirror B would take a few moments, while the time of passage of a photon through the apparatus is of the order of 10^{-10} seconds! However, equivalent delayed-choice experiments using electronic switches at the position of mirror B have been performed. We will continue discussing the experiment depicted in Fig. 1, because thinking about it makes the issues clear.

The puzzle of delayed-choice experiments is one of many quantum mechanical puzzles. In the sequel we will comment about two other puzzling aspects of quantum mechanics:

1. *The violations of Bell's inequalities*: they *seem* to show the existence of superluminal propagation of influences.
2. *The riddle of the collapse of quantum states*: in spite of strenuous effort for the past 70 years, there is no convincing mechanism for the choice involved in a collapse.

The fact that things seem strange often indicates the need for a paradigm shift. It is a characteristic of a paradigm shift that what was inexplicable in the old paradigm becomes simple and obvious in the new. Consider, for example, the fact that, if one keeps traveling east, one comes back to the point of origin. This fact, which is puzzling in a flat earth paradigm, indicates the need to shift to the paradigm of a spherical earth. Can we see the direction of a paradigm shift that will make the seemingly strange feature of delayed-choice experiments simple and obvious?

2   Whitehead's Philosophy

2.1   A Historical Perspective

During the twenties, when quantum mechanics was being discovered, Alfred North Whitehead introduced, in his book *Process and Reality* (Whitehead

1979), a revolutionary paradigm which he called "The Philosophy of Organism" and is now usually called "Process Philosophy". As we shall see, this philosophy is in remarkable agreement with the findings of quantum mechanics.

Whitehead, working at Harvard University, was aware only of "the old quantum mechanics" of 1900–1915. He was unaware of the discoveries of Heisenberg, Born, Jordan, and Schrödinger. This makes the agreement between his paradigm and the findings of quantum mechanics all the more remarkable.

Strangely, the confluence between Whitehead's "Process Philosophy" and quantum mechanics was noticed only in the sixties. The seminal paper is by J.M. Burgers (1963). Among the others who worked on the relationship between Whitehead's philosophy and quantum mechanics are A. Shimony (1965), H.P. Stapp (1979), and myself. My paper "Whiteheadean Approach to Bell's Correlations" (Malin 1988) contains references to earlier work.

2.2   Some of the Main Ideas

As the name "Philosophy of Organism" indicates, Whitehead's paradigm is one of an alive universe, one in which creativity and experience are fundamental to the universe and to everything in it. This may sound like cheap new age stuff. In fact, Whitehead was one of the most rigorous thinkers of this century. Before he turned to philosophy he was a mathematician, a co-author (with Bertrand Russell) of *Principia Mathematica*. The rigorous thinking of an outstanding mathematician is evident in Whitehead's magnum opus, *Process and Reality*.

One basic idea of Whitehead's philosophy is this: The fundamental building blocks of the universe, the "atoms of reality", are not enduring objects, but what Whitehead calls "actual entities" or "throbs of experience". "Actual entities" can be characterized as follows:

1. They are neither purely subjects nor purely objects; they have both subjective and objective characteristics, as we will see.
2. They do not endure in time: They flash in and out of existence in spacetime. The apparent existence of enduring objects is due to many collections of actual entities coming one after the other in quick succession, like frames in a movie.
3. Each actual entity is a nexus of relationships with all the other actual entities.
4. An actual entity is the process of its own self-creation.
5. This self-creation involves accommodating and integrating within itself (comprehending, or "prehending" in Whitehead's terminology) all the previous actual entities as "settled facts" that cannot be changed, and all the future actual entities as potentialities. This process of self-creation

involves a sequence of phases, which are delineated and analyzed in detail in *Process and Reality*.

An example: Listening to an orchestra playing a symphony involves, at each moment, accommodating the sounds produced by the orchestra. This accommodation depends, in turn, on many collections of past actual entities, such as previous knowledge and training in music, associations with the symphony, etc. Notice that at each instant there is one experience.

6. The end product of the process is one new actual entity, one "throb of experience". The fundamental building blocks of the universe are, then, elementary experiences; we do not live in a "universe of objects", but in a "universe of experience".

7. Subjectively, i.e., for itself, an actual entity is a "pulse of experience". The end of the process of self-creation is called "the satisfaction of the actual entity". Its subjective existence is momentary. Objectively, i.e., for other, future actual entities, it is "a settled fact": The fact that it did happen cannot be erased. "The end of ... [its] private life – its 'perishing' – is the beginning of its public career." (Lowe 1951, p. 404)

8. The point that is crucial for our purposes is this: *The process of self-creation of an actual entity is not a process in time; it is rather an atemporal process leading to the momentary appearance of the completed actual entity in space-time.* Quoting Whitehead (1979, p. 283): "[In the process of self-creation which is an actual entity] the generic passage from phase to phase is not in physical time ... the generic process is not the temporal succession ... Each phase in the generic process presupposes the entire quantum."

3   Atemporal Processes

Let us pause to examine the idea of an "atemporal process". This idea is strange for us, because we are used to the Newtonian paradigm of a clockwork universe, in which everything happens in space-time. However, the idea of an "atemporal process" is not an invention of Whitehead's; it goes back at least to Plato.

All of Plato's *Dialogues* are based on the distinction between "that which always is and has no becoming and ... [that which] is always becoming and never is", that is, the distinction between "the world of immutable being" and "the world of generation". The universe consists of both. Plato's account of its creation, the *Timaeus*, is an account of creating, first, "the world of immutable being", and second, "the world of generation".

"The world of immutable being" has nothing to do with time. It just is, not only unchanging, but also, by its very nature, unchangeable. The concepts of time and change do not apply to it. Equality, Sameness, Difference, Beauty, Justice, and all the other "Forms" are of this nature. The Soul of the Universe, too, is eternal. And yet Plato describes, in considerable length, the process by which the Soul of the Universe is created! Clearly, he does not mean creation as a process in time. Indeed, we read about the creation of Soul, Forms, and of much else besides, before we get to the *creation of time*.

Processes that are outside time are central to Plato's thinking. As far as the account of creation goes, they are the essence of the story – a confinement of the account of creation within time would be unthinkable. But this notion of atemporal processes is necessary not only for the big story of the creation of the universe; atemporal processes are, according to Plato, all around us. They are the essence of any process of "becoming". A beautiful object becomes beautiful through its "participation" in the Form of Beauty. This participation is not a process in time. It is rather an atemporal process that leads to the appearance of a beautiful object in space and time.

One person who was deeply influenced by Plato in general and the *Timaeus* in particular was Heisenberg (1972, p. 8):

> "Quite often it happened that, after spending the whole night on guard in the telephone exchange, I was free for a day, and in order to catch up with my neglected school work I would retire to the roof of the Training College with a Greek school edition of Plato's Dialogues. There, lying in the wide gutter, and warmed by the rays of the early morning sun, I could pursue my studies in peace ..."

4   Whitehead's Philosophy and Quantum Mechanics

According to Heisenberg's interpretation of quantum mechanics, quantum states represent potentialities, some of which are actualized in acts of measurement. So, the quantum mechanical universe is an interplay of potentialities and actualities. A measurement brings about, through the process of the "collapse of a quantum state", what John Wheeler called an "elementary quantum event". Such an event is a "flash of actual existence"; as soon as it appears out of the background of potentialities, it disappears back into the background of potentialities.

This concept of elementary quantum events is remarkably similar to Whitehead's concept of actual entities: According to both quantum mechanics and Whitehead's philosophy, reality is not made of enduring objects, but of "flashes of existence" that disappear as soon as they appear. The main difference between actual entities and elementary quantum events is that the latter have no subjective aspect – they are completely objective. This is due to the fact that quantum mechanics, like all of our science, is based on what Schrödinger called the "Principle of Objectivation" (Schrödinger 1958, p. 37f):

> "By this [i.e., by the 'principle of objectivation'] I mean what is also frequently called the 'hypothesis of the real world' around us. I maintain that it amounts to a certain simplification which we adopt in

order to master the infinitely intricate problem of nature. Without being aware of it and without being rigorously systematic about it, we exclude the Subject of Cognizance from the domain of nature that we endeavor to understand. We step with our own person back into the part of an onlooker who does not belong to the world, which by this very procedure becomes an objective world."

Schrödinger's point is similar to what Whitehead called, in his book *Science and the Modern World* (Whitehead 1967, Chaps. III and IV), the "fallacy of misplaced concreteness": When I look at a tree, the statement "there is a tree out there" is an *abstraction*. The *concrete* fact is that "*I see* a tree out there."

Indeed, the first statement, "there is a tree out there," may not even be true. When I say it, I may be hallucinating. In contrast, the second statement, "*I see* a tree out there," is valid even if I am dreaming or hallucinating. Schrödinger's and Whitehead's point is that we are so used to mistake the abstract for the concrete, that the fundamental fact of experiencing is excluded from scientific analysis. With this exclusion the universe becomes, for science, a lifeless object. *The standard view of the universe as inanimate is a characteristic of the scientific method, not of the universe.*

Let us return now to the comparison between Whitehead's actual entities and Wheeler's elementary quantum events: Because of the limitations of Western science, i.e., its adherence to the principle of objectivation, there is no way for any subjective aspect of elementary quantum events – if they have such an aspect – to appear in their quantum mechanical description. We can summarize the situation by saying that *if we consider elementary quantum events as objectivized actual entities, the agreement between Whitehead and quantum mechanics is perfect.*

This conclusion opens the door to the following insight about the collapse of quantum states: *The collapse is an atemporal process, whose final outcome is the (momentary) appearance of an elementary quantum event in space-time.* It may well be, then, that we cannot discover a mechanism for collapse because there is no mechanism, no process *in time* that corresponds to a collapse.

5 The Potential and the Actual

The final piece we need in order to solve the puzzle of the delayed-choice experiments is an understanding of the nature of potentialities. *While actual events take place in space-time, potential events do not*; potential events only *refer* to space-time locations.

In Whitehead's system the only real items are actual entities. Potentialities are like thoughts in the mind of God (who is an actual entity). They are aspects of God that refer to space-time locations, but they do not happen in space-time.[1] In the context of the delayed-choice experiment, this means: Since the photon is an elementary quantum event only when measured and a field of potentialities at all other times, *nothing happens in space-time when it supposedly enters the apparatus.*

Finally, a comment about the role of the speed of light limit: The speed of light limit refers to the passage of *signals* from one actual entity to another, not to the relationship between different aspects of one process of transition from potential to actual. Transitions from the potential to the actual have a holistic character that *seems* like a superluminal connection. This is not in contradiction to special relativity, because it cannot be used to transmit information faster than light. This holistic character is a feature of Einstein-Podolsky-Rosen correlations.

6 A Whiteheadean View of Delayed-Choice Experiments

How does thinking about delayed-choice experiments in Whitehead's philosophy differ from the usual way of thinking about them?

The usual way of thinking: We envision a photon approaching the apparatus, and (loosely speaking) having to make a decision as it enters the apparatus. Will it make a choice between path (a) and path (b), or will it stay in a state of superposition of the two paths? If it stays in a superposition, it is hard to see how this can be changed later on, when the mounting of the mirror is changed while the photon is already inside the apparatus.

The Whiteheadian way of thinking: According to Heisenberg's interpretation of quantum mechanics, and according to Whitehead's understanding of the ontological status of potentialities, as long as no measurements are made, nothing is happening in space-time. There are only potentialities for events to happen, should a measurement be made, and potentialities do not happen in space-time. It follows that the statement "a photon is about to enter the apparatus, and this entails a decision between a choice of path and a superposition" does not correspond to anything in reality.

The place of decision is the place of measurement. If the mirror is mounted on a spring, and path (a) is chosen for the photon, an actual entity or an elementary quantum event representing a photon will be created there. The self-creation of this photon will have to accommodate all the previous actual entities including the properties of the mirrors in the apparatus.

Because the speed of light limit does not apply to the self-creation of a single photon event, a change of the mounting of the mirror along path (a) to a spring mounting leads to an instantaneous change in the state of the photon from a superposition of paths (a) and (b) to either a detection by

---

[1] Whitehead's concept of the actual entity "God" is fascinating; its discussion is, however, beyond the scope of this paper.

the vibration of the mirror (if path (a) is chosen) or a non-detection at the mirror (if path (b) is chosen). The puzzle that comes out of the usual mode of thinking simply does not arise.

7   The Significance of Quantum States

The holistic, seemingly "faster-than-light" feature will be demystified further by considering the following question: What is the significance of the abrupt change of the quantum state of the photon when the mounting of the mirror is changed? The answer to this question depends on the answer to a more general question: What is the significance of quantum states? Do they represent ontological information (i.e., information about how the system is) or epistemological information (i.e., information about our knowledge of the system)?

There are strong arguments against the ontological interpretation:

1. The collapse occurs on hypersurfaces with $t =$ const. Since these surfaces are observer dependent, whatever occurs on them cannot be an ontological description of the system.
2. Schrödinger's cat loses its mystery when the ontological interpretation is abandoned: The superposition of "cat alive" and "cat dead" then does not refer to the cat, but to the knowledge about it.

There are also strong arguments against the epistemological interpretation: It is too subjective. If someone makes mistakes in reading the results of a measurement, does the quantum state of the system change?

I suggest that the significance of quantum states is neither purely ontological nor purely epistemological. It is rather a synthesis of these two interpretations, namely: *A quantum state of a system at time t represents the knowledge that is available about the system at time t*, regardless of whether anyone actually knows it. Applying this to the delayed-choice experiment, the change in the mounting of the mirror brings about a change in the knowledge that is available about the photon, it says nothing about the photon itself.

8   Conclusions

Let us summarize:

1. The seemingly weird aspects of quantum mechanics, such as delayed-choice experiments, indicate the need for a paradigm shift. Whitehead's "Process Philosophy" is a strong candidate. Within its framework, with the addition of the understanding that the speed of light limit applies to propagation of signals and not to transitions from the potential to the actual, these seemingly weird aspects have simple, straightforward explanations.

2. *Space and time apply to the ordering of things and events in the actual world.* A potentiality is neither a thing nor an event, hence it does not exist in space-time. Therefore the process whereby an electron or a photon appears as an "elementary quantum event" in space-time is not temporal. Just like Plato's "becoming", it is an atemporal process whose final outcome is the appearance of an event in space-time.

3. The significance of quantum states: A quantum state of a system at time t represents the knowledge that is available about the system at time t, regardless of whether anyone actually knows it.

Acknowledgments: This work was supported by the Colgate Research Council.

References

Burgers J.M. (1963): The measuring process in quantum theory. *Rev. Mod. Phys.* **35**, 145–150.

Gilmore R. (1995): *Alice in Quantumland* (Springer, New York).

Heisenberg W. (1972): *Physics and Beyond* (Harper Torchbooks, New York).

Lowe V. (1951): Introduction to Alfred North Whitehead. In *Classic American Philosophers*, ed. by M.H. Fisch (Appelton-Century-Crofts, New York), 395–417.

Malin S. (1988): A Whiteheadian approach to Bell's correlations. *Found. Phys.* **18**, 1035–1044.

Schrödinger E. (1958): *Mind and Matter* (Cambridge University Press, Cambridge).

Shimony A. (1965): Quantum physics and the philosophy of Whitehead. In *Boston Studies in the Philosophy of Science, Vol. II*, ed. by R.S. Cohen and M.W. Wartofsky (Humanities Press, New York), 307–330.

Stapp H.P. (1979): Whiteheadian approach to quantum theory and the generalized Bell's theorem. *Found. Phys.* **9**, 1–25.

Wheeler J.A. (1978): The "past" and the double-slit "delayed-choice" experiment. In *Mathematical Foundations of Quantum Theory*, ed. by A.R. Marlow (Academic, New York), 9–48.

Whitehead A.N. (1967): *Science and the Modern World* (The Free Press, New York).

Whitehead A.N. (1979): *Process and Reality*, corrected edition, ed. by D.R. Griffin and D.W. Sherburne (The Free Press, New York and London).

The Deconstruction of Time and the Emergence of Temporality

Eva Ruhnau[1,2]

[1] Research Center Jülich, D–52425 Jülich, Germany
[2] Institut für Medizinische Psychologie, Ludwig-Maximilians-Universität,
   Goethestr. 31, D–80336 München, Germany

1   Introduction

Time is an enigmatic concept. Although seemingly nowhere to be found and
the least tangible of anything, time is the most basic and undeniable aspect
of experience. The list of words describing this enigma is comprehensive and
often confusing. Therefore, to begin with, I want to clarify my use of the
words Time, Temporality, and Now. By Time I do not mean the general
concept, but rather a very specific idea: the mathematical-physical concept,
linear-successive real-valued time measured by clocks and strongly tied to
facticity. Concerning the Now, I do not correlate Now with the present, as is
customarily done. Embedded in the paradigm of the Cartesian distinction between res extensa and res cogitans, the extended and the knowing substance,
the concepts of Time and Now usually reflect the Cartesian division between
matter and mind. In this paradigm, the Now occurs in physical time only as
a transition point between earlier-later, past and future, whereas in mental
time, the phenomenon of an experienced Now occurs, which is itself transient.
In this Cartesian view, the Now belongs to subjective consciousness and has
no place in objective physics.

   This peculiarity of the Now was the topic of a discussion between Einstein and Carnap. "Einstein said that the problem of the Now worried him
seriously. He explained that the experience of the Now means something special for man, something essentially different from the past and the future,
but that this important difference does not and cannot occur within physics.
That this experience cannot be grasped by science seemed to him a matter of
painful but inevitable resignation" (Carnap 1963). There are two important
points here. First, with respect to the Now Einstein referred to human experience. Second, he assumed that past and future are captured within physics.
This, however, would need a genuine derivation of temporal irreversibility
within the physical framework. Such a derivation does not exist (see Görnitz
et al. 1990), although several attempts have been made claiming such an
achievement. As already pointed out, I do not associate the Now with the
human mind. The Now is considered here as absolute nontemporality in the
sense that there exists neither succession nor a measure of duration. What

is associated with the human mind – though not exclusively with it – is Temporality.

Temporality encompasses earlier-later distinctions as a unified experience. The "atom" (the present), the elementary building block of Temporality, embraces immutability together with change; immutability because of the individuality, the closure, of the process itself, and change because of the earlier-later differentiation. Within the atom of Temporality, there is no subject-object separation. Such a separation occurs only retrospectively when experience has turned into observation. Observation (the explicit foundation of science) together with the structure of the present leads to the comprehension of past-present-future. Temporality is regarded as necessarily experiential and not – like in a Kantian sense – as a precondition of the possibility of experience. On the contrary, experience is seen as a precondition of the possibility of Temporality. In this way, the three concepts Time, Temporality, and Now are divided into two classes, one tied to experience, and one to abstraction. Time and Now are considered as purely abstract notions forming two poles with respect to a content-and-context-free iteration of Temporality. (To distinguish my definitions of these concepts from common usage, capital letters are used.)

This short introductory characterization is worked out in the following way: First, the concept of mathematical-physical Time and its dependence on facticity are discussed. Next, the Cartesian method as a program for the production of "facts" is discussed, and the limitations of this paradigm are demonstrated. This provides the basis for a deeper analysis and deconstruction of mathematical-physical Time. I choose a Derrida-like method of deconstruction and not, for example, a Hegelian analysis, because I do not want to contrast a thesis or a concept with its antithesis and produce a synthesis. The idea is rather to search for a concept which is "inter" two opposing concepts, uniting and differentiating them to create one new nondual concept from which the former ones derive as complementary.

To give a brief outline of deconstruction let me first introduce the distinction between ontic and epistemic arguments. The ontic realm is that of an abstract world. The objects of this world are, for example, the Kantian "thing in itself", Platonic ideas, Fregean referents, and metaphysical concepts like consciousness or Being. The epistemic realm is the realm of our knowledge of the ontic world. This knowledge depends on language – the universe of discourse we choose and have at our disposal. (Detailed descriptions of the ontic/epistemic distinction and its relevance to quantum theory can be found in Atmanspacher 1994, Primas 1983.) An object of the ontic world is termed in Derrida's language as "transcendental signified" or "presence". Derrida's principal thesis is a denial of "presence".

Deconstruction (Derrida 1978) proceeds as follows. Let us assume that we have a candidate for a "presence"; let us call it P. It is then shown that P draws its meaning from its opposite, non-P. This pair (P, non-P)

represents a generating dichotomy for the universe of discourse, the language we use. However, the two elements are in a hierarchical relation, P is dominant (privileged) since P creates the metaphysics. In the next step, the privilege is reversed; non-P becomes the dominant member of the pair. The final step is called "displacement". In their achieved equality-relation, the two concepts open up a space between them in which we find a new concept. This new concept unites and differentiates the pair, but is not reducible to either.

In the context of this paper, deconstruction is carried out in the following way. Deconstruction of the Cartesian paradigm with respect to Time leads to non-Time, i.e., a concept of the Now as nontemporal. The reversal, taking the Now as dominant provides the basis for displacement. In the open space between Time and Now, Temporality is located. Temporality unites and differentiates the two concepts and is neither reducible to Time nor to the Now.

2 Time

"Time might be taken as typical of the kind of stuff of which we imagine the physical world to be built. Physics has no direct concern with that feeling of "becoming" in our consciousness which we regard as inherently belonging to the nature of time, and it treats time merely as a symbol; but equally matter and all else that is in the physical world have been reduced to a shadowy symbolism" (Eddington 1929).

Physicists are usually not concerned with the meaning of the mathematical formulae and symbols they use. What is relevant is much less than meaning, it is an interpretative rule only, a mapping between observed phenomena and mathematical symbols. Basically, a physical theory is a formalism for the prediction of results of completed measurements. An attribute of a system defined by a description of an apparatus to measure it is called an observable. Predictions of the numerical results of observables are tested by the factual results of experiments. A completed measurement is the registration of the final outcome of a yes/no inquiry, not the superposition of possible results. The truth values of propositions referring to observables are determined by measurements.

The logic of measurement and observation is the classical two-valued logic. Therefore, the compatibility of scientific results with their classification by binary logic is not contingent because, in the realm of the exact sciences, facticity and two-valued logic are mutually dependent. To test the truth or falsehood of predictions, measurements have to be made. Measurements produce facts. Predictions are about possible future events. Facts are constituted in the present and are, retrospectively, described as past events with respect to instants which have already passed.

Mathematically, time is modeled as a (strictly) ordered set T. Referring to the elements of T as instants and interpreting their ordering as earlier-later relation leads to (physical) Time as a set of points in a linear-successive order. Furthermore, this set is considered to form a continuum – the continuum of the real numbers. This is based on the idea that between two Time points there is always a third Time point. However, in measuring temporal data we do not observe points of Time; what we observe are, for example, the factual positions of the hands of clocks. In other words, we measure time by observing facts. Therefore, taking the linear continuum as the mathematical model of time is adequate for the following definition of Time: Continuous Time is the abstract structure of unlimited observability.

3   Descartes' Method and Its Limitations

The progress and success of modern (classical) science with its enormous production of facts, the reducibility of experience to observability and facticity, is based on the Cartesian cut between a material and a mental world, between the object of knowledge and the knower. Descartes sets up four methodological principles (Descartes 1637/1969):

1. Never accept anything as true which is not evidently true.
2. Divide every problem under investigation into as many parts as possible and necessary to reach a solution easily.
3. Think in the adequate order, i.e., begin with the most simple things and achieve – step by step – knowledge of the composite.
4. Establish complete lists and general surveys to be sure not to forget anything.

In other words, Descartes' methodology is concerned with truth, reduction, hierarchical composition and completeness to do proper justice to facts. To predict knowledge about the objective world, one has to draw correct inferences from already known facts, i.e., classical logic has to be used.

But what is really true? Descartes is sceptical about everything until he grasps his unique and certain truth: Cogito, ergo sum. This is the first principle, the definite pillar of his philosophy. He then deduces two substances, res extensa and res cogitans. This substance dualism is the basis of the mind-body-problem, of the relation between mind and matter. In the following, I want to demonstrate briefly how Descartes' systematic approach – the basis of modern science and modern thinking – fails in its absoluteness.

Descartes' philosophy is not a single and unique event; it is the culmination of a long historical tradition. The question "what is true?" arises in this explicit form first in Athens. In this Greek polis, 2500 years ago, we encounter a historically unique situation, a factual conflict of interests in a density never experienced before. The interests of peasants, tradesmen, slaves, aristocracy – all these interests clash at one place. How could one differentiate among all these "universes", what is true? The order of nature – physis – is no longer sufficient when faced with this sudden accumulation of diversity. A human truth, a human law – nomos – has to be defined and set up. The sophists appear teaching rhetorics to supply partial truths with the appearance of the whole truth. Rules are needed to deal with these partial truths, to dispute them properly and to push them through. At this time, logic is developed, the theory of correct inferences with its three fundamental principles:

- the Law of Identity,
- the Law of Forbidden Contradiction,
- the Law of Excluded Middle (Tertium Non Datur).

As a consequence, the truth of common living and acting is disintegrated into the question of right and wrong within contexts, within universes of discourse which are defined and determined by needs, prejudices, interests, etc. In the course of history, it is then forgotten that there are many interests, many universes of discourse, many truths, and one seeks the one unique truth.

However, something proves to be a problem, and this something is "time". The proposition "it is raining tomorrow", is it true or false today? With respect to the Tertium Non Datur, it has to be either true or false, there is no third possibility. Aristotle, the inventor of logic, had already realized this critical point of his own system of logical rules. Propositions concerning the future do not come under the Tertium Non Datur. This is somehow a rent within the secure edifice of truth and logic.

Propositions concerning the future are neither true nor false. How can one deal with this uncertainty? If there exist rules and laws, then within the range of validity of these laws predictions about the future are possible. If we do not just encounter phenomena, but if we know how they are composed, how they are made up, i.e., if we know their regularities, we have a chance to gain control of the uncertainty about the future. Prediction and production are better and easier the simpler, the more reduced, something is. This is the crucial point to understand the great importance and effectiveness of the Cartesian program. These methods lead to an enormous production of factual knowledge and to the conviction that complete knowledge can be achieved. However, Gödel's incompleteness theorem should cast doubt upon such a belief in completeness. It leads to the consequence that a theory which is universal in the sense that a proof of its generating hypotheses can be carried out within the theory itself is not possible in a logically consistent way. Gödel's results are based on self-referentiality. The ban of self-referentiality like in the Russell-Whitehead approach (Whitehead and Russell 1910) leads to the unfolding of an infinite hierarchy. This can be demonstrated, for example, with the paradox of the knower.

It can be shown (Grim 1988) that a knowledge for which the following simple sentences are valid is logically inconsistent:

- If something is true, it is so.
- The sentence above is true.
- If B follows from A, and A is true, B is true as well.

A way out of this dilemma of logical inconsistency is to give up the demand of any coherent notion of omniscience. Necessarily, truth then becomes a relative concept, it is related to a specific universe of discourse. Such a universe of discourse is based on hypotheses being necessary to define this specific universe. However, the truth of the hypotheses cannot be derived within this same universe of discourse. For such a proof of truth another universe of discourse containing the former one is necessary. Thus, this opens an infinite regress; a whole hierarchy of universes of discourse and relative truths unfold. The quest for the one and only truth or the search of modern natural sciences for the single, unique, true, complete, grand unified theory leads into a void. Metaphorically, it seems that the different partial truths of the Greek polis have emerged again. However, they display themselves no longer in the original form as conflicts of interests, but as a well-orderd hierarchy of knowledge.

What remains so far from Descartes' program is the interplay of reduction and hierarchical composition having lost much of its attraction because of the loss of omniscience. In the following, I want to show that even the reduction of knowledge cannot be carried through rigorously.

Again, the problem is time. The familiar time of the clocks, the objective time of physics is expressed as a real-valued parameter, a continuous Time. States, events, objects, and perceptions are ordered successively along this temporal line. Such a temporal continuum can be divided into parts which are infinitely further divisible. However, what would happen if time in a dynamic discrete form were necessary for the existence, i.e., facticity of events, objects, percepts? If this were the case, limits to the reducibility to facts would be the consequence. These limits of reduction by the existence of discrete dynamic entities are most clearly seen in the study of the functioning brain. One has to consider the so-called binding problem, i.e., the linking of spatially distributed activities of the brain into one perceptual and action unit. Neurophysiological and psychophysical experiments and theoretical considerations suggest that the functioning brain creates discrete operating entities within two temporal domains (for more details see Pöppel 1988, 1996, and this volume; Ruhnau 1994, 1995):

- A high-frequency mechanism in the 30-msec range provides the basis for binding operations. Within such 30-msec windows, temporal relations between spatially distributed activities are undefined, i.e., these windows are internally atemporal zones. Only in external time are these discrete units represented as quanta of duration lasting approximately 30 msec.
- A low-frequency mechanism in the range of 3 sec exists within which atemporal zones of 30-msec external duration are automatically linked together. This pre-semantic process is the syntactic basis of the experienced present.

The brain itself creates a dynamic clock. To be more precise, one should probably say that the system "brain" is this clock – this temporal Gestalt.

Considered this way, the brain is a measuring rod; measuring itself leads to limits of temporal reducibility. The functioning brain cannot arbitrarily measure itself accurately. To be what it is (at least with respect to certain functional domains), atemporal windows are necessary, i.e., windows where the concept earlier-later is abandoned.

This argument may seem quite restricted and even inadequate because one could argue that a psychological, a subjective concept of time, is introduced. In contrast, the linear-successive objective Time does not contain any concept of an atemporal zone. It is a continuum being infinitely divisible, and its constituents are point-like. However, within the Cartesian program, the paradigm of reduction is not restricted only to physics, it applies to all scientific theories. In this view, biology, neurosciences, psychology, etc., have to be reduced to physics. Furthermore, physics as a theory has to be reduced to its basic constituents. In this universal claim, reductionism fails. Interestingly, there is an increasing accumulation of hints that the concept of Time is no longer adequate with respect to fundamental theoretical problems in physics itself. I dare to predict that even in physics concepts of atemporal zones in the sense of non-observability will turn out to be necessary (for more details see the next sections and Ruhnau and Pöppel 1991).

4   The Now

It has been demonstrated how linear-successive Time relies on facticity. To make an observation, a completed measurement is necessary. A completed measurement changes possibilities, i.e., future potential facts into actual facts. Facticity and the concept of "past" are mutually dependent. The measurement itself is considered to be point-like with respect to Time. (Concerning modifications of this last point of view, see considerations of the phenomenon of decoherence, for example, Zurek 1991.) A fact obeys the laws of classical logic. It keeps its identity in Time (because of the order relation, there are no temporal loops). As the definite value of an observable it does not contradict itself and is not a superposition of possible results.

In short, linear-successive Time is based on:

- the rules of classical logic,
- the point-like structure of the present,
- the distinction between past and future,
- facts.

However, the entries of this list are not independent of each other and their status has to be differentiated. Some of them are assumptions of the formalism (the rules of logic, point-like elements and an order relation); others are interpretations needed to complete the formalism providing a physical theory which can be tested (the distinction between past and future and the reduction of observability to facticity). The former are inherent to the formalism; the latter cannot be derived within the formalism.

Following the idea of deconstruction, reversal of the basis of Time leads to assumptions which form the basis of non-Time, i.e., the Now (in my terminology). Therefore, the Now is based on:

- the negation of the rules of classical logic,
- a non point-like structure, i.e. an extension,
- no distinction between past and future,
- possibilities.

Interestingly, such concepts of Now reveal themselves with great clarity in the realm of quantum physics, especially in connection with delayed-choice experiments, which were introduced as a generalization of EPR-experiments (see also Malin, this volume; Ruhnau and Pöppel 1991). Let us consider an electromagnetic (light) wave (the intensity of the light source can be as low as one single photon at a time), which is split by a half-silvered mirror into two beams of equal intensity. These beams are then reflected by two other mirrors to a crossing point. Now one can choose. Choice 1: two counters are located past the crossing point checking the route an arriving photon has come. Choice 2: another mirror is inserted at the crossing point creating destructive interference of the two beams on one side and constructive interference on the other. This results in a beam of equal intensity as the incoming beam, measured by a third counter. In the latter case, every photon entering the whole measuring apparatus is registered in this counter. Choice 1 exhibits the particle-like nature of light, choice 2 its wave-like nature.

However, the decision to choose version 1 or 2 of the experiment can be delayed until the object (photon/wave) has already passed the splitting point. Sticking to the usual space-time description, one runs into a problem. In choice 1, the particle picture seems to differentiate the way the quantum object has come; in choice 2, the wave picture seems to demonstrate that the quantum object has travelled both paths.

Therefore, the conclusion seems to be that present choice influences past dynamics. But this contradicts the concept of causality. Advocating causality, one is not able to provide states of both beams multiplied by some complex numbers. However, this does not imply that the quantum object travels both ways; it only states that it is in a superposition of the two states, being a superposition of possibilities, not of facts. Being registered with a measuring device (1 or 2), the system collapses from its superposition of possibilities into one definite factual state influenced by the mode of measurement the observer has chosen. The measurement defines the individual process (Bohr 1935). The observed phenomenon does not represent an independent reality. It may be more adequate to say it presents itself in the context of the experiment.

In this interpretation, there is no problem with causality; one is no longer allowed to visualize a definite space-time behavior of the quantum object. The disintegration of the outer temporal order within the system would generate a different individual process. The experiment demonstrates that there are situations in which space-time description and causality cannot be fulfilled together. Space-time description and causality are complementary aspects (Bohr 1958).

In the following, I interpret the delayed-choice experiment with respect to the reversal of Time. The observer chooses what to measure – particle or wave aspect. However, the concepts of "particle" or "wave" have an identical meaning only for the observer who keeps the identity of meaning in his mind. The cause of the fallacy of common sense is the extrapolation of the observation (particle or wave) at a certain time t back in time, ascribing the fact (identity) generated at t to the quantum object for instances earlier than t (as seen by the observer). One attached identity over time leads to a certain space-time behaviour. Another attached identity leads to another space-time behaviour. In delayed-choice, this leads to a conflict with causality. In this case, the problem of causality is a problem for the observer, not for the quantum object. Identity defined by the observer has no trans-successive meaning for the object and is not constitutive for the identity of the object. If the observer wants to see the fact "particle" in the course of (his) Time, continuous monitoring of the fact, i.e., continuous observation is necessary.

In this context, for the quantum system neither the Law of Identity, nor the Law of Contradicton, nor the Tertium Non Datur are fulfilled. The delayed-choice experiment also exhibits clearly the mutual dependence of these laws. (However, it is not the place here for more detailed investigation and interpretation of this interdependence.) Furthermore, the experiment reveals nonlocality, i.e., some kind of extension. It is a spatio-temporal nonlocality, not an exclusively temporal nonlocality. Isolation of temporal nonlocalities is a topic too detailed for this paper. One of the first papers dealing with temporal nonlocality explicitly is the paper by Paz and Mahler (1993). Temporal nonlocality is the absence of an order relation in the sense of an earlier-later interpretation. It is an immediate consequence of the absence of an earlier-later relation that makes a distinction between past and future impossible. It is worth mentioning that the attribution of possibilities instead of facts to the quantum system before the completion of a measurement brings with it the interpretation that possibilities are in a certain way "earlier" than facts. But a careful analysis shows that this being "earlier" is based on the establishment of Time itself. Only retrospectively, in Time, are possibilities earlier than facts.

5   Time and Now: The Substance-View of Reality

The Cartesian dualism, the division of reality into res extensa and res cogitans, is itself based on the substance-view of reality. In describing reality, we face the problem that linguistic categories are inherently dualistic. The modelling of reality requires at least two concepts, a conceptual pair to generate the language we use. One of the most fundamental – or even the most fundamental – of these generating dichotomies is the pair "permanence and change". To perceive change, something has to be invariant. Vice versa, permanence needs a background of change to be observed.

Within the conceptual realm of Western philosophy, the complementarity of permanence and change is turned into a hierarchy. Beginning with Parmenides, permanence is chosen as dominant. Change becomes a secondary concept to be considered as an illusion superimposed on the static factual world. In this substance-view, the dynamics of experienced reality is represented as the duality between an object (matter) that is observed and a subject (mind) that observes it. Experience as experience – in its genuine non-duality – can not be described adequately within this hierarchical reduction. Experience is objectified into the dualism of perceived object and perceiving subject. Time – neither belonging to the object nor to the subject – appears as a necessary background to objectify experience in this substantive-dualistic picture. Metaphorically, within time the concept of change escapes its fixation as a pure, timeless object.

However, this separation of time and objects makes time itself accessible for objectification. Again, time becomes objectified in a dualistic way. It is described as Time, the pure counting of successive (f)actual states, and as Now, a single instant within the stream of Time. Like space, Time is considered to be a container. But to be a container, something must be contained within it, i.e., objects. However, for objects to be "in" Time, they must be nontemporal. This bifurcation between Time and objects poses the problem of how nontemporal objects are ordered in a sequential, causal way. To put it differently, not the succession of perceptions and observations is the problem, but the perception of the succession. As a "solution" to this dilemma, one has to hypothesize something as nontemporal which, nevertheless, is responsible for the perception of succession. The objectified subject as something permanent and unchanging is constructed. Such a subject in its constructed nontemporal identity is able to produce trans-successive unity.

In this picture, the creation of unity is not connected to something temporal itself; it is the performance of a nontemporal subject immersed in an absolute linear Newtonian Time which flows smoothly, regardless of occurring event and of the existence of a unifying subject. In this view, there is no extended present – present is a durationless instant, the nontemporal Now. However, in the objectified reality of the natural sciences, there is no place for a subject. Therefore, in the subject-object substance dualism, the unifying power of the subject is delegated to the Now.

The dualistic splitting of time (as the result of the subordination of reality to objective categories of permanence) into Time as the counting of successive (f)actual states and the Now as unifying power can be traced back to Aristotle (Aristoteles 1987). In his books "Physics" we find the following famous passage: "For this is time: that which is counted in the movement which we encounter within the horizon of the earlier and later" (Physik IV, 220a). Furthermore, for him time and the now condition each other. There would be no time without now and no now without time. For him it is essential that through the now time is continuous as well as separated. The now or present separating past and future is a point. But is this point always the same, or are there many nows? If the present were always the same, there would be no time. If there were many presents, one present point would have to perish, but where? It cannot perish in itself. Therefore, Aristotle concludes that time cannot consist of present points. Instead, time is the relation of such points of presence. In its function to establish continuity of time, the now is itself a unity. In its function to constitute time via separating moments, it is itself a diversity. The now as the unity of a difference generates continuity and perceivability of time.

To repeat my thesis: The substance/permanence based reduction of reality leads to the subject-object duality, and, furthermore, to a dualistic description of substantialized time as

– Time: the counting of successive (f)actual states,
– the Now as a unifying power.

Again, problems arise if one of these dualistic aspects is taken to be universal. Taking the aspect of Time as universal, facticity has to be derived in order to achieve a universal theory. As can be seen most clearly in quantum theory, this is not the case without further assumptions (see Primas 1983). Accepting the Now as universal, a notion of completeness, of closure, necessary for the unity of objects should be derived without further assumptions. However, this notion of closure is either a designation from "outside" or the effect of an active principle "inside".

6   Unity in a Nontemporal World

Let me map this again to some philosophical considerations. Aristotle came to see things as a unity of two components, namely matter and form. Matter, in itself and pure, was entirely formless and inert; form had the character of an active principle. This active principle represented in form was responsible for the generation of unity.

Newton made a fundamental distinction between the phenomena and the "things themselves" and explained their relation as that of effect and cause. Phenomena are to be described, while the things themselves provide explantions for the phenomena. Science is concerned with the phenomena. In this way, Newton reduced the inherent active principle of Aristotle to a cause for the phenomena described by science. The inherent generating principle for unity was transformed into an external cause in the world of the objective. But this leads immediately to the question of how the unity of objects can be achieved? What makes objects distinct?

Leibniz found it incomprehensible how things which are distinguished only by number and are entirely identical in all other respects can be described at all. There must be a sufficient reason for making a distinction between things. He introduced two ontological principles: the Principle of the Identity of Indiscernibles, i.e., things are identical if no difference is discernible, and the Principle of Sufficient Reason, i.e., there must be a sufficient reason for things to be as they are in the distinction from other things as they are. Hence, each thing which is distinct from another thing must have an intrinsic difference with respect to that thing.

Leibniz generalized this into a universal principle of individuality. Every thing, the nonliving included, must have a distinct internal principle which makes it distinct from others. Organic things all have a life-principle which makes them subjects with perceptions. However, perception does not mean a representation of something from the outside. It denotes an internal drive which shows itself by the degree of activity of the subject calling forth its own perceptions when isolated from any outer influence. These substantial life-principles he called monads. Bodies or objects, in contradiction to monads or subjects, are seen to be nothing but composites of monades. Objects have no individual character and are considered to be phenomenal. They have extension because of the multiplicity of the monads they contain. Monads themselves are extensionless. Each object is an actual infinity of (point-)monads and is infinitely divisible (Leibniz 1714/1956).

Leibniz and, in a quite similar way, Whitehead (see Klose, this volume; Malin, this volume; Whitehead 1929/1978) both take an action principle seriously. Without such an "internal" action principle, unity can be conceived as a composition seen from "outside" only. In this context, it is interesting to consider the debate about the foundations of mathematics in this century. It is a debate between the discrete and the continuum, and a similar problem arises. This can be demonstrated by Russell's paradox.

Set theory as the most general foundation of analysis is, in a certain way, incomplete. Russell's paradox is about the set of all sets which do not contain themselves. The question whether this set belongs to itself or not leads to a paradox. Two points of view have to be distinguished here. One is to refer to a set; the other is to substantialize it, i.e., to make it available as an element for another set. It is not at all clear in what sense a collection of things is, indeed, a whole, a new entity. The theory of types, which was developed as a way out of this paradox, led to the idea that the notion of a set is incomplete. Something beyond the set itself is needed to determine whether the set referred to actually denotes a (meaningful) entity. Therefore, the logical problem is solved, but the problem of what defines an entity remains.

7   Temporality: The Dynamic Unity of Permanence and Change

As mentioned above, we learn from the working brain that time in a dynamic, discrete form is necessary to solve the binding problem, to achieve the unity of percepts. A clear distinction between "non-temporality" and "atemporality" has to be made here. With respect to the brain, the creation of zones which limit the perception of a before-after relation are important. However, in external Time, these discrete units are represented as quanta of duration of approximately 30 msec. Therefore, these zones are called atemporal to distinguish them from the non-temporal Now points.

Discreteness and unity are condensed and expressed by the hypothesis that the brain creates these atemporal zones and is structured by them. The atemporal zones form the logistic basis for binding operations – a kind of a syntax. Activations from different regions of the brain and from different sensory modalities are "collected" within these zones. Thus, via such elementary integration units, spatial segregation of functional representation can be overcome, thus solving the so-called "binding problem".

One could speculate that windows of atemporality in the observational sense are necessary preconditions of any kind of integration capacity to achieve changes which can be observed. To obtain earlier-later propositions, i.e., to get temporal propositions we have to observe change. In a Gedanken experiment called quantum Zeno effect (see, for example, Omnès 1992), it is possible to construct a quantum system which is extremely unstable, i.e., it has a high probability to decay. Now the system is observed. This can be accomplished with sophisticated laser techniques. The crucial point is to check within shorter and shorter Time intervals whether the decay has already happened. It can be deduced from quantum theory that a quantum system whose change is measured arbitrarily precise, i.e., a system which is under constant observation cannot decay. In such a situation, it is not possible for the system to undergo any change, it is frozen in Time. This indicates that atemporal windows within which no earlier-later relation can be measured or observed may be important even in physics.

With regard to the brain, it is assumed that the neurophysiological correlate of elementary integration units are synchronized relaxation oscillations of neurons based on re-entrant connections between cortical areas. However, if synchronized oscillations are postulated as a solution of the binding problem, the question of their functional effectivity arises. If we do not want to assume the existence of a homunculus in the brain observing synchronization, the mass activity generated by the synchronized activities of neurons has to be functionally effective. What could be an appropriate experimental paradigm to study the functional relevance of coherent oscillatory phenomena? In contrast to recent work on synchronized oscillations (Eckhorn and Reitboeck 1989; Engel et al., this volume; Gray and Singer 1989), all data supporting the proposed model of elementary integration units are mainly results of reaction-time studies (Pöppel 1988, 1994, and this volume; Pöppel et al. 1990). This suggests that the appropriate paradigm might be to take into account complete sensory-motor cycles. If coherent oscillations are functionally effective, this effectiveness should be relevant on the efferent, the output side. In the last decades (also as a result of the subject-object paradigm of experience), significant experimental and theoretical prejudices in favour of cognition and the sensory part of awareness have been manifest. However, this makes the homunculus unavoidable. In the context of the cognitive paradigm, the neuronal pattern of activity per se cannot reflect a holistic aspect, but necessarily and immediately leads to the question "holism for whom or what?". Correction of the cognitive prejudice and appropriate consideration of the motor aspect can solve this problem. The homunculus dissolves in sensory-motor integration. In brief: perception and motion constitute an inseparable unity.

At the phenomenological level, we usually observe (!) observation and action as separate. It seems that in the flow of action no observer can constitute itself. Observation has to interrupt, to break the flow to establish objects. However, turning attention away from objects and towards units of action could provide us with a different picture.

The atemporal zones of 30 msec external duration can be considered to form a necessary basis for the constitution of the perception of primordial events. Next, successive atemporal zones should be combined to achieve the transfer from the atemporal zones to the experienced present. Such a binding process is actually observed (Pöppel 1988). It is characterized by the fact that it automatically links atemporal windows together to intervals of approximately 2-3 seconds, forming the syntactic basis of the experienced present. The continuity of experience is the result of semantic linkages of these pre-semantic 3-sec islands of present. On the basis of these experimental results and hypotheses, I shall briefly outline a conceptual frame to describe the Temporality of experience.

The class of activity changes within one elementary integration unit (EIU) can be characterized as an equivalence class (in the mathematical sense). This equivalence relation is a formal expression of the partition of the set of activity changes into mutually exclusive EIUs. The perception of a stimulus is always correlated with a state of activity of the brain. To be more precise, it is correlated with quite a number of such states. In general, talking about the state of a system denotes the present condition of the system, i.e., states correspond to Time points. If a state lasts for a while, it corresponds to several Time points, i.e., to an interval. But this is the description in external Time. To distinguish this duration in external Time from the "internal extension" of elementary integration units, the concept of Gestalt can be introduced, based on the description of atemporal zones as equivalence classes. A concept of homogeneity (with respect to the atemporal zones) and a concept of concatenation (with respect to the islands of present) have to be defined mathematically. Also in the latter case, the linkage (to a chain) is essential for the functioning of the system, not the mapping onto a Time axis which is external to the system. In other words, a dynamic time is created by the system itself. The syntax necessary for the concatenation of such atomic events is given by the extension of the atemporal zone.

Usually the concept of Gestalt is related to spatially extended and delimited entities. The changing activities belonging to one equivalence class are extended with respect to external Time, delimited with respect to the before-after relation, and identified with respect to an internal dynamics. In a certain sense, these activities generate a temporal Gestalt. A temporal Gestalt defined in such a way differs from a permanent state in that it is not just a simple succession of Time points. The actual window as an entirety is constitutive for a temporal Gestalt. It is an interesting mathematical question, whether one can describe states starting with a process structure, i.e., not to presuppose Time points, not to use the topological space of the real numbers. On the basis of corresponding recent results (for details see Ruhnau 1996), one can define the transformation of processes to process germs as direct limits of a certain filter structure. It is possible to show that there is generally no bijection between the set of processes in a certain interval and the set of families of process germs. In other words, processes are different from changes of states.

The binding together of activities to classes of atemporality is an expression of the internal dynamics of the brain. The concept of homogeneity is connected to atemporality; the concept of concatenation is connected to the before-after relation. The linkage within the present of 3 seconds may be a necessary prerequisite for the experience of causality. But causal connections grasp more than a syntactic structure only. The meaning of the entities which are to be connected turns out to be essential. To establish meaning, a certain context is necessary. However, within the experienced present percepts, memories, feelings, and volitional acts form a unity; only retrospectively are they conceptually disconnected on the qualitatively operating level of reflexion, only retrospectively is this unity analyzed within a context – it is observed.

Brains are open systems embedded into a specific environment. Permanent observation or measurement would correspond to a continuous transposition of incoming (afferent) into outgoing (efferent) patterns of activity. A flow of motion which would yield the goal-oriented behavior of the organism as a totality as impossible. Therefore, the coordinated action of the whole organism requires zones free of observation or atemporal zones. In this case, observation is defined as functional effectivity of the appropriate neuronal coherences.

To summarize: The syntax of atemporal zones and their sequentiality within the experienced present are based on re-entrant structures and relaxation oscillations. The creation of unity is connected to a "value" system of the brain which consists of its own motor acts. The continuity of experience is a construct. In general, all this points to the following conclusion: We should not exclusively capture reality as facticity, Now, and Time, but also as dynamics creating temporal Gestalt and Temporality.

References

Aristoteles (1987): *Physik, Bücher I–IV* (Meiner, Hamburg).

Atmanspacher H. (1994): Is the ontic/epistemic-distinction sufficient to represent quantum systems exhaustively? In *Symposium on the Foundations of Modern Physics 1994*, ed. by K.V. Laurikainen, C. Montonen, and K. Sunnarborg (Editions Frontières, Gif-sur-Yvette), 15–32.

Bohr N. (1935): Can quantum mechanical description of physical reality be considered complete? *Phys. Rev.* **48**, 689–702.

Bohr N. (1958): Quantum physics and philosophy, causality and complementarity. In *Philosophy in the Mid-Century: A Survey*, ed. by R. Klibansky (La Nuova Italia Editrice, Florence), 308–314.

Carnap R. (1963): Autobiography. In *The Philosophy of Rudolf Carnap*, ed. by P.A. Schilpp (Open Court, La Salle), p. 37.

Derrida J. (1978): *Writing and Difference*. Routledge and Kegan Paul, London).

Descartes R. (1637/1969): *Discours de la méthode pour bien conduire sa raison, et chercher la verité dans les sciences (Von der Methode des richtigen Vernunftgebrauchs und der wissenschaftlichen Forschung)* (Meiner, Hamburg).

Eckhorn R. and Reitboeck H.J. (1989): Stimulus-specific synchronizations in cat visual cortex and their possible role in visual pattern recognition. In *Synergetics of Cognition*, ed. by H. Haken and M. Stadler (Springer, Berlin), 99–111.

Eddington A. (1929): *Science and the Unseen World* (London), p. 11.

Görnitz Th., Ruhnau E., and von Weizsäcker C.F. (1991): Temporal asymmetry as precondition of experience – the foundation of the arrow of time. *Int. J. Theor. Phys.* **31**, 37–46.

Gray C.M. and Singer W. (1989): Stimulus-specific neuronal oscillations in orientation columns of cat visual cortex. *Proc. Natl. Acad. Sci. USA* **86**, 1698–1702.

Grim P. (1988): Truth, omniscience, and the knower. *Phil. Stud.* **54**, 9–41.

Leibniz G.W. (1714/1956): *Monadologie* (Meiner, Hamburg).

Newton I. (1687): *Philosophiae Naturalis Principia Mathematica* (London), translated by A. Cajori (University of California Press, Berkeley 1934).

Omnès R. (1992): Consistent interpretations of quantum mechanics. *Rev. Mod. Phys.* **64**, 339–382.

Paz J.P. and Mahler G. (1993): Proposed test for temporal Bell inequalities. *Phys. Rev. Lett.* **71**, 3235–3239.

Pöppel E. (1988): *Mindworks: Time and Conscious Experience* (Harcourt Brace Jovanovich, New York).

Pöppel E. (1994): Temporal mechanisms in perception. *Int. Rev. Neurobiology* **37**, 123–129.

Pöppel E. (1996): Reconstruction of subjective time on the basis of hierarchically organized processing systems. In *Time, Internal Clocks, and Movement*, ed by M.A. Pastor and J. Artieda (Elsevier, Amsterdam), 165–185.

Pöppel E., Ruhnau E., Schill K., and von Steinbüchel N. (1990): A hypothesis concerning timing in the brain. In *Synergetics of Cognition*, ed. by H. Haken and M. Stadler (Springer, Berlin), 144–149.

Primas H. (1983): *Chemistry, Quantum Mechanics, and Reductionism* (Springer, Berlin).

Ruhnau E. (1994): The Now – a hidden window to dynamics. In *Inside Versus Outside*, ed. by H. Atmanspacher and G.J. Dalenoort (Springer, Berlin), 293–308.

Ruhnau E. (1995): Time Gestalt and the Observer. Reflexions on the tertium datur of consciousness. In *Conscious Experience*, ed. by T. Metzinger (Imprint Academic, Thorverton), 165–184.

Ruhnau E. (1996): Processes and states. In preparation.

Ruhnau E. and Pöppel E. (1991): Adirectional temporal zones in quantum physics and brain physiology. *Int. J. Theor. Phys.* **30**, 1083–1090.

Whitehead A.N. (1929/1978): *Process and Reality* (Free Press, New York).

Whitehead A.N. and Russell B. (1910–1913): *Principia Mathematica, Vol. I–III* (Cambridge University Press, Cambridge).

Zurek W.H. (1991): Decoherence and the transition from quantum to classical. *Physics Today*, October, 36–44.

Self-Reference and Time According to Spencer-Brown

Andreas Kull

Max-Planck-Institut für extraterrestrische Physik, D–85740 Garching, Germany

1   Introduction

Self-reference is ubiquitous. It occurs in any comprehensive description of the universe. It is found in neuroscience where the brain studies itself, in attempts to describe the biological characteristics of life (Maturana and Varela 1980), in mathematics when dealing with questions of provability (Gödel 1931), in sociology when describing the society from within (Luhmann 1992), and in the physics of quantum cosmology (Wheeler 1983). However, self-reference is irritating, too. It leads to paradoxes and antinomies, to propositions equivalent to their own negation. The difficulties related to self-reference seem to be rooted in language. Thus, despite its ubiquity, self-reference in everyday life is neither a topic of practical interest nor does it present a problem. But why is this the way it is?

An attempt to clarify the problem of self-reference on a fundamental level has been made by George Spencer-Brown (1969). In his book *Laws of Form*, Spencer-Brown presents a calculus dealing with self-reference without running into inevitable antinomies. The main feature of interest here is the occurrence of a tight connection of self-referential and temporal structures leading to the emergence of time out of self-reference.

2   Laws of Form

The basic idea of the *Laws of Form* is that any description presupposes a fundamental distinction which is found, e.g., in the difference between a description and its object, the described, as well as in the difference between the observer and the observed. The *Laws of Form* develop a formal system called the "calculus of indications", whose fundamentals are not only in agreement with but also are restricted to this presupposition. As a consequence, the calculus may be considered as underlying every attempt to describe any universe. As far as a universe is inaccessible without a description and every description presupposes a distinction,

"... a universe comes into being when a space is severed ...".[1]

---

[1] Spencer-Brown (1969), a note on the mathematical approach, first sentence.

In analogy, the development of the calculus starts with an injunction to make a distinction, the first distinction. The laws governing the form of this first as well as every other distinction, are the topic of the *Laws of Form*. These laws are based on a difference between the descriptive and injunctive aspect of indications as expressed in the two axioms (Spencer-Brown 1969, Chap. 1):

Axiom 1. The law of calling
The value of a call made again is the value of the call.
That is to say, if a name is called and then called again, the value indicated by the two calls taken together is the value indicated by one of them.
That is to say, for any name, to recall is to call.

Axiom 2. The law of crossing
The value of a crossing made again is not the value of the crossing.
That is to say, if it is intended to cross a boundary, and then it is intended to cross it again, the value indicated by the two intentions taken together is the value indicated by none of them.
That is to say, for any boundary, to recross is not to cross.

The attempt undertaken in the *Laws of Form* goes beyond the concepts of truth and logic. It is intended to reach out for the common ground of both, the logic and its object, the structure of any universe. Thus, it provides a foundation for a theory of structures in general – including self-referential structures by the capability of the calculus of indications to deal with self-reference. The manifestations of self-referential structures recognized within the calculus leads to a deeper insight into the problem of self-reference, whether related to logic, language, or to the observer himself.

3   The Calculus of Indications

3.1   Distinction and Indication

The starting point of the *calculus of indications*[2] is given by the *idea of distinction* and the *idea of indication*. Both concepts are related by *the form* of the distinction, which itself is considered as the fundamental structure. The form represents the distinction and its subject but without a reference to either side of the distinction. However, there is no factual distinction without an indication of one side of the distinction, as well as there are no factual sides without a distinction. Thus, the act of making a distinction involves both the idea of indication and the idea of distinction in a complementary way. When referring to *the first distinction* the reader is advised to make, the form becomes realized by the act of making this first distinction. The

---

[2] For exact definitions of terms printed in italic, c.f. Spencer-Brown (1969).

indicated part of the first distinction is known as the *marked state*, marked by *the mark* ⌐. In a schematic way, this may be represented as shown in Fig. 1.

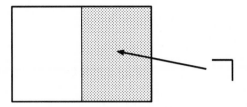

Fig. 1. Schematic representation of the first distinction, the marked state and its indication.

It is essential that the mark itself is considered as distinction. It has an outside and an inside as indicated in Fig. 2. The two-fold meaning of ⌐ allows a definition of a calculus without introducing the ordinary operator-operand distinction.

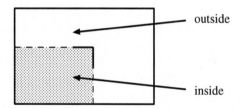

Fig. 2. The inside and the outside of the mark.

The *initials* of the calculus of indications are

$$\overline{\overline{}}\,\overline{} = \overline{} \quad , \tag{1}$$

$$\overline{\overline{}} = \quad , \tag{2}$$

where the blank   on the right-hand side of (2) is associated with the *unmarked state*. The initials (1) and (2) refer to the two-fold possibility of establishing or making a distinction by either a descriptive or an injunctive approach to indicate the marked state. The descriptive approach indicates the marked state by *calling* it by its name ⌐ while the injunctive approach indicates the marked state by interpreting ⌐ as an injunction to *cross* the boundary towards it. The initials (1) and (2) reflect, in a formal way, the two different aspects of repetitive applications of indications as expressed in the two axioms of the *Laws of Form*. The two-fold structure of the indication lies at the very heart of the calculus of indications. It is intended as a comprehensive criticism to sentence 7 of Wittgenstein's *Tractatus* (Wittgenstein 1984):

"Wovon man nicht sprechen kann, darüber muss man schweigen"

which may be interpreted to refer only to the descriptive aspect of indications.

3.2   The Primary Arithmetic

The *primary arithmetic* exclusively deals with the first distinction made by the reader following the injunction at the starting point of the development of the Calculus of Indications. The primary arithmetic is limited to the part of the calculus directly and finitely generated from the initials (1) and (2). As its numerical counterpart, the primary arithmetic deals with methods, means, and theorems of calculations. However, it does not refer to numbers but to the marked and unmarked state.

It follows from the initials (1) and (2) that every expression within the primary arithmetic is either equivalent to $\rceil$ or   . An example of an arithmetic calculation demonstrating the equivalence of a complex expression with $\rceil$ is

$$\overline{\overline{\rceil\;\overline{\rceil\,\rceil}}\;\rceil} = \overline{\rceil\;\overline{\rceil\,\rceil}}\;\rceil = \overline{\overline{\rceil\,\rceil}}\;\rceil = \rceil\;\rceil = \rceil\,.$$

(3)

The demonstration involves first (1), subsequently two times (2), and finally again (1). The two central theorems of the primary arithmetic are

$$\overline{\overline{p}\;\;p} = \qquad ,$$

(4)

$$\overline{\overline{pr}\;\;\overline{qr}} = \overline{\overline{p}\;\;\overline{q}}\;\;r \quad ,$$

(5)

where p, q and r represent expressions of known value either equivalent to $\rceil$ or   .

3.3   The Primary Algebra

The *primary algebra* considers the theorems (4) and (5) taken out of the context of the primary arithmetic as its initials. Within the algebra, p, q and r are variables indicating expressions of unknown value. Thus, the primary algebra is a calculus for the primary arithmetic studying the operations performed on $\rceil$ and   , i.e., the operations defined on the ground of the two-fold meaning of $\rceil$ as operator and operand by the initials of the calculus of indications (1) and (2). Although both the representations of the primary arithmetic and the primary algebra make use of the same mark $\daleth$, its interpretation changes with the context. The descriptive and injunctive aspect of the arithmetic $\daleth$ in the context of the primary algebra is reduced to the merely descriptive nature of the algebraic $\daleth$, indicating the injunctive aspect of the arithmetic $\daleth$.

The theorems of the primary algebra may be classified as either algebraic theorems with an arithmetic or without arithmetic counterpart. Theorems of the last type are purely algebraic. They involve expressions without arithmetic representation as, e.g., the infinite expression

$$\cdots \overline{\overline{\overline{}}} \, . \tag{6}$$

3.4   Re-Entry

The key feature of the primary algebra is the possibility of re-entering an algebraic expression in itself. The *re-entry* process explicitly reveals the self-referential structure of the calculus of indications and transforms infinite expressions like (6) into a finite form. For instance, expression (6) may be considered to be generated by re-entering it in itself according to

$$f1 \;=\; \overline{f1} \,. \tag{7}$$

$$=\; \cdots \overline{\overline{}} \,. \tag{8}$$

However, the same expression (6) may also result from

$$f2 \;=\; \overline{\overline{f2}} \tag{9}$$

$$=\; \cdots \overline{\overline{\overline{}}} \,, \tag{10}$$

and thus

$$f1 \,,\, f2 \;=\; \cdots \overline{\overline{\overline{}}} \,. \tag{11}$$

In the context of the primary algebra, (7) and (9) are equivalent. In the context of the primary arithmetic they are not. Arithmetically, (9) is fulfilled for $f2 = \daleth$, as well as for $f2 = \;$. But (7) is neither fulfilled for $f1 = \daleth$ nor for $f1 = \;$. As a consequence, (7) represents arithmetically an inequality. While (9) may be considered as a tautology, (7) may be considered as a paradox. Since (7) has no solution in the primary arithmetic, the connection of the primary algebra and the primary arithmetic is lost as a consequence of the re-entry process.

3.5   Self-Reference and Time

By the loss of connection of the primary algebra and the primary arithmetic, the calculus of indications becomes inconsistent. There are two common possibilities to resolve this situation. First, (9) could be excluded from the calculus. This, however, would result in an incompleteness. The second possibility is to include (9) as an additional axiom.[3]

Spencer-Brown's attempt to solve the problem considers the state described by (9) as an *imaginary state*, alternating between ⌐ and   . Thus, if there is no other state besides[4] ⌐ and   , and if (9) is considered to have a solution, this solution has to be regarded as invoking *time* by its oscillations.[5]

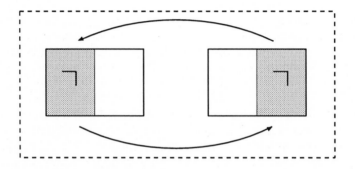

Fig. 3. Schematic representation of an imaginary state oscillating in time. This state is considered to be the solution of (9).

The primary algebra contains a variety of other self-referential expressions besides (6). By referring to the imaginary state they are all in close connection with the notion of time. It is interesting to note as a side remark that whenever potentially infinite expressions occur in finite form, antinomies are circumvented by the introduction of time. Self-reference, time, and finiteness thus seem to belong to one side, while infiniteness belongs to another.

To summarize, the self-reference introduced into the calculus by the re-entry process leads to the paradox of (9), canceling the connection of the primary algebra and the primary arithmetic. Restoring this connection without changing the domain of the primary arithmetic leads to the inclusion of time.

---

[3] This line of thinking has been explored, e.g., by Varela (1975).

[4] The term "besides" may be taken literally here. The only relation the calculus of indications introduced so far is the inside-outside relation associated with the mark or the marked-unmarked relation of the first distinction. Both are of the instantaneous, timeless type of side-by-side relations.

[5] Including oscillations brings up a relation of the type before-after besides the side-by-side relation.

4   The Emergence of Time from the Self-Reference of Distinctions

The fundamental starting point of the calculus of indications is the distinction inherent in any description. As a consequence, the calculus of indications may be considered as the essence of every description. The self-referential structures encountered within the calculus of indications thus correspond to the self-reference occurring ubiquitously in comprehensive attempts to describe, e.g., a universe including observers. The calculus of indications relates self-reference and time insofar as a consistent picture of self-reference restoring the connection between primary arithmetic and primary algebra cannot be conceived outside time. In analogy, any description involving self-referential structures invokes time.

However, this is only one side of the coin. The other side is encountered when contemplating that, to be self-consistent, the development of the calculus of indications must be based on the calculus itself. Taking the attempt of Spencer-Brown most seriously, we retrieve the observer, i.e., ourselves dealing with the calculus of indications, at the level of the calculus. All the attempts an observer makes to deal with the universe and even with its description take place in the context of the calculus of indications. Thus the structures found within the calculus of indications do not only comply with the structures of any description but they *are* these structures. As a consequence, the observer recovers itself within the calculus and from within the calculus. In this sense:

"We see now that the first distinction ... and the observer are ... identical."[6]

From this point of view the connection between self-reference and time takes on another quality. Time is not explicitly included in the calculus of indications. Its occurrence is related to the calculus in the same way other findings are, like the position of the observer. Self-reference gives rise to time, or time emerges out of self-reference. Equation (9) is seen as transcending its space-like context into temporality. Therefore, time appears as a necessary consequence of any description dealing with self-referential structures. In particular, time is a consequence of every attempt to provide a comprehensive description of the universe from within. Thus, time in this sense is not related to the universe itself but to the attempt to describe it. In other words, the more or less consistent day-to-day picture we have of ourselves as a part of the universe is necessarily temporal. Its temporality emerges out of considering ourselves as a part (the distinction) of the universe and our being conscious of this consideration (the self-reference).

---

[6] Spencer-Brown (1969), last chapter, last sentence.

5   Summary and Conclusions

At its arithmetic level, the calculus of indications deals with the first distinction. At the algebraic level, the arithmetic itself becomes the subject. Thus, the topic of the primary algebra is different from its arithmetic predecessor. Making these distinctions is accompanied by a shift of level. The arithmetic level is immediately related to the first distinction. The primary algebra, obtained by contemplating the arithmetics, leaves this basic level. Its topic continues to be the first distinction, mediated by the primary arithmetic.

As long as only finite, non self-referential algebraic expressions are considered, there is a direct representation of the primary arithmetic by the algebra. The loss of connection between primary arithmetic and primary algebra occurs by allowing algebraic expressions to be self-referential. As a consequence, introducing self-reference does not affect the immediate, arithmetic level of the calculus of indications but, roughly speaking, the way in which the primary arithmetic is contemplated. Compared to the common relation between algebraic and arithmetic structures, the relation between the primary arithmetic and the primary algebra of the calculus of indications is reversed. The primary arithmetic is indeed primary to the algebra. Changing the domain of the primary arithmetic to restore its connection to the algebra would, thus, result, in the context of the *Laws of Form*, in a change of the primary algebra. To restore the connection and to retrieve a consistent algebraic contemplation of the primary arithmetic that allows self-reference, the calculus of indications introduces time.

In much the same way as self-reference, time is linked to the primary algebra. Time is not mediated by the primary arithmetic. Thus, it is not to be found at the level of the first distinction. Rather it is, by its algebraic nature, related to the description of it. By extending the context, time appears to be related to the description of the universe but not to the universe itself. As a consequence, time then appears to emerge from the attempt to describe the universe from within.

One of the key features of the notion of time according to the *Laws of Form* is its algebraic nature. Since the primary algebra contemplates the injunctive aspect of the primary arithmetic, time, also not immediately related to the arithmetic level, reflects its operational aspect. Such a concept of time is clearly non-ontological. Time is seen to emerge from self-referent algebraic expressions contemplating the arithmetic, operational aspect of the calculus of indications. Accordingly, in its last consequence time originates from self-referential distinctions and indications.

There is a wide variety of conclusions applicable to systems including self-referential structures (Maturana and Varela 1980, Luhmann 1992, Wheeler 1983). First of all, time appears directly related to self-reference. Second, time may be considered to emerge from self-reference. The calculus of indications provides a strong hint that any description of a system excluding its observer does necessarily lead to an atemporal picture tending to antinomies. However, antinomies may be taken to indicate the implicit presence of time in a description of a system. In the context of quantum mechanics, this underlines the tight connection conjectured between the measurement problem and the time problem.

The calculus of indications is intended to provide a basis for any kind of description. Therefore, it is an interesting speculation that the problem of time as it appears in physical theories might become more transparent by reformulating the foundations of quantum mechanics in terms of the *Laws of Form*. However, as a word of caution, one should be aware that a possible connection between the calculus of indications and physical theories is far from obvious.

Another, more modest challenge is the comparison of other time concepts to that of the *Laws of Form*. Kant's concept of time (Kant 1989), e.g., is strictly non-ontological. It also possesses a kind of self-referential structure since time is considered to be the form of inner perception. Thus, the time concept of the *Laws of Form* shares, roughly speaking, at least two features with the Kantian concept of time.

Acknowledgments: The author thanks Prof. P. Reisinger for valuable discussions.

References

Gödel K. (1931): Über formal unentscheidbare Sätze der Principia Mathematica und verwandter Systeme. *Monatshefte für Mathematik und Physik* **38**, 173–198.

Kant I. (1989): *Kritik der reinen Vernunft* (Reclam, Stuttgart).

Luhmann N. (1992): *Die Wissenschaft der Gesellschaft* (Suhrkamp, Frankfurt).

Maturana H.R., Varela F.J. (1980): *Autopoesis and Cognition: The Realization of the Living* (Reidel, Dordrecht).

Spencer-Brown G. (1969): *Laws of Form* (George Allen and Unwin, London).

Varela F.J. (1975): A calculus for self-reference. *Int. J. General Systems* **2**, 5–24.

Wheeler J.A. (1983): Law without law. In *Quantum Theory and Measurement*, ed. by J.A. Wheeler and W.H. Zurek (Princeton University Press, Princeton), 182–213.

Wittgenstein L. (1984): *Tractatus logico-philosophicus* (Suhrkamp, Frankfurt).

Time and Information

Holger Lyre

Institute of Philosophy, Ruhr-Universität Bochum, D–44780 Bochum, Germany

Abstract. The idea of this paper is to propose distinguishability and temporality as basic principles of empirical knowledge. On the basis of these principles, the outline of an abstract theory of information (ATI) is developed. The assumption that time and information are always interwoven and the circular structure of a "complete concept of information" are the most important characteristics. The suggestion is made that an empirical science like physics must have the structure of an ATI. Several reasons are given why Weizsäcker's quantum theory of ur alternatives could be considered as the concrete program for the realization of such an ATI-like structure.

1 Introduction

The intention of this paper is to present an outline of an abstract theory of information (ATI), which is mainly based on the principles of distinguishability and temporality. These principles are motivated in Sect. 2. I should emphasize from the beginning that the proposed structure does not claim to be a closed systematic theory. It is not even a theory in the sense that it is presented in mathematical form. I try to present the theoretical framework of ATI as a clearly arranged verbalization of the intended structure (Sect. 3).

A further assumption is that the structure of ATI must essentially be described as an elementary circle of cognition. Knowledge is based on distinctions, but to give such a description of knowledge primary distinctions, on which the description itself is based, have to be presupposed. Thus, in Sect. 3 propositions are presented which contain terms like "circularity" and "self-applicability" as keywords.

Although I already mentioned that the present attempt to develop an ATI will be formulated verbally, there exists a preliminary mathematical framework for such an ATI in terms of the so-called *quantum theory of ur alternatives* (in short: *ur-theory*), based on the physical and philosophical ideas of Carl Friedrich von Weizsäcker (Weizsäcker 1985, Chaps. 8–10). As a physical program, ur-theory also uses the ideas of circularity and self-applicability and is not formulated in a rigorous axiomatic manner. I shall point out the conceptual essence of ur-theory, the concept of *information in time*, in Sect. 4.

2   Distinguishability and Temporality as Preconditions of Experience

Let us start with the following assertion: *Knowledge is based on the possibility of distinction.* This remark reflects the fact that *distinguishability* – the possibility of making distinctions – presumably represents a precondition to the cognitive abilities of any rational living being. Without the possibility of distinction, we would not be able to perceive, to form concepts, or to speak. We would not even be able to live. Therefore, distinguishability could be considered as a basic principle of knowledge. But this principle is so rudimentary, self-evident, and obvious that we usually don't even realize it. It seems to be a tacit presupposition to the possibility of knowledge.

Let us, next, consider science in general. *Science can be characterized by conceptual knowledge.* We presuppose the possibility of forming concepts when we perform science. Strictly speaking, science *is* conceptual. I will not assert that all existing knowledge is conceptual, but, in my opinion, scientific knowledge at least has to be conceptual – and concepts are based on the possibility of distinction.

Here, the domain of *empirical* science will be discussed. Therefore, in addition to distinguishability as a precondition of science in general, we have to look for the presuppositions of empirical science, i.e., for the preconditions of experience. As von Weizsäcker has pointed out, experience can be understood as *learning from the facts of the past for the possibilities of the future* (Weizsäcker 1985, S. 25). In that sense, the concept of experience presupposes the difference between the past, which is factual, and the future, which is potential (Weizsäcker, this volume). According to von Weizsäcker, this can be called *the structure of time* or *temporality.* Hence, temporality seems to be a precondition of empirical knowledge as distinguishability is a precondition of knowledge in general.

3   Outline of an Abstract Theory of Information

I shall use a transcendental starting point to form a systematic framework of some propositions and terminological introductions as a possible and abstract foundational basis of empirical science, in which *information* appears to be the central concept. The reason for this is that the "complete concept of information" – as I call it – is to be derived from the two principles of distinguishability and temporality alone. Thus, a rough outline of an abstract theory of information shall be presented. Here "abstract" means that information is neither to be understood as necessarily related to the existence of concrete energy or matter nor as an entity which is strictly independent of energy or matter. Moreover, information is neither to be understood as a purely subjective nor as a purely objective quantity, but as a general concept, which possibly covers anything which is, in principle, empirically knowable.

I start with distinguishability as the first of the two supposed preconditions of experience and introduce it as a basic principle of ATI.

Basic Principle 1 (Principle of Distinguishability)
Conceptual knowledge presupposes distinguishability. Thus, distinctions are possible as far as conceptual knowledge is possible.

To quantify distinguishability, a new term is introduced:

Terminological Introduction 1 (Binarity)
The simplest distinction possible, i.e., minimal distinguishability, is called a binarity.

Indeed, there always exists the possibility of considering a smallest kind of distinguishability. "Small" in this context means small in a conceptual or logical sense. Binarities refer to simple distinctions like yes/no, black/white, $+/-$, or true/false. It seems to be possible to have such a kind of conceptual or logical "atomism".
I now turn to the terminological introductions of subject and object:

Terminological Introduction 2 (Subject)
Subjects (S) make distinctions. Distinguishabilities exist for subjects.

Terminological Introduction 3 (Object)
Objects (O) are constituted by possible distinctions. Distinguishabilities exist on objects.

A binarity is characterized as the possibility of a distinction of S made on O. Moreover, the essential point is that S and O themselves can in principle be represented by distinguishabilities.
This leads to a first central proposition:

Proposition 1 (Self-Applicability of Distinguishability)
The principle of distinguishability is necessarily self-applicable.

What does "self-applicable" mean? One single binarity is hardly imaginable, because by stating one binarity several other binarities must have been stated before. This is primarily related to the fact that any binarity presupposes an indefinitely large number of other binarities.
    Now the considerations above have to be "translated" into an information-theoretical representation.[1] First, a *preliminary* introduction of information will be given.

Terminological Introduction 4 (Information)
Information is a measure of distinguishability. Its unit is called a bit. *One bit is the amount of information of one binarity.*

---

[1] Some of the following ideas are inspired by von Weizsäcker's considerations on the concept of information (Weizsäcker 1971, Chap. III,5; Weizsäcker 1985, Chap. 5).

This preliminary terminological introduction characterizes information as a measure of distinguishability, but in the light of the self-applicability of distinguishability (Proposition 1) the following aspects of information have to be distinguished:

Terminological Introduction 5 (Syntactics)
The syntactic aspect of information is related to the occurrence of distinguishabilities.

Terminological Introduction 6 (Semantics)
The semantic aspect of information is related to those distinguishabilities, which are a presupposition of other distinguishabilities and their meaning.

Terminological Introduction 7 (Semantic Levels)
Any number of distinguishabilities, i.e., any accummulation of bits, can, in principle, be called a semantic level.

Terminological Introduction 8 (Pragmatics)
The pragmatic aspect of information is related to those distinguishabilities which occur as a consequence of previous distinguishabilities.

It is important to emphasize that the three terminological introductions 5, 6, and 8 only describe certain *aspects* of information, but that they necessarily belong together regarding a *complete concept of information.*

Proposition 2 (Meaning)
The meaning of information has to be conceived as just that information, which occurs in connection with or as a consequence of information according to terminological introduction 4. Meaning presupposes the difference between semantic levels.

My assertion is that meaning can, in principle, be reduced to the level of distinguishabilities. Moreover, it can also be seen from Proposition 1 that there is a close connection between the semantic aspect of information and the pragmatic aspect. On the one hand, information presupposes information, i.e., the meaning of just a single bit presupposes certain semantic levels, which must again be described by information or, in other words, as an accumulation of bits. On the other hand, information leads to new information, i.e., the meaning of a single bit corresponds to the pragmatic effect of the bit on the creation of new information.

Proposition 3 (Non-Separability of Semantics and Pragmatics)
The semantic and the pragmatic aspect of information are non-separable, i.e., the necessary presupposition of semantics for the meaning of information must simultaneously be considered as its own pragmatic consequence.

Any information I_a presupposes information I_b, which represents the semantics. But the same is also valid for I_b. Let I_c be the information invested in the semantics of I_b; then the semantics of I_c will be I_d and so on. Suppose there is only a finite number of bits in the world – this describes an inherent circular structure.

Proposition 4 (Inherent Circularity of Information)
The concept of information refers to an inherent circular structure, i.e., each semantics presupposes other semantics.

I suppose that this is *the* elementary circle of cognition. Presumably, science is not possible in a less complicated and simultaneously non-contradictory way. In such a system, there will be essentially no "starting point".
Let us now introduce temporality as a second basic principle.

Basic Principle 2 (Principle of Temporality)
Empirical knowledge presupposes temporality. The difference between the past and the future is a precondition of experience.

The following proposition is a direct consequence of the two basic principles.

Proposition 5 (Distinguishability and Temporality Are Interwoven)
Distinguishability and temporality are always interwoven. Any temporal transition can be looked upon as a change of distinguishabilities. Distinctions which are made in the past lead to distinguishabilities in the future.

On the one hand, temporality always represents itself as a change of distinguishabilities or, in other words, of information in time. On the other hand, there is a clear difference between distinguishabilities (distinctions which are possible in the future) and distinctions (of the past). Moreover, there is a temporal quality as regards the semantic and the pragmatic aspects of information. The former is correlated to the pre-existence of bits, whereas the latter is correlated to their creation.

Now, an information-theoretical representation of temporality will be given.

Terminological Introduction 9 (Potential Information)
Distinguishabilities of the future are called potential information.

Terminological Introduction 10 (Actual Information)
Distinctions of the past are called actual information.

Terminological Introduction 11 (Flow of Information)
The flow of information has to be regarded as the transition of potential to actual information.

Thus, the flow of information gives a general representation not only of temporality but also of the fact that distinguishability and temporality are interwoven. This does also imply that information is an intrinsically dynamical concept.

Now, the so-called *complete concept of information* can be introduced.

Terminological Introduction 12 (Complete Concept of Information)
The abstract concept, which is given by the basic principles, terminological introductions, and propositions above will be called the complete concept of information.

Let us try to discuss a meta-theoretical point of view. According to ATI, information is based on preconditions of experience. As far as I can see, the consequence is that information can neither be understood as a purely subjective nor as a purely objective entity. Subject and object belong to the complete concept of information (Terminological Introductions 2, 3). Thus, subject and object are not separated a priori. At the same time, subject and object can, in principle, be described in terms of information itself – as semantic levels. This refers to an objective description of both, whereas their rigorous separation is, for fundamental reasons, not possible (Propositions 3, 4). Since information is based on distinguishability as a precondition of experience, i.e., because of the transcendental status of information, its clear characterization as a purely subjective or as a purely objective entity fails.

Proposition 6 (The Transcendental Status of Information)
Information can be considered as an entity beyond any rigorous separation of subject and object because it is based on the preconditions of experience.

4   Ur-Theory as a Quantum Theory of Information

I suggest now that quantum theory has to be considered as a general theory of information. This is reasonable because of the common interpretation of the quantum theoretical wavefunction ψ as a *catalogue of information*. According to quantum theory, ψ represents the complete physical description of an object. Anything which is in principle empirically knowable for an observer can be expressed in terms of information.

In the last decade, the special concept of *quantum information* has been developed. Charles Bennett and several other authors use it to describe new applications of quantum theory, such as quantum teleportation, quantum cryptography, and quantum computation (see Bennett (1995) for an overview). Instead of classical information theory, quantum information theory deals with *quantum bits* – sometimes called *qubits*. A qubit represents a two-dimensional state space – the simplest object in quantum theory. Of course, qubits have the same strange properties that all quantum objects have. First, they exist in superposition states with interferences. Second, in an n-qubit-state, the qubits can be entangled, i.e., they are not independent. Third, they are not "clonable", i.e., they can only be teleported, but cannot be copied or multiplied.

If we combine the two concepts of the ψ-function as a catalogue of information and of qubits as the simplest objects in quantum theory, we are fairly close to the starting point of von Weizsäcker's quantum theory of ur-alternatives, which was mentioned in the beginning. My suggestion is that ur-theory has to be considered as a rigorously developed quantum theory of information. The basic assumption of ur-theory, the ur-hypothesis, is that every physical object may be embedded in a tensor product of simple ur-alternatives (urs), which are, without this basic assumption, the same as qubits. Evidently, the ur-hypothesis is mathematically trivial, but it becomes physically plausible when we consider the essential symmetry group of urs, which is

$$SU(2) \otimes U(1) \sim \mathbf{S}^3 \times \mathbf{S}^1 \rightarrow \mathbf{S}^3 \times \mathbb{R}^+. \tag{1}$$

As, according to the ur-hypothesis, any empirical object must be build up from urs, the symmetry properties of urs have to be the symmetry properties of empirical objects, which are essentially the space-time-symmetry properties. Thus, the group $SU(2)$, conceived as a homogeneous space of the group itself, is in ur-theory understood as a model of our three-dimensional position space, i.e, $\mathbf{S}^3$ gives an approximation for a model of our cosmos. For that reason, ur-theory explains why we find all empirical objects in a three-dimensional space! Therefore, ur-alternatives refer primarily to spatial alternatives (Lyre 1995).

The topic of this paper is the information-theoretical aspect of the theory. According to the outline of ATI given above, ur-theory can be understood as a quantum theory of information only if it has the structure of ATI. To demonstrate this in the realm of ur-alternatives, we give the following terminological introduction:

Terminological Introduction 13 (Alternatives)
Alternatives can be regarded as logical representations of distinguishabilities.

In the light of an ATI, urs represent *1 bit of quantum theoretically treated potential information*. Thus, as quantum objects, urs have to be indistinguishable. As illustrated in the previous section, all information depends on the existence of prior information, as well as on the creation of new information, i.e., information is only information in the combination of its semantic and pragmatic aspect (Terminological Introductions 6, 8). A quantum theory of information must take this into account. If urs have to be considered as information in the complete sense, their indistinguishability must be dropped. This could presumably be achieved by Weizsäcker's concept of *multiple quantization* (Weizsäcker 1985). In this framework one starts with one ur, and quantization leads to a many-ur-theory. It turns out that urs obey a general parabose statistics. The parabose quantization decomposes the Hilbert space of n urs into all possible symmetry classes of the permutation group S_n. The effect of this kind of "class-generation" is like forming the simplest kind of semantic levels or, in other words, "concepts". At the next step of quantization, elementary particle states can be gained. This is the semantic level at which the standard model begins. By multiple quantization starting with one ur, the semantic hierarchy of evolution can in principle be described (Lyre 1996).

The preceding considerations mainly apply to the aspects of distinguishability (Basic Principle 1) and semantics (Terminological Introduction 6); we shall now consider the aspects of temporality (Proposition 2) and pragmatics (Terminological Introduction 8), keeping in mind that both are strongly connected (Propositions 3, 5). At the second level of multiple quantization, one gets a many-ur-theory, where the states of urs are created and annihilated, i.e., the number of urs always changes. Because of the ur-theoretical derivation of space-time structure from the spin-structure of the space of ur-spinors, even the spatio-temporal translations and rotations, i.e., the Poincaré transformations, are described by ur-operators. A Lorentz-boost, for instance, changes the number of urs (Görnitz et al. 1992). In this sense, there is always a flow of information – a pragmatic change.

The total number of urs in the universe, which can be estimated to $N \approx 10^{120}$, represents the maximum amount of potential information in the world – its "total potentiality". The number of urs is increasing, which means that the number of possibilities is growing. This can be regarded as the irreversible flow of information expressed by the second law of thermodynamics. However, N is presumably not a precise number because it is highly improbable that the universe is in an *eigenstate* of the number-operator of the total number of urs.

5   Concluding Remarks

To summarize: Based on distinguishability and temporality as preconditions of experience, an abstract theory of information (ATI) has been sketched. The central statements of ATI are:

1. Time and information are interwoven (Proposition 5).
2. The concept of information refers to an inherent circular structure (Proposition 4).
3. Information should be understood in terms of the complete concept of information (Terminological Introduction 12).
4. Information is neither a purely subjective nor a purely objective entity (Proposition 6).

I further suggested that *information in time* conceptually covers, at least in principle, the domain of empirical knowledge. Related to this, quantum theory can be understood as a general theory of information and ur-theory could be the consequent and concrete quantum theoretical realization of the structure of an ATI. Thus, position space is a consequence of information in time. Position space itself, therefore, is not a precondition of experience, as Immanuel Kant thought (Kant 1971), but it is the representation of information in time. Nevertheless, Kant's basic idea proves to be fruitful: empirical knowledge depends on the preconditions of experience.

Acknowledgments: I should like to thank Prof. M. Drieschner, Prof. Th. Görnitz, and Prof. C.F. von Weizsäcker for helpful discussions. I am grateful to *Wissen und Verantwortung – Verein zur Carl Friedrich von Weizsäcker-Stiftung e. V.* for financial support.

References

Bennett C.H. (1995): Quantum information and computation. *Physics Today*, October 1995, 24–30.

Görnitz T., Graudenz D., and von Weizsäcker C.F. (1992): Quantum field theory of binary alternatives. *Int. J. Theor. Phys.* **31**, 1929–1959.

Kant I. (1971): *Kritik der reinen Vernunft* (Meiner, Hamburg).

Lyre H. (1995): The quantum theory of ur objects as a theory of information. *Int. J. Theor. Phys.* **34**, 1541–1552.

Lyre H. (1996): Multiple quantization and the concept of information. *Int. J. Theor. Phys.* **35**, 2263–2269.

von Weizsäcker C.F. (1971): *Die Einheit der Natur* (Hanser, München).

von Weizsäcker C.F. (1985): *Aufbau der Physik* (Hanser, München).

Time – Empirical Mathematics – Quantum Theory

Carl Friedrich von Weizsäcker

Maximilianstrasse 14c, D–82319 Starnberg, Germany

1   The Question

Quantum theory has stayed essentially unchanged in its basic mathematical structure since Heisenberg, Schrödinger, Dirac (1925–1926), and its codification by John von Neumann (1932). Its success in describing and predicting experience is so far unprecedented in the history of physics. It comprises all known basic ranges of physics (with the possible, but perhaps not final exception of general relativity), and no experimental result has been found that would recognizably contradict it. However, its interpretation has remained controversial to the present day. Quantum theory is highly abstract, and concepts derived from earlier, concrete experience are evidently inadequate to interpret it.

How can we understand or even deduce that such an abstract theory should be strictly valid for all known experience? As a working hypothesis (not as a dogma!) I shall use the idea of Kant that a concept or a statement is valid for all experience if it poses a precondition for the possibility of experience. Can we find such preconditions? Here we must first ask the question: what do we mean by "experience"?

2   Time

Experience takes place in time. An experienced person is one who was learned something in the past that can be expected to be valid in the future. Basic for time is the progress of the present from the past into the future. Past events may be known as *facts*. We cannot change them. Future events are only known to us as *possibilities*; we can expect or even rationally try to predict them as consequences of past facts.

My working hypothesis is that quantum theory contains or even is an abstract mathematical formulation of the structure of time. Whatever we describe from experience will already presuppose the triple structure of facts, events, and possibilities. We shall first analyze the mathematics presupposed in physics in these terms, and then we shall proceed to the foundation of physics. How this approach is connected with the subjective elements in the experience of time will very briefly be considered in the final section.

3 Natural Numbers

1, 2, 3, 4, 5, 6, 7, ...

Here I have quoted the beginning of L.E.J. Brouwer's first treatise. For him, the original intuition of counting is more clear and more certain than any possible expression in language. The reader must be reminded of this before the author begins to speak (see Weizsäcker 1992, p. 111).

Counting is done in time. When we have, for example, reached number 7, then the earlier numbers from 1 until 6 are quotable facts. The succeeding numbers, however, are possibilities which are indicated by the points after 7. We can continue to count and thus translate further possibilities into quotable facts. The field of possibilities, however, remains open. We cannot explicitly name all numbers as facts.

In this way, the natural numbers represent the structure of time with factual past and possible future.

4 Groups

In the realm of natural numbers, we can define an *operation*, called *addition*. Beginning with a number m we can count through the succeeding n numbers. We thereby reach a number which we call $p = m + n$. This operation has some understandable properties; first of all commutativity: $m + n$ is the same number as $n + m$. We can perform the addition of any two natural numbers, and we will always reach another well-defined inatural number.

We can also define the "inverse"of addition, called *subtraction*. If $m + n = p$, then we write $m = p - n$. Starting from p we count through n, going backwards. Yet the operation $p - n$ will only lead us to another natural number m if p is larger than n. The operation $+n$ and $-n$ we call "integers" ("whole numbers"). "Whole numbers" are not an enlargement of natural numbers, but operations on natural numbers.[1]

A class of operations without their inverse we call a *semigroup*, a class with the inverse operations we call a *group*. The operations of a group can be applied subsequent to each other as often as we like. The integers form a group. Their operations in time can be so described: natural numbers, starting from one moment of time, go into the future; integers, however, go into the future as well as into the past.

Time, however, is continuous. When counting, there will always pass a period of time between the words which express two different numbers. We do not know empirically whether we might count with natural numbers as far as we like ("infinitely") into the future, or similarly with negative numbers into the past. This is why I have used the term "empirical mathematics" in the title of this paper. If mathematics is to describe the structure of time, we must ask what possible limitations we assume of time in past and future.

---

[1] Görnitz et al. 1992; see also Weizsäcker 1992, pp. 145–152.

5   Continuity

In describing time as continuous, we must ask how we understand a mathematical continuum.

Aristotle defined a continuum as something that can be infinitely divided into parts similar to itself, i.e,, into other (smaller) continua. Here the possibility of an operation called *division* is presupposed. A mathematical model of this operation can be found in the group of multiplications of natural numbers. If one adds the number m to itself n times, then we call this operation a *multiplication* $m \cdot n$. Again this operation is commutative: $m \cdot n = n \cdot m = q$. The inverse operation is called *division*. Choose any pair m and q; define $m \cdot n = q$ and $n = q/m$. In general, n will not be a natural (or whole) number, and we call it a fraction or a rational number.

Ancient Greek mathematics tried to apply this concept of division to geometric distances. If two finite straight lines, measured in a given unit of length, have the lengths m and n, their length ratio will be m/n. One of the most amazing and enlightening discoveries of Greek mathematics was the fact that there are length ratios which cannot be described by rational numbers. The classical example is the diagonal of a square. If the side of a square has the length a, the Pythagorean theorem gives the length d of its diagonal as $d^2 = 2a^2$. Therefore, $d = \sqrt{2}\,a$. $\sqrt{2}$ turns out not be a rational number; the equation $m/n = \sqrt{2}$ has no strict rational solution. The Greeks did not describe length ratios as "numbers". The concept of "rational numbers" is modern.

$\sqrt{2}$ is an "irrational number". In the 19th century it was discovered that irrational numbers cannot always be defined by algebraic operations. Sometimes an infinite series of rational numbers is required; e and π are examples. We call the entire set of rational numbers and irrational numbers *real numbers*. They are a subset of the set of "complex numbers", which contain square roots of negative numbers.

In analogy to the question whether we can count indefinitely in time, we must ask whether we can indefinitely divide a continuum in time. Aristotle discussed this question, starting from Zeno's paradox of Achilles and the tortoise. Achilles, who goes faster than the tortoise, will indeed reach the place where the tortoise was when he started to pursue it; but by then it will have reached a further place. This must happen in infinitely many moments of time until Achilles catches up with the tortoise. He will indeed have gone through a finite length in finite time. However, according to Aristotle, since every act of counting needs a finite period of time, one would have to use an infinite amount of time to count all these places and moments. Thus, continuity cannot be copied in a description of real human actions.

Mathematicians in the late 19th century (especially Cantor) tried to solve the problem by defining continuity not by human actions but as an infinite "class of points", thus presupposing the mathematical existence of really existing objects. But what does "mathematical existence" mean here? Russell's paradox already showed during Cantor's lifetime that this naive concept of a class leads to logical paradoxes. The present axiomatic theory of classes avoids those paradoxes of which we have become aware. According to Gödel, the strict proof of its non-contradictory nature cannot be given by a calculus in time.

6   Transition to Quantum Theory

I now dare to hypothesize that the structure of a continuum which is possible in real time is approached more closely by quantum theory than by traditional mathematics.

Planck's quantum hypothesis turned out to solve a problem which had appeared to be insoluble in classical physics under precise analysis – the thermodynamics of a continuum.[2] If the law of equal distribution is valid in thermodynamical equilibrium, i.e., all degrees of freedom have the same average value of energy, then a mechanical continuum, due to its infinite number of degrees of freedom, must have temperature zero if it is to have a finite total energy.

In final quantum theory, the continuous-state functions are probability amplitudes. That means that they describe possibilities in the future. The eigenvalues of operators which represent measurable quantities are discrete if the eigenfunctions are square-integrable; thus, the eigenvalues describe possible facts. We shall now ask whether this formalism can be justified by a consistent theory of representations of time.

7   Real Quantum Theory
as an Approximate Representation of Time

I shall not presuppose traditional quantum theory as basic. I shall, rather, try to justify it as a representation of time. In this way, complex wave functions will only constitute a secondary step, and a special case.

First, let us consider the concept of a *free process*. A process should describe the progress of time which can be measured by a clock. A process shall be called "free" if it is not causally influenced by another process. "Causality" is to be understood as an interaction of processes.

In a first approach the progress of time shall be naively described by the additive group of real numbers. The time variable is called t. A free process has an attribute (e.g., the position of the pointer of a clock), which we call x and which should be proportional to t:

$$x = vt, \tag{1}$$

where v is called the velocity of the process.

---

[2] Weizsäcker 1985, pp. 287–295: "The impossibility of classical physics".

Then we consider processes with interactions. The states of two such processes will be described by a real vector in two dimensions,

$$\mathbf{x} = \begin{pmatrix} x_1 \\ x_2 \end{pmatrix}. \tag{2}$$

This process with interaction shall also be described by a group with real numbers, now in a two-dimensional vector space. Here the group is described by products of matrices. If both processes were free, the generator of their group would be

$$g = \begin{pmatrix} v_1 & 0 \\ 0 & v_2 \end{pmatrix}, \tag{3}$$

and the group element at time t,

$$\mathbf{x}(t) = \begin{pmatrix} e^{v_1 t} & 0 \\ 0 & e^{v_2 t} \end{pmatrix} \begin{pmatrix} x_1^0 \\ x_2^0 \end{pmatrix} = \begin{pmatrix} x_1^0\, e^{v_1 t} \\ x_2^0\, e^{v_2 t} \end{pmatrix}. \tag{4}$$

In this multiplicative representation of groups the additive description (1) goes into the exponent.

The group has four linearly independent generators

$$a' = \begin{pmatrix} 1 & 0 \\ 0 & 1 \end{pmatrix},\ b' = \begin{pmatrix} 1 & 0 \\ 0 & -1 \end{pmatrix},\ c' = \begin{pmatrix} 0 & 1 \\ 1 & 0 \end{pmatrix},\ d' = \begin{pmatrix} 0 & 1 \\ -1 & 0 \end{pmatrix}. \tag{5}$$

We write the representations produced by these four generators as functions of time,

$$a(t) = \begin{pmatrix} e^t & 0 \\ 0 & e^t \end{pmatrix},\ b(t) = \begin{pmatrix} e^t & 0 \\ 0 & -e^t \end{pmatrix},\ c(t) = \begin{pmatrix} K & S \\ S & K \end{pmatrix},\ d(t) = \begin{pmatrix} k & s \\ -s & k \end{pmatrix}. \tag{6}$$

with

$$K = \cosh t,\ S = \sinh t,\ k = \cos t,\ s = \sin t. \tag{7}$$

In the space of these four generators, we can produce four other basic elements which are interesting for the interpretation:

$$e' = \frac{1}{2}(a' + b') = \begin{pmatrix} 1 & 0 \\ 0 & 0 \end{pmatrix},\ f' = \frac{1}{2}(a' - b') = \begin{pmatrix} 0 & 0 \\ 0 & 1 \end{pmatrix},$$

$$g' = \frac{1}{2}(c' + d') = \begin{pmatrix} 0 & 1 \\ 0 & 0 \end{pmatrix},\ h' = \frac{1}{2}(c' - d') = \begin{pmatrix} 0 & 0 \\ 1 & 0 \end{pmatrix}. \tag{8}$$

They give four representations

$$e(t) = \begin{pmatrix} e^t & 0 \\ 0 & 1 \end{pmatrix},\ f(t) = \begin{pmatrix} 1 & 0 \\ 0 & e^t \end{pmatrix},\ g(t) = \begin{pmatrix} 1 & t \\ 0 & 1 \end{pmatrix},\ h(t) = \begin{pmatrix} 1 & 0 \\ t & 1 \end{pmatrix}. \tag{9}$$

In $g(t)$ and $h(t)$ we do not find components t rather than e^t. This is because the matrices g' and h' are nilpotent,

$$g'^2 = h'^2 = 0. \tag{10}$$

Formally, $g(t)$ might, for example, describe Newton's law of inertia,

$$\mathbf{x} = g(t) = \begin{pmatrix} x_1 \\ x_2 \end{pmatrix} = \begin{pmatrix} x_0 + pt \\ p \end{pmatrix}, \tag{11}$$

where p would be the momentum and $x_0 + pt$ the position.

8   Complex Quantum Theory as a Representation of Objects

The group $d(t)$ in (6) is orthogonal. If we write the real two-vector as a complex number,

$$u = x_1 + ix_2, \tag{12}$$

then u will be transformed unitarily by $d(t)$. Thus $d(t)$ keeps invariant the norm of the real vector $\mathbf{x}$. In the language of the preceding section, this can be interpreted as an interaction of two processes: according to the matrix d', the process x_2 changes the process x_1, proportionally to itself, i.e., to x_2. If x_2 is positive, then x_1 becomes larger; if x_2 is negative, then x_1 becomes smaller. On the other hand, the process x_1 changes the process x_2 at the same time. These two counteracting influences produce a rotation of x_1 and x_2 through the possible values of cos and sin, the sum of their squares remaining constant. We now have a complex number with imaginary time dependence,

$$u = u_0 \, i\omega t. \tag{13}$$

In this manner, real quantum theory contains complex quantum theory as a submanifold.

Through the interaction of two processes which, if they were separate, would individually have to tend towards infinity, we obtain a process which returns into itself like a circular motion. This might be understood as the basis of the concept of an *object* which stays identical to itself.

We cannot exclude that the non-complex parts of real quantum theory also play a role in the dynamics of objects, or of the entire universe. H. Saller has pointed out that the interaction of elementary particles contains "ghosts" which cannot be normalized and can be represented by nilpotent operators (like g' or h').

9   Finite Alternatives

During the past years we have constructed an abstract quantum theory by postulates for empirically decidable alternatives.[3] Let A_n describe n mutually exclusive states or n statements about those states. "Mutually exclusive" here

---

[3] Weizsäcker 1985, Chapter 8; see also Drieschner (1970), Drieschner (1979), and Drieschner et al. (1987).

means two things: (1) if one of the statements is true, then all other $(n-1)$ statements are false; (2) if $(n-1)$ statements are false, then the remaining one is true. According to one of our postulates, there exist possible states or statements such that, when the statement is true, each of the n statements in the alternative has a well-defined precise probability, and the sum of these probabilities is equal to one.

This postulate can be formally deduced from the complex quantum theory of objects (cf. Sect. 8) if in (8) we consider $|u_0|^2$ as the probability of the state. A_n comprises n such states, and when all of them are constant, A_n describes a stationary quantum state of the object. If, for example, the object is an angular momentum $j \cdot h$ then $n = 2j + 1$; for $j = 0$ we have $n = 1$, for $j = 1/2$ we have $n = 2$, etc. The number of real dimensions in which A_n can be described is $2n$. It is possible to introduce time-dependent states in this formalism.

10 Binary Original Alternatives and the Space-Time Continuum

Every finite alternative A_n can be decided by a finite number of yes/no decisions. The complex vector space of an A_n can be described by tensor products of two-dimensional vector spaces. These two-dimensional spaces have the structure of the symmetry group $U(2)$ if unitary, and $GL(2,C)$ if linear. I call such a binary alternative an original alternative or, in German, "Ur-Alternative", abbreviated "Ur". I also use the term "ur" in English.

In an earlier paper (Weizsäcker et al. 1958), it was shown in detail that the laws of free states of objects, if they obey the general abstract complex quantum theory, can always be mathematically described in the $(3+1)$–dimensional space-time of special relativity. The general unitary group $U(n)$ in n complex dimensions can always be described as the product of a sufficient number of $U(2)$ groups. But $SU(2)$ is (with one ambiguity) isomorphic to the real, three-dimensional rotation group $SO(3)$, and $SL(2,C)$ to $SO(3,1)$, the Lorenz group. This is described in detail elsewhere.[4]

As stated in Sect. 7, we have so far described the progress of time and the group parameters naively by the additive group of real numbers. (This applies to all papers I have quoted so far). This might, however, be too simple for a final approach to continuity.

11 Free Particles

The simplest real object might be a free particle. How do we have to describe it mathematically?

---

[4] Weizsäcker 1985, Chaps. 9 and 10 (especially 9.3); Castell 1975.

As a basis for its state space we may choose the eigenstates of its momentum operator. The mathematical problems which we described earlier appear here also. A momentum state in an infinite position space, hence also in Minkowski space, is not square-integrable. We are, however, not obliged to describe the space-time-continuum as a Minkowski space. If we presume that particles consist of "urs", we must first distinguish "urs" and "anti-urs", with time-dependence according to $e^{-i\omega t}$ and $e^{i\omega t}$ for a free state.

Using products of production and annihilation operators of urs and anti-urs, we can then produce the group $SU(2,2)$ or $SO(4,2)$.[5] This is the conformal group of special relativity which contains as subgroups the Poincaré group of Minkowski space, as well as the group $SO(4,1)$ of de Sitter space with an expanding spherical position space and the group $SO(3,2)$ of "anti-de Sitter space" with a hyperbolic position space and cyclic time. With other operators, we can produce other world models. Thus, an expanding spherical Einstein universe does always contain a finite, but steadily increasing number of urs at any time.

T. Görnitz and I have made detailed calculations for free momentum states in a Minkowski space with a Poincaré group. Such a state is an infinite sum of terms, each of which contains $2N$ or (with spin) $2N + 1$ urs as factors, and $0 \leq N < \infty$. The Poincaré group implies that every well-defined particle has a fixed value of spin and rest mass, independent of the momentum value. The numerical value of every summand depends on the momentum. For $N \to \infty$ it goes to zero, but the sum diverges. The factor is maximal close to the momentum value of the state if the momentum of a single ur is defined as one. For experimentally observed momenta of actual particles, we can – without profound contradictions between theory and experiment – change these factors such that the sum converges. In an expanding Einstein universe, momentum states strictly adhere to this property. They contain a finite, but time-dependent number of urs.

12   Interaction

The work reviewed here aims at a theory of the interactions, and hence of the systematics of elementary particles. We have not yet arrived there but I shall describe the idea. As an example, I choose the quantum theory of the hydrogen atom.

Naively one says that the (light) hydrogen atom consists of two elementary particles: proton and electron. Less naively, we must necessarily add the electromagnetic field. In the ground state of the atom this is just the Coulomb field. If the atom changes its state within its discrete spectrum, photons are emitted or absorbed. Concerning the continuous part of the spectrum, an indefinite number of photons participates in these processes. Therefore, one

---

[5] Castell 1975; Weizsäcker 1985, p. 407.

can say that the hydrogen atom consists of one proton, one electron, and the electromagnetic field.

Moreover, at some distance from the atom into which the photons go or from which they come, the real world also contains other atoms. The idea of a "free" atom is always only an approximation. And the "rest of the world" is not only spatially extended. The protons consist of quarks. The quark-gluon field is confined on the outside, but it is hugely complex within small distances.

If we try to describe collisions between "free" particles, a strict calculation would finally lead to an incalculable theory of the universe. On the other hand, in strong collisions the individuality of the colliding partners is not always conserved; Hagedorn calls such processes a "fire ball". Today we possess very good approximations, e.g., quantum electrodynamics. Here, too, we still make assumptions about the dynamics and interactions of particles and fields which come from the empirical classical field theory. The question is whether we might be able to deduce these assumptions from fundamental abstract quantum theory.

A plausible way toward such a goal would be to describe all particles and fields as composed of urs. But the ur is the unit of information in relation to the entire universe. Present estimates of the radius of the universe imply (Weizsäcker 1985, p. 474) that one might think of a proton as composed of something like 10^{40} urs. From so many bits one could construct an immense complexity of different combined states of many particles. Here the first question is why there exist particles at all, i.e., arrangements of many urs in identical states.

In the "theory of urs" this question has not yet been resolved. I have some unfinished presumptions which I shall not discuss here. The remaining sections of this paper deal with problems which must be clarified before we can approach this question.

13 Quantum Theory of the Continuum

We can now attempt to correct the preceding considerations. I have so far described the continuum "naively" by the so-called real numbers. Let us reconsider the structure of time.

We can know the past as far as we know *facts* from it. Describable facts can be *counted*. Past facts known to us are a discrete manifold. For the future, we can only estimate *possibilities*. In classical physics and in traditional quantum theory we describe them by *probabilities* or even by *complex functions*. These concepts are known as *continuous*. However, the *present* only consists of discrete *events* out of such a continuum, events which later remain known as facts. How does our prognostic vision of the future connect these possible discrete events with the continuum, with which we conventionally describe possibilities?

Here I refer to earlier results in the history of quantum field theory. Schrödinger's wave function ψ was interpreted by Born and Heisenberg using $|\psi|^2$ as a probability density for the classical observable (e.g., position or momentum) on which ψ was assumed to depend in the chosen representation. The wave provides a continuous measure of possibility for discrete facts. In classical physics, position and momentum were also continuous variables. In quantum theory, one had to describe them as matrices, i.e., as operators acting on ψ, so that the classical orbit should itself not appear as a continuum of facts (cf. Sect. 6 on the origin of quantum theory). Consequently, a dilemma was created when one tried to apply this way of thinking to the electromagnetic field. One the one hand, electric and magnetic field strengths were considered to be classically measurable. If we wanted to apply quantum theory to them, we had to describe them as operators, themselves acting in a Hilbert space (which, in first approximation, was non-separable). In consistent quantum electrodynamics one had, on the other hand, to describe current and density of matter equally as fields (as functions of ψ), whose interaction with the electromagnetic field made it necessary to introduce ψ as another operator in a higher level Hilbert space. This procedure was called "second quantization".

This term, however, appeared arbitrary or paradoxical, especially in the Copenhagen effort to clearly understand the relationship between quantum theory and classical physics. When I was Heisenberg's student, he forbade me to use this term. But the procedure was successful in quantum field theory. Later I tried to interpret it as a consistent application of the probability concept (Weizsäcker 1985, pp. 306–309). I defined probability as the expectation value of the relative frequency of events in an ensemble. So one could meaningfully define an ensemble of ensembles: relative frequency of relative frequencies. In the present paper I have discussed the procedure in steps by defining rational numbers through division (Sect. 5). "Real numbers" are only ideal limiting cases of such a procedure of division.

This consideration leads to the concept of "multiple quantization". In quantum field theory, the stages which separate the steps of quantization are primarily defined by the classical laws in the first stage. In a systematic construction of the interaction theory, they would have to follow from the still incomplete attempt to relate urs to particles.

I conclude this section by looking back into the history of geometry in the field of mathematics. Since the Greeks, we have possessed the concept of Euclidean three-dimensional space. Since the 19th century, we have perceived the possibility of many different geometries. In the present attempt to reconstruct quantum theory we apply different geometries, depending on the states of the objects, and formalised in the theory of urs. I have already indicated the corresponding different cosmological models. But the free choice in the distribution of urs will also admit different geometries in the limit of small distances. In this sense, I speak of "empirical mathematics".

14   Events

So far I have described quantum theoretical models only as continuous representations of a continuous progression of time. However, traditional quantum theory also presents completely different features. The probability of an event leads to a fact with a ψ-function that is completely different from the ψ out of which the probability was calculated. One occasionally speaks of a *collapse* of the wave function to indicate this situation. We know mathematically that in an immediate repetition of a measurement the probability of an identical result will be extremely close to one. In this sense, the created fact is irreversible.

This connection of the probability of events with the absolute square of the wave function is traditionally understood as the *indeterminism* of quantum theory – the violation of a strict law of causality. But this poses a new problem: *When do events happen?* An answer which has convinced me is: only when, according to the time-dependence of the wave function, an irreversible event must happen. Twentyfour years ago I developed a formulation of this answer which I called the "Trieste theory", due to the place where it was presented first. It is connected with the theory of the *measuring process*.

The traditional statement is: when we measure a quantity x of an object, then the probability of finding precisely a value x' is $|\psi(x')|^2$. Here an irreversible process occurs in the measuring instrument: a *fact* is established. Some interpreters of quantum theory concluded that the state of the object is a consequence of an act of our consciousness. This would pose very difficult philosophical questions, which I shall touch upon in the following section. Niels Bohr advocated the thesis that for every measuring process a description of the result by classical physics must be possible. Then, he argued, the measured result is objectively present, independent of whether anybody perceived it immediately or even whether anybody at any time was or will be aware of it. "Classical" here probably means only: not affected by the quantum theoretical uncertainty of the measured result. This then means: events take place objectively, independent of whether they are observed or not. They must have a property which is identical for any observer. In other words: they always must have been created as an *irreversible* event. To interpret this concept more precisely, I would have to go into more detail in statistical thermodynamics.

In the "Trieste theory", I made the further statement: as long as no irreversible process happens, one might maintain the continuous evolution of any event without a chance of being refuted, and, therefore, without a chance of defending it. In the present paper I do not focus on this issue. But the question remains: which conditions must be fulfilled for a process to be irreversible? I first give two reconcilable examples.

1. A. Schlüter suggests the following model: let a free particle have two possible states, a ground state and a metastable excited state. When the particle

relaxes from the excited state to the ground state, a photon is emitted. Since the particle is free, the photon will disappear and never return. This is, of course, only a very probable result, because there are no completely free particles, i.e., particles without a surrounding universe which contains objects. Irreversibility is always only a highly probable feature – never absolutely certain.

2. In Sect. 7, we defined a state

$$\mathbf{x} = \mathbf{x}_0 \, e^{-\alpha t} \tag{14}$$

where $\mathbf{x}$ monotonically decreases toward zero.

Both models describe something like a free object which, however, can only arise through a preceding interaction. If we consistently describe this coexistence of interacting objects and their subsequent separation (which can happen by force-free motion), and, finally, the irreversible "relaxation", then there is no special moment in time at which the "event" comes into existence. A "moment of occurence of the event" can only be seen macroscopically.

What, then, is the origin of the unpredictability of an event, of the probabilistic indeterminism? Omnès (1994) made me aware of an interpretation which has been the object of lively discussion for nearly two decades. Observable objects generally consist of many atoms. And even single atoms move under the influence of a complicated surrounding. Yet the Schrödinger equation by which we describe the object contains only a few observable quantities. However, the "background" or "environment" of unobserved quantities can strongly influence the measurable quantities. Therefore, even with a motion which has been observed for a time, the unobserved motion of the environment can substantially influence the interaction with the measuring instrument. This is compatible with the assumption that the temporal development of all quantities is fully determined by an initial state which is never completely known to us. The remaining ignorance is sufficient for making the measured result unpredictable.

If we introduce this consideration into the theory of urs, we expect this unpredictability even for free particles. Hence, one might accept the idea that the "collapse" of the known wave function is only a superficial effect of the unknown structure of the complete wave function. This might, if at all, only be testable by the predictions of a consistent theory of interaction.

15   Self-Knowledge

In this last section I can only roughly indicate the basic philosophical problems which, in my opinion, underly the unresolved debate about the interpretation of quantum theory. Niels Bohr saw a fundamental difference between classical physics and quantum theory in the fact that classical physics describes pure objects while, in his view, subject and object are not separated in quantum theory.

The quantum state defines the probabilities of possible results of measurement. A probability is an aspect of human knowledge. If ψ is changed deterministically without any measurement (according to Schrödinger's equation), but indeterministically by measurement, then the action of the measuring subject is an essential part of the natural evolution of the object.

I admit that, at the end of the preceding section, I refered to the possibility that the process, if it could be completely described according to quantum theory, might in fact be deterministic. Indeterminism might only be the consequence of our lack of knowledge of initial conditions. I have, however, attempted to describe quantum theory and the corresponding mathematics as a representation of time and to describe time as the basic precondition of experience. The difference between past facts and future possibilities corresponds to the possibility that subjects can acquire knowledge. This becomes evident in the theory of the continuum (Sect. 13). We cannot avoid the dual question, what is the role of physics as subjective knowledge of nature, and what is the role of subjects as parts of nature.

Here I want to emphasize the historical and philosophical role of the quite modern distinction between the concepts of subject and object. Aristotle defined physics as the doctrine concerning those motions which have their cause in the essence of the moved object. A stone falls and a flame ascends because both are drawn toward their natural destination: stones to the earth, the flame to the sky. Living organisms move similarly, and by similar reasons they produce offspring. Aristotle refered to these moving principles as psyche, or soul. For him they represented the essence and the cause of motion of what we today call subjective.

The modern distinction of subject and object was postulated and discussed most strikingly be Descartes. He distinguished *res extensa*, the extended substance which obeys geometry, hence mathematics (by which it can be understood in its causalities), from *res cogitans*, the thinking substance which knows itself and can conceive the mathematics, thus explaining the causalities of the extended substance. In the discussion of this distinction, Thomas Hobbes wrote to Descartes that he could not see why an extended substance should not be able to think. Descartes interpreted this objection as an escape into doubt. He searched for a certainty which would not admit doubt. A detailed discussion of his arguments would exceed the scope this paper.

Since Newton, classical physics has presupposed that there is an objective space in which bodies (and, in later versions, also fields) can be moved. These theories encouraged the development of classical mathematics of the continuum and its interpretation as an a priori truth. This view has convinced me of the thermodynamic impossibility of classical field physics in quantum theory, and of the problematic classical concept of the continuum in the paradoxes of fundamental mathematics. Therefore I have attmepted to formulate an abstract quantum theory with countable alternatives and a group theoretical concept of probability. The next question was to which realities we might apply this abstract theory. Countable alternatives and their connection by probabilities can be found for bodies in space. However, they may equally be found for psychic processes in self-perception and in the perception of other human beings or of other living beings which we consider to be comprehensible. If quantum theory is a representation of time, we should not be surprised by this consequence: we know objects and subjects in time.

I emphasized above that single free objects or alternatives do not strictly exist. An object with absolutely no interaction could not be perceived by us, hence would not appear in an empirically testable system of physics. Freedom from interaction is always an approximative assumption. However, if we wanted to include all interactions, we would have to invent a model of the complete universe. This model would describe the world like a free object within the world. This means that logic and mathematics as we use them in physics can essentially not describe reality as a whole.

With this caution mark we may nevertheless conclude: in the quantum theory of original alternatives, the space of Newton or of Minkowski-Einstein is the natural way to put everything that is accessible by such a theory into a mathematical order. Nothing prohibits concluding the same for psychical alternatives. Nothing contradicts the hypothesis that the thinking substance, if we describe it approximately in mathematical structure, would also be observable as an extended substance: the brain, a psychical reality, considered from outside. The universe: God as seen with the eyes of a natural scientist with a modern European cultural perspective. This is the extent of what we can conclude from quantum theory. It is satisfactory that the theory does not exclude more profound questions.

References

Castell L. (1975): Quantum theory of simple alternatives. In *Quantum Theory and the Structure of Time and Space II*, ed. by L. Castell, M. Drieschner, and C.F. von Weizsäcker (Hanser, München), 147–162.

Drieschner M. (1970): Quantum Mechanics as a General Theory of Objective Prediction (Dissertation Universität Hamburg).

Drieschner M. (1979): *Voraussage – Wahrscheinlichkeit – Objekt* (Springer, Berlin).

Drieschner M., Görnitz T., Weizsäcker C.F. von (1987): Reconstructioon of abstract quantum theory. *Int. J. Theor. Phys.* **27**, 289–306.

Görnitz T., Ruhnau E., Weizsäcker C.F. von (1992): Temporal asymmetry as precondition of experience – the foundation of time. *Int. J. Theor. Phys.* **31**, 37–46.

Omnès R. (1994): *The Interpretation of Quantum Mechanics* (Princeton University Press, Princeton).

Weizsäcker C.F. von (1985): *Aufbau der Physik* (Hanser, München).

Weizsäcker C.F. von (1992): *Zeit und Wissen* (Hanser, München).

Weizsäcker C.F. von, Scheibe E., Süßmann G. (1958): Komplementarität und Logik III. Mehrfache Quantelung. *Z. Naturforsch.* **13a**, 705–721.

Part II

Cognition and Time

The Brain's Way to Create "Nowness"

Ernst Pöppel[1,2]

[1] Institut für Medizinische Psychologie, Ludwig-Maximilians-Universität,
  Goethestrasse 31, D–80336 München, Germany
[2] Research Center Jülich, D–52425 Jülich, Germany

1 Introduction

Continuity of time appears to be obvious to the naive observer, but it is a misleading idea. This concept probably goes back to Isaac Newton and his overwhelming influence on psychophysics as well as basic physics. One underlying idea in psychophysics is that subjective reality is a direct reflection of objective reality. By understanding the transformation rules between objective and subjective reality as expressed in psychophysical laws, one can reconstruct subjective reality (e.g., Stevens 1951). Applying the same thesis to subjective time, we can go back directly to Newton, who stated in the beginning of his *Philosophiae Naturalis Principia Mathematica* under *Definitions*:

> "*Absolute, true, and mathematical time, of itself and from its own nature, flows equably without relation to anything external,* and by another name is called duration: relative, apparent, and common time, is some sensible and external (whether accurate or unequable) measure of duration by the means of motion, which is commonly used instead of true time; such as an hour, a day, a month, a year."

The central statement (in italics) refers to the general nature of time, continuity being one of the main features of physical processes as observed on the macroscopic level ("time flows equably"). Presumably on this basis, subjective time has erroneously also been considered a continuous phenomenon. However, if one analyzes subjective phenomena in different temporal domains, one is impressed by the great number of experimental observations that indicate discontinuous information processing. It will be argued that the apparent continuity of time is a secondary phenomenon – actually an illusion – which is only made possible by discrete information processing on different temporal levels.

Experimental evidence suggests the existence of at least two processing systems employing discrete time sampling. These presumably independent processing systems are hierarchically linked with one another. Both these systems are fundamental for any perceptual act, for cognitive control, and for motor planning. First, I shall present observations that suggest the existence of a high-frequency processing system generating discrete time quanta in the domain of approximately 30 milliseconds. Then I shall address a low-frequency processing system, which is operative in the domain of 2 to 3 seconds. Evidence for the two temporal processing domains derives mainly from psychophysical, neuropsychological, and neurophysiological experiments. Thus, arguments are anchored on a broad empirical basis, taking into consideration qualitatively different experimental paradigms. In addressing these two processing systems, a remark of caution is, however, necessary: I do not refer to physical constants but to temporal domains of information processing; on both levels substantial intra- and inter-individual variability can be seen.

2 The Neuronal Event Machine

Evidence for a distinct high-frequency mechanism derives, for instance, from studies on *temporal order* thresholds (Hirsh and Sherrick 1961; von Steinbüchel 1995). If subjects are asked to indicate the sequence of two sensory stimuli, temporal order thresholds in the domain of approximately 30 ms are observed independent of the sensory modality. To measure auditory order thresholds, two click stimuli are presented binaurally to a subject. If the stimuli are presented simultaneously, the subject will fuse the stimuli perceptually so that only one stimulus is heard in the center of the head. A delay of one stimulus against the other results in hearing the two clicks separately in each ear if the interval between the two exceeds 2 to 3 milliseconds. Although the subject hears two clicks and might even know they are no longer simultaneous, he will not be able to indicate their temporal order correctly. The delay between the two clicks must exceed approximately 30 ms before a subject can reliably indicate the correct sequence. Similar threshold values are observed both for the visual and the tactile modality in analogous experimental situations.

The similarity of temporal order thresholds in the different sensory systems is particularly noteworthy because it indicates that the temporal order threshold is probably based on a neuronal process different from that responsible for the temporal fusion threshold. I suggest that fusion thresholds are a direct reflection of the peripheral transduction processes of the different sensory systems, which are characterized by different time constants. The transduction process in the auditory system is short (up to 1 ms) compared to the transduction processes in the visual or tactile system. Because of this temporal difference in transduction, the auditory fusion threshold (a few milliseconds) is considerably shorter than the critical flicker-fusion threshold in the visual modality (tens of milliseconds) (Pöppel et al. 1975). It is obvious that temporal order thresholds must be longer than fusion thresholds because independent representations of stimuli are a necessary condition to define their sequence. Interestingly, such an independent representation is not sufficient to allow the mental construction of temporal order.

Order thresholds indicate directly that temporal processing is discontinuous. Different physical stimuli which are processed within a *temporal window* of approximately 30 ms are treated as *co-temporal*, i.e., a temporal relationship with respect to the before-after dimension cannot be established for such distinct stimuli. Information gathered within a temporal window of 30 ms is treated as *a-temporal*, i.e., there is no temporal continuity defined and definable for stimuli that follow each other within such intervals. *Time* in the Newtonian sense (Ruhnau and Pöppel 1991) does not exist on an experiential level for intervals shorter than approximately 30 milliseconds. This statement does not imply that the central nervous system cannot process information on time scales shorter than 30 ms (for example, the localization of objects in auditory space requires a much higher temporal resolution). However, to establish distinct *events* that are related to each other such that their temporal order can be indicated, the shortest temporal interval is observed in the domain of approximately 30 milliseconds.

Before turning to a discussion of mechanisms (in particular neuronal oscillations, which are thought to be essential for structuring subjective time in a discrete fashion), some other observations from psychophysical experiments supporting the notion of temporally distinct processing stages may be appropriate. Interesting evidence comes from a certain class of experiments on reaction time. It has been shown that, under *stationary conditions*, response distributions of reaction time often show multimodal characteristics (Harter and White 1968; Pöppel 1968, 1970). This has been observed in choice reaction time, when subjects have to selectively press a response button when either a visual or an auditory stimulus is presented. The distinct modes in the response distributions are separated by 30 to 40 ms (Ilmberger 1986; Jokeit 1990). It has been hypothesized that these distinct modes represent successive and discrete decision-making steps being processed in central neuronal populations (Pöppel 1994).

Multimodal response distributions have also been observed in measurements of response latencies in saccadic and pursuit eye movements. If a moving visual stimulus initiates pursuit eye movements, the latency of such eye movements has a strong tendency to be multimodally distributed with 30 to 40 ms intervals between neighbouring modes (Pöppel and Logothetis 1986). Similar observations on multimodal latency distributions have been made on saccadic eye movements for human subjects (Frost and Pöppel 1976; Ruhnau and Haase 1993) and other primates (Fuchs 1967).

It should be pointed out that multimodalities as referred to here are not always observed in standard experimental situations. There are many reasons why this is so. They can all be summarized under one heading: *stationarity*. Practically all psychological variables are strongly influenced by fatigue, lack of concentration, learning, diurnal variation, etc. These modulating variables are a rich source of *non-stationarity* in experimental observations. As the distinct temporal effects referred to above are in the millisecond domain, these non-systematic effects might easily shadow the temporal processing proper unless *stationary* conditions are meticulously controlled. Other *non-stationarities* may result from inappropriate statistics often employed in such experiments (Pöppel et al. 1990a; Ruhnau and Pöppel 1996).

The multimodal response distributions of reaction time for eye movements and the observations of temporal order thresholds can be explained on the basis of *excitability cycles* or *neuronal oscillations* (Pöppel 1970; Pöppel et al. 1990b,c). According to this model, if a supra-threshold stimulus comes into contact with the sensory surface, a neuronal oscillation with a period of approximately 30 ms is initiated. This oscillation is (technically speaking) thought to be a relaxation oscillation and not a pendulum oscillation (Wever 1965). Relaxation oscillations can be triggered almost instantaneously by external stimuli, or their phase can be reset with only a minor delay. Furthermore, relaxation oscillations fade out after several periods unless they are retriggered. This fading can be observed experimentally in a progressive reduction in response amplitude. In the extreme case, only two or three successive periods are observed because of high damping constants characterizing the system. In addition, the period of relaxation oscillations is not stable as in pendulum oscillations, i.e., the period can vary depending on the special properties of the system or due to environmental conditions. Thus, if I refer to processing periods of 30 ms, this does not mean a fixed number like a physical constant (see above), but indicates a temporal domain in the range of 20–50 ms.

These processing periods set up by neuronal oscillations represent temporal quanta within which – as mentioned before – information is treated as *co-temporal* and, therefore, *a-temporal*. I believe that this interpretation is nicely exemplified by the experiments on temporal order threshold. The domain of *perceptual successiveness* can only be entered beyond 30 ms, since only after approximately 30 ms is it possible to determine temporal order. This means that information within this temporal limit is still in the domain of *non-successiveness*. This interpretation is supported by choice reaction-time studies. In a choice situation, neuronal processes of the brain apparently have to go through discrete temporal phases within which new constellations are set up as reflected in *decisions*. These are then fed into a motor response system. Sternberg (1966) observed that the exhaustive scanning process through short-term memory is discontinuous, with approximate step durations of 30 to 40 ms. Thus, experiments from very different experimental domains appear to support the notion of neuronal processing in this temporal domain.

Up to now, I have mentioned observations from psychophysical experiments, but there are also other observations deriving from neurophysiological experiments supporting the notion of discrete temporal processing. Experiments on the auditory evoked potential demonstrate that the midlatency response shows a clear oscillatory component in the first 100 ms after stimulus onset (Galambos et al. 1981; Mäkelä and Hari 1987). After an initial phase of brain-stem activity, an oscillation sets in with high interindividual stability. Similar oscillatory responses can be seen in the other sensory modalities on the level of single cells (Gray et al. 1989; Murthy and Fetz 1992, Podvigin and Pöppel 1994). It has been suggested that oscillations in the 40-Hz domain are essential for cognitive mechanisms (Llinas and Ribary 1993), and it has been demonstrated that oscillatory components in the midlatency region of the auditory evoked potential are sensitive markers for wakefulness in contrast to general anesthesia (Madler and Pöppel 1987). When patients receive general anesthetics, one can still observe the brain-stem response, but the mid-latency oscillatory activity disappears. If oscillatory activity is preserved under anesthesia, patients still can process sensory information, although often only implicitly (Schwender et al. 1994).

It can be hypothesized that general anesthetics selectively destroy the *temporal coherence* within neuronal populations. It is as if a *temporal glue* between neurons were lacking, which is a necessary condition for reliably processing sensory information. Patients going through this physiological state when the oscillatory responses have disappeared have been subjectively in a phase of *a-temporality*. Such patients often ask, after an operation: "When will the operation start?" – which suggests that, unlike sleep, the brain is processing no information at all. Nothing has happened, and no information has been made available to central processing mechanisms of the brain.

This overview provides some factual information about discrete time sampling in the 40-Hz domain. There also exist theoretical considerations about the practicability and perhaps even the biological necessity of processing neuronal information discontinuously. I wish to present two such arguments in support of this conclusion, one derived from sensory analyses, and one based on the mode of functional representation.

As already mentioned, transduction processes in different sensory systems are characterized by different time constants. This biophysical fact leads to a logistic problem for the brain if information from different sensory channels is to be integrated. An object in space might, for instance, be defined by visual and auditory information. As transduction in the auditory system takes much less time than in the visual system, central availability of auditory information is earlier than of visual information. A difference of this kind would, however, only apply to nearby objects, when the time the sound-wave takes from its source can be neglected. At a distance of approximately 10 meters, the transduction time in the retina corresponds to the time a sound-wave takes (Pöppel et al. 1990b). Beyond this *horizon of simultaneity*, the central availability of visual stimuli will be prior to that of auditory stimuli.

For an object moving in space and changing its distance in relation to the observer, the central availability of stimuli in the two sensory modalities is highly unpredictable. Both the physics (light vs. sound velocity) and the biophysics (different transduction times) result in permanent shifts of central temporal availability in the two sensory systems. With respect to sensory integration allowing central fusion of two sources of information, such unpredictability could be overcome logistically by introducing temporally neutral zones within which information is treated as *co-temporaneous*. Such zones could be set up by neuronal oscillations, one period representing an interval within which information from various sources is collected indiscriminately from its source.

An important argument for such a collective mechanism comes from the demands of the motor system. To program a movement, information from different sensory channels often has to be integrated. This can be done most efficiently if temporally neutral zones are introduced in which information coming from various sources is integrated, defining a distinct physiological state. Especially ballistic movements, whose neuronal program has been completed before the beginning of the movement, would profit from such a mechanism.

A qualitatively similar argument suggesting the necessity of discrete processing states comes from observations on the mode of functional localization. One essential conclusion that can be drawn on the basis of anatomical studies and neurological observations is that elementary functions are locally represented. Anatomical studies of the visual system, for instance, have clearly demonstrated a spatial segregation of functions (Zeki 1978). Similarly, neuropsychological observations strongly support the notion of a spatially distributed representation of elementary functions (e.g., Pöppel 1989). It has to be stressed that each functional state is characterized by the simultaneous activity of different neuronal modules, as has been demonstrated in PET-studies on pain perception (Talbot et al. 1991). Distributed and simultaneous activities require a strict temporal regimen to guarantee functional competence. As in the co-ordination of sensory input, system states could serve a useful function in temporal co-ordination of distributed activities, and such system states could be implemented by neuronal oscillations. Simulation studies based on theoretical considerations have actually demonstrated the essential role oscillations could play in functional co-ordination (Sporns et al. 1989; Tononi et al. 1992).

Neuronal oscillations in the 40-Hz domain may actually represent spatial binding of distributed activities. In this case, oscillations represent binding, i.e., equal frequency and identical phase of an oscillatory activity can be conceived of as the expression of binding of spatially distributed activities (Gray et al. 1989). The model suggested here is different: Neuronal oscillations in the 40-Hz domain are triggered by exogenous and endogenous events and provide a temporal framework within which binding operations are implemented by distinct neuronal algorithms. These oscillations thereby represent a necessary condition for binding and not the binding operation itself.

3   The Neuronal STOBCON Machine

While the high-frequency mechanism discussed above is thought to organize distributed neuronal activities and to implement "primordial events", a low-frequency mechanism appears to integrate successive events within a temporal window of approximately 2 to 3 seconds. One of the most convincing results on temporal integration comes from studies on the reproduction of visual and auditory stimuli of different durations (Pöppel 1978; Elbert et al. 1991). If visual or auditory stimuli of different durations are presented, subjects reproduce these stimuli reliably up to intervals of approximately 3 seconds. Longer stimuli are reproduced shorter and with much higher variability. Often stimuli of up to about 3 seconds are reproduced slightly longer than the stimulus, probably due to a reaction-time component. The longer reproduction of short intervals and the shorter reproduction of long intervals characterizes a transition from over- to underestimation. The transition point has been referred to as *indifference point*.

The observation of a veridical reproduction of temporal intervals up to approximately 3 seconds suggests the existence of a specific temporal integration mechanism in this domain. Further experimental evidence suggests that this temporal integration is *automatic* and *presemantic*, i.e., it is not driven by content, but occurs prior to a semantic evaluation of the information processed. The undistorted reproduction of temporal intervals within a defined temporal window is presumably just one expression of the well-known time-order error (Köhler 1923), which can be observed for all prothetic continua in the different sensory systems (Stevens 1951; Pöppel 1978). If two stimuli are to be compared with respect to subjective intensity, they have to be presented within a *temporal window* of up to approximately 3 seconds. If the temporal interval between the two stimuli is longer than this critical interval, the second stimulus will be overestimated with respect to intensity compared to the first stimulus. The brain seems to provide an operative basis in which successive events are represented reliably and can, therefore, be compared appropriately only within a limited time span.

Another expression for the temporal integration up to approximately 3 seconds comes from psychophysical experiments studying Weber's law in time perception (Getty 1975). It turns out that Weber's law is valid only up to intervals of 2 to 3 seconds if temporal intervals are to be compared. For longer intervals, the difference between temporal intervals has to be increased substantially to allow their distinct perception. This also demonstrates that temporal processing up to approximately 3 seconds obeys mechanisms that are qualitatively different from those for longer intervals.

There is another class of observations supporting the idea of automatic temporal integration in which introspective co-operation of the subjects is required. One such class includes experiments on the spontaneous alteration rates in perception using ambiguous figures or sound sequences as stimuli. If stimuli that can be perceived under two perspectives or with two different meanings are presented, there is an automatic change of perceptual content after an average of 3 seconds. The Necker cube, which can be seen under two different perspectives, has been used for visual stimulation, and a phoneme sequence (like CU-BA-CU-BA, where one hears either CUBA or BACU) has been used for auditory stimulation (e.g., Radilova et al. 1990). The spontaneous alteration rates in the visual and the auditory modality can be interpreted as follows: after an *exhaust period*, attentional mechanisms may be automatically elicited that open sensory channels for new information. If the information of the perceptual stimulus is simply the other alternative of an ambiguous figure, this alternate perspective or meaning will gain control in conscious representation. It is as if the brain asks, "What is new in the world?" every 3 seconds or so, and, under such unusual stimulus conditions, the temporal *eigen-operations* of the brain are unmasked.

Neurological observations have shown that brain lesions often result in a slowing down of processing (e.g., Pöppel et al. 1975). It could be demonstrated that this effect is also true for the alteration rate in binocular rivalry (Pöppel et al. 1978). In one such case, an interesting observation could be made with respect to the dynamics of the alteration process itself. Because of the extreme deceleration of neuronal processing, the patient was able to describe how the vertical grating, when dominant in his perception, was gradually pushed away by a slow movement of the horizontal grating, as if a travelling wave were carrying the horizontal grating. Similarly, when the horizontal grating was dominant, it was gradually removed by the vertical grating until it became dominant. Observations like these fortify the hypothesis that the neuronal shift mechanism resulting in a new percept is not an irregular process which occurs in randomly selected positions in the neuronal net. Rather, neurons in local networks interact with each other and their synchronized activity spreads sequentially in a topographically organized way (like in a crystallization process) to gain new territory.

Additional evidence for an automatic temporal integration process comes from a simple experiment in which a subject is required to mentally structure auditory sequences. If one listens to the beats of a metronome, one is automatically drawn into the perceptual habit of accenting each second or third beat, thereby structuring the continuous metronome beats subjectively into a rhythm. By positing a subjective accent to every second beat, for instance, two successive beats are perceived as a unit (Szelag et al. 1996). It turns out that two beats cannot lie further apart than 2 to 3 seconds to allow subjective accentuation. Beyond this interval, it is no longer possible to mentally connect the second to the first beat, i.e., the first beat has then already disappeared into a perceptually not directly available past. This is one reason why one can refer to such integration intervals as *subjective present* (Pöppel 1988; Edelman 1989).

A temporal grouping for an auditory chain of stimuli as suggested by the metronome experiment can also be observed in the temporal segmentation of spontaneous speech. Independent of age or of the language spoken, speakers have the tendency to construct closed verbal utterances of durations up to about 2 to 3 seconds (Kowal et al. 1975; Pöppel, 1988; Vollrath et al, 1992). The expression of such a linguistic unit is usually followed by a pause, presumably used to mentally prepare the next linguistic unit and load it into an output system. It is important to stress that temporal segmentation as observed in spontaneous speech does not necessarily apply to reading out loud. Written language is characterized by a different syntactic structure, and the reader apparently uses different neuronal processes from the speech domain to transport verbal utterances.

It may not be accidental that, in many languages, a line in classical verse most often has a duration of approximately 3 seconds, as if poets had an implicit knowledge of the temporal segmentation process governing our brain (Turner and Pöppel 1988; see also Kien and Kemp 1994).

It has also been demonstrated that *working memory* is similarly limited to a few seconds if rehearsal is not allowed. In classical experiments on short-term memory (Peterson and Peterson 1959), sequences of letters were presented, and the subject's task was to reproduce them correctly. If the subject is asked to perform another task immediately after stimulus presentation, the access to the presented information prior to the disruptive task is severely disturbed. Only within a temporal window of approximately 3 seconds can information be retained veridically. Temporal segmentation in the domain of approximately 3 seconds has also been observed in studies of movements, as demonstrated by experiments on sensory-motor synchronization. A subject is requested to synchronize finger taps to a regular sequence of short clicks, which are presented by earphone. It has been shown that click occurrence is anticipated with a motor response by some tens of milliseconds when the interstimulus interval is, for instance, 700 milliseconds (e.g., Radil et al. 1991; Vos et al. 1995).

This *negative asynchrony* (as the anticipatory response has been referred to) has not been resolved yet, but the following hypothesis seems plausible to explain the effect. The subject attempts to synchronize the anticipated perceptual representation of the next auditory stimulus with the anticipated movement corresponding to this stimulus. Both these systems, the auditory and the tactile or muscle-spindle system (which informs the perceptual center about the accuracy of the movement), are characterized by different delays between the periphery (i.e., the beginning of the transduction process) and the perceptual center. Because of this temporal difference, a *phenomenal synchrony* of the anticipated events must be *asynchronous* on the observer level. Thus, it is actually misleading to refer to *stimulus anticipation*; we actually observe the expression of proper internal synchronization masked by the different delays in different sensory systems.

Negative asynchrony is, however, only observed for interstimulus intervals up to approximately 3 seconds (Mates et al. 1994). A subject can program a properly synchronized motor response corresponding to a sensory event only within this temporal window. If the next sensory event lies too far in the future, the subject is no longer capable of programming the corresponding response. If the interstimulus interval is too long, synchronization, if attempted, shows an extremely high variability. Subjects often try to synchronize their movements with the sensory stimuli by simply reacting to them, which would express *positive asynchrony*. In this situation, we are no longer dealing with active anticipation, but with passive reaction to the sensory event.

Temporal segmentation in the domain of 3 seconds has also been observed in studies on the duration of spontaneous intentional acts. It could be shown that there is a strong tendency of such movements to be structurally embedded in a temporal window of approximately 3 seconds duration. Homologous movements have been studied in different cultures and it was found that, independent of the state of acculturation, the duration of such movement patterns is identical (Schleidt et al. 1987). Cultures studied were those of the Yanomani Indians, Trobriand Islanders, Kalahari Bushmen, and Europeans. As a cultural transfer between these groups is extremely unlikely, one is forced to conclude that a universal temporal constant dominates a certain class of human motor behavior.

It has been demonstrated recently that typical movement patterns of different mammalian species, like scratching, also have a strong tendency to last 3 seconds on average. This observation suggests a universal temporal mechanism transcending human behavior (Gerstner and Fazio 1995). It can be argued that, for all higher mammals, movement patterns are controlled by a homologous neuronal mechanism that automatically integrates information up to approximately 3 seconds.

In a recent study on *mismatch negativity* (Sams et al. 1993) using the technology of magnetencephalography (MEG), it was observed that interstimulus intervals of 3 seconds elicited by far the greatest amplitudes of the MMN (magnetic mismatch negativity). The mismatch negativity is thought to be an expression of increased cortical activity. In the study mentioned, the MMN was recorded for the auditory cortex. An explanation for the resulting preference for approximately 3 seconds is based on the idea that the auditory system has to open up its channels to pick up new information. It is approximately every 3 seconds that the "closed mind" opens up for new information coming from the environment.

4   Concluding Remarks

The above-mentioned experiments on temporal segmentation of 3 seconds duration cover perception, memory, cognition, and movement control. The different experiments suggest that temporal segmentation is an underlying principle of all higher data-generating mechanisms. Evidence from neuropsychology suggests that the different temporal segmentation processes might each be implemented in the corresponding neuronal domains. Nevertheless, we must conclude that there is a general temporal segmentation mechanism which is *automatic* and *presemantic*. It provides an operative basis for percepts, memories, and volitional control. Because of the omnipresence of this mechanism, the single states of 3-second segments could be referred to as "states of being conscious" (STOBCON), accompanied by the feeling of "nowness". Each STOBCON represents a mental island of activity distinctly separate from the temporally neighboring ones.

How, then, does temporal continuity arise? It is suggested that each STOBCON, each island of "nowness", being implemented in a 3-second window, is semantically linked with the previous and with the following one. The conservation of exogenous phenomena results in a hysteresis of meaning. Thus, the continuity of experience is an *illusion*: actually, distinct STOBCONs follow each other. Continuity arises because of a specific mechanism linking the contents of each temporal window of "nowness" to the next one.

However, temporal segmentation in the two domains described, namely in the domain of 30 milliseconds and in the domain of 3 seconds, is a necessary prerequisite for the construction of subjective continuity. Without this continuity, the brain would be lost in a jungle of unrelated pieces of information. Paradoxically, continuity is made possible by discontinuous information processing in the brain. Its individual steps are inseparable; neither an event nor its meaning can be divided into temporal parts. But the argument can also be turned around: Meaning arises because discrete temporal sampling phases are implemented on a processing level. "Nowness" creates meaning.

Acknowledgements: Experiments referred to were mainly supported by DFG and BMFT/BMBF. This paper is based on a text "Reconstruction of subjective time on the basis of hierarchically organized processing systems" which appeared in: M.A. Pastor and J. Artieda (eds.): *Time, Internal Clocks, and Movement* (Elsevier, Amsterdam 1996), 165–185.

References

Edelman G.M. (1989): *The Remembered Present: A Biological Theory of Consciousness* (Basic Books, New York).

Elbert T., Ulrich R., Rockstroh B., and Lutzenberger W. (1991): The processing of temporal intervals reflected by CNV-like brain potentials. *Psychophysiology* **28**, 648–655.

Frost D. and Pöppel E. (1976): Different programming modes of human saccadic eye movements as a function of stimulus eccentricity: Indications of a functional subdivision of the visual field. *Biol. Cybernetics* **23**, 39–48.

Fuchs A.F. (1967): Saccadic and smooth pursuit eye movements in the monkey. *J. Physiol.* **191**, 609–631.

Galambos R., Makeig S., and Talmachoff P.J. (1981): A 40-Hz auditory potential recorded from the human scalp. *Proc. Natl. Acad. Sci. USA* **78**, 2643–2647.

Gerstner G.E. and Fazio V.A. (1995): Evidence of a universal perceptual unit in mammals. *Ethology* **101**, 89–100.

Getty D.J. (1975): Discrimination of short temporal intervals: a comparison of two models. *Perception and Psychophysics* **18**, 1–8.

Gray C., König P., Engel A.K., and Singer W. (1989): Oscillatory responses in cat visual cortex exhibit inter-columnar synchronization which reflects global stimulus properties. *Nature* **338**, 334–337.

Harter R. and White C.T. (1968): Periodicity within reaction time distributions and electromyograms. *Quart. J. Exp. Psychol.* **20**, 157–166.

Hirsh I.J. and Sherrick C.E. (1961): Perceived order in different sense modalities. *J. Exp. Psychol.* **62**, 423–432.

Ilmberger J. (1986): Auditory excitability cycles in choice reaction time and order threshold. *Naturwissenschaften* **73**, 743–744.

Jokeit H. (1990): Analysis of periodicities in human reaction times. *Naturwissenschaften* **77**, 289–291.

Kien J. and Kemp A. (1994): Is speech temporally segmented? Comparison with temporal segmentation in behavior. *Brain and Language* **46**, 662–682.

Köhler W. (1923): Zur Theorie des Sukzessivvergleichs und der Zeitfehler. *Psychol. Forschung* **4**, 115–175.

Kowal S., O'Connell D.C., and Sabin E.J. (1975): Development of temporal patterning and vocal hesitations in spontaneous narratives. *J. Psycholinguistic Res.* **4**, 195–207.

Llinas R. and Ribary U. (1993): Coherent 40-Hz oscillation characterizes dream state in humans. *Proc. Natl. Acad. Sci. USA* **90**, 2078–2081.

Madler C. and Pöppel E. (1987): Auditory evoked potentials indicate the loss of neuronal oscillations during general anaesthesia. *Naturwissenschaften* **74**, 42–43.

Mäkelä J.P. and Hari R. (1987): Evidence for cortical origin of the 40 Hz auditory evoked response in man. *Electroencephal. Clin. Neurophysiol.* **66**, 539–546.

Mates J., Müller U., Radil T., and Pöppel E. (1994): Temporal integration in sensorimotor synchronization. *J. Cogn. Neurosci.* **6**, 332–340.

Murthy V.N. and Fetz E.E. (1992): Coherent 25- to 35-Hz oscillations in the sensorimotor cortex of awake behaving monkeys. *Proc. Natl. Acad. Sci. USA* **89**, 5670–5674.

Newton I. (1687): *Philosophiae Naturalis Principia Mathematica* (London).

Peterson L.B. and Peterson M.J. (1959): Short-term retention of individual items. *J. Exp. Psychol.* **58**, 193–198.

Podvigin N.D. and Pöppel E. (1994): Characteristics and functional meaning of oscillatory processes in the retina and lateral geniculate body. *Sensory Systems* **8**, 97–103.

Pöppel E. (1968): Oszillatorische Komponenten in Reaktionszeiten. *Naturwissenschaften* **55**, 449–450.

Pöppel E. (1970): Excitability cycles in central intermittency. *Psychol. Forsch.* **34**, 1–9.

Pöppel E. (1978): Time Perception. In *Handbook of Sensory Physiology, Vol. 8: Perception*, ed. by R. Held, H.W. Leibowitz, and H.-L. Teuber (Springer, Berlin), 713–729.

Pöppel E. (1988): *Mindworks. Time and Conscious Experience* (Harcourt Brace Jovanovich, Boston).

Pöppel E. (1989): Taxonomy of the subjective: An evolutionary perspective. In *Neuropsychology of Visual Perception*, ed. by J.W. Brown (Erlbaum, Hillsdale), 219–232.

Pöppel E. (1994): Temporal mechanisms in perception. *Int. Rev. Neurobiol.* **37**, 123–129.

Pöppel E., Brinkmann R., von Cramon D., and Singer W. (1978): Association and dissociation of visual functions in a case of bilateral occipital lobe infarction. *Arch. Psychiat. Nervenkr.* **225**, 1–21.

Pöppel E., von Cramon D., and Backmund H. (1975): Eccentricity-specific dissociation of visual functions in patients with lesions of the central visual pathways. *Nature* **256**, 489–490.

Pöppel E. and Logothetis N. (1986): Neuronal oscillations in the brain. Discontinuous initiations of pursuit eye movements indicate a 30-Hz temporal framework for visual information processing. *Naturwissenschaften* **73**, 267–268.

Pöppel E., Schill K., and von Steinbüchel N. (1990a): Multistable states in intrahemispheric learning of a sensorimotor task. *Neuroreport* **1**, 69–72.

Pöppel E., Schill K., and von Steinbüchel, N. (1990b): Sensory integration within temporally neutral system states: a hypothesis. *Naturwissenschaften* **77**, 89–91.

Pöppel E., Ruhnau E., Schill K., and von Steinbüchel N. (1990c): A hypothesis concerning timing in the brain. In *Synergetics of Cognition*, ed. by H. Haken and M. Stadler (Springer, Berlin), 144–149.

Radil T., Mates J., Ilmberger J., and Pöppel E. (1990): Stimulus anticipation in following rhythmic acoustical patterns by tapping. *Experientia* **46**, 762–763.

Radilova J., Pöppel E., and Ilmberger J. (1990): Auditory reversal timing. *Act. Nerv. Super.* **32**, 137–138.

Ruhnau E. and Haase V.G. (1993): Parallel distributed processing and integration by oscillations. *Behav. Brain Sci.* **16**, 587–588.

Ruhnau E. and Pöppel E. (1991): Adirectional temporal zones in quantum physics and brain physiology. *Int. J. Theor. Phys.* **30**, 1093–1090.

Ruhnau E. and Pöppel E. (1996): On a relationship between the prediction paradox in logic and the statistics of interstimulus intervals in biological experiments. *Naturwissenschaften*, submitted.

Sams M., Hari R., Rif J., and Knuutila J. (1993): The human auditory sensory memory trace persists about 10 sec: neuromagnetic evidence. *J. Cogn. Neurosci.* **5**, 363–370.

Schleidt M., Eibl-Eibesfeldt I., and Pöppel E. (1987): A universal constant in temporal segmentation of human short-term behaviour. *Naturwissenschaften* **74**, 289–290.

Schwender D., Madler C., Klasing S., Peter K., and Pöppel E. (1994): Anaesthetic control of 40-Hz brain activity and implicit memory. *Consciousness and Cognition* **3**, 129–147.

Sporns O., Gally J.A., Reeke G.N., and Edelman G.M. (1989): Reentrant signalling among simulated neuronal groups leads to coherency in their oscillatory activity. *Proc. Natl. Acad. Sci. USA* **86**, 7265–7269.

von Steinbüchel N. (1995): Temporal system states in speech processing. In *Supercomputing in Brain Research: From Tomography to Neural Networks*, ed. by H.J. Herrmann, D.E. Wolf, and E. Pöppel (World Scientific, Singapore), 75–81.

Sternberg S. (1966): High-speed scanning in human memory. *Science* **153**, 652–654.

Stevens S.S. (1951): Mathematics, measurement, and psychophysics. In *Handbook of Experimental Psychology*, ed. by S.S. Stevens (Wiley, New York), 1–49.

Szelag E., von Steinbüchel N., Reiser M., de Langen E.G., and Pöppel E. (1996): Temporal constraints in processing nonverbal rhythmic patterns. *Acta Neurobiologiae Experimentalis* **56**, 215–225.

Talbot J.D., Marett S., Evans A.C., Meyer E., Bushnell M.C., and Duncan G.H. (1991): Multiple representations of pain in human cerebral cortex. *Science* **251**, 1355–1357.

Tononi G., Sporns O., and Edelman G.M. (1992): Reentry and the problem of integrating multiple cortical areas: Simulation of dynamic integration in the visual system. *Cereb. Cortex* **2**, 310–335.

Turner F. and Pöppel E. (1988): Metered poetry, the brain, and time. In *Beauty and the Brain. Biological Aspects of Aesthetics*, ed. by I. Rentschler, B. Herzberger, and D. Epstein (Birkhäuser, Basel), 71–90.

Vollrath M., Kazenwadel J., and Krüger H.-P. (1992): A universal constant in temporal segmentation of human speech. *Naturwissenschaften* **79**, 479–480.

Vos P., Mates J., and van Kruysbergen N.W. (1995): The perceptual center of a stimulus as the cue for synchronization to a metronome: Evidence from asynchronies. *Quart. J. Exp. Psychol.* **48A**, 1024–1040.

Wever R. (1965): Pendulum versus relaxation oscillation. In *Circadian Clocks*, ed. by J. Aschoff (North Holland, Amsterdam), 74–83.

Zeki S. (1978): Functional specialisation in the visual cortex of the rhesus monkey. *Nature* **274**, 423–428.

Temporal Integration of the Brain as Studied with the Metronome Paradigm

Elzbieta Szelag

Department of Neurophysiology, Nencki Institute of Experimental Biology, 3 Pasteur Street, PL–02-093 Warsaw, Poland

1 Introduction

It is well known that our subjective experience of the passage of time is influenced by what we are doing. More than 70 years ago, Axel (1925) theorized that the level of behavioral activity determines the experience of time. The subjective flow of time is influenced by the amount of information processed or by the mental content. It is commonly believed that time appears to pass rapidly if one is engaged in a higher order of behavioral activity. In contrast, a low level of behavioral activity results in the impression that time is dragging. Retrospectively, this impression may be turned around; a rapid passage of time leads to a long subjective duration, a slow passage of time to a short subjective duration. Considerable support for such observations comes from experimental studies on reproduction of temporal intervals (Fraisse 1984). Although several experimental studies have investigated the effect of the information content of presented stimuli on the reproduction of time intervals, the influence of such a content on presemantic temporal integration (PTI) has not been studied before (e.g., Pöppel 1994 and this volume; Ruhnau 1995). To gain better insight into mechanisms underlying PTI, we report here some observations concerning the integration of temporal information contained in continuous beats of a metronome.

2 Information Content and Temporal Integration

In a series of experiments, sequences of metronome beats at nine different frequencies (from one up to five beats/s) have been presented to a number of subjects. The subjects were asked to listen to the equally-spaced beats and to mentally accentuate every second, third, fourth, ... etc. beat to create a subjective rhythm. Obviously, this accentuated rhythm existed only in the subjects' minds, and not objectively. In one condition, subjects reported verbally how many beats they could integrate into a unit. In a second condition, they pressed a button at the beginning and at the end of a sequence they could mentally integrate. For verbal responses, the measured integration interval length (MIIL) was defined as the number of reported beats times the temporal distance between two successive beats for a particular metronome frequency. For nonverbal responses, reaction times between the first and the second button pressing were measured.

The results indicate an integration process, i.e., temporally separated successive beats are mentally connected with each other into larger perceptual units. Thus, the separate beats in exposed sequences lose their dominance at the perceptual level and are organized into a higher-order structure dominating the serial order. One can anticipate that, to follow the present integration task, at least three strategies are possible (see Fig. 1). First, if the subjects had integrated information only by time (i.e., in a constant period of 2, 3 s, ... etc.), MIIL would be constant and independent of presented frequency. Second, if the subjects had integrated information only by number (i.e., connecting them mentally by always counting the constant number of beats: 2, 3, ... etc.), MIIL would be strongly dependent on presented frequency. The first strategy reflects a global, holistic process, whereas the second one is predominantly semantic and reflects serial ordering or serial labelling. Third, if one looks at Fig. 1 it appears that in the present experiment subjects apply a combination of the two strategies mentioned. Thus, they integrate the information partly by time and partly by number. Moreover, it can be seen that MIIL significantly depends on the presented metronome frequency.

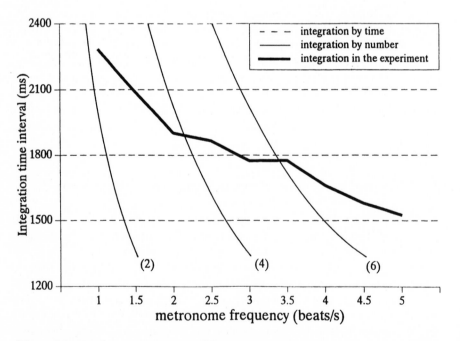

Fig. 1. Measured integration interval length for various metronome frequencies using three different strategies (verbal response).

The upper time limit of MIIL was approximately 2.3 s and was characteristic for the lowest frequency. This observation corresponds with the hypothesis that the temporal extent of the integration process is limited to approximately 2-3 s (Pöppel 1988, 1994). The integration of temporal information can apparently not exceed this time interval. When beats were separated by intervals longer than 1 s, an accentuated rhythm could not be perceived, and only isolated beats were heard. However, the lower limit of this integration was approximately 1.5 s and was characteristic for the two highest frequencies. Thus, the binding mechanism can also integrate the information in intervals shorter than 3 s if necessary. This relationship was predicted by Pöppel et al. (1991, p. 65): "... the temporal binding process is closed up to approximately 3 s and open for shorter intervals ...".

It is worth to be noted that the temporal limitation of PTI is related to the number of events within a given time interval. An interesting relationship was found: if the information content of the presented stimuli was high (beats presented with high frequencies), MIIL was found to be shorter; if the mental content was low (beats presented with low frequencies), the duration of this period appeared to be significantly longer. If one looks at time limits of PTI using nonverbal responses (Fig. 2), qualitatively similar results are observed. As shown in Fig. 2, a general tendency for shorter MIIL for higher metronome

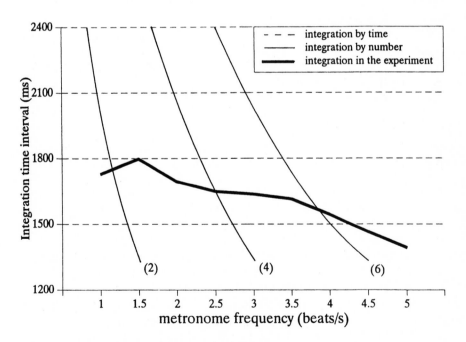

Fig. 2. Measured integration interval length for various metronome frequencies using three different strategies (nonverbal response).

frequencies is also found, although, because of the different type of responses to be applied, higher individual variability is observed. The results show that the temporal limitation of PTI is related to how much information can be processed in a given time interval.

This result can be explained by referring to the experience of boredom (Pöppel 1978, 1988). According to this hypothesis, people usually become bored during a low level of behavioral activity accompanied by a lack of events. In contrast, a higher level of behavioral activity and time filled by many events gives an impression of time passing quickly. Accordingly, one of the major findings in the present experiment is that this hypothesis is valid not only for experience of duration and reproduction of temporal intervals, as mentioned earlier (e.g., Fraisse 1984, Nichelli 1993), but also for the temporal extent of PTI. More events taking place in a given interval are integrated faster, and when a person finds himself in a situation of high informational content (no boredom), time appears to pass more rapidly.

3   Lack of Hemispheric Differences in Temporal Integration

According to common belief, differentiated hemispheric information processing may be based on temporal factors (Efron 1991). It has been suggested that the left cerebral hemisphere, rather than specializing in speech per se, mediates the underlying temporal information in the speech signal. Experimental results and theoretical considerations clearly indicate that PTI in the domain of 3 s plays a significant role in speech perception and production – typical left-hemispheric functions. On this basis, one may wonder whether there is any hemispheric asymmetry concerning the temporal limits of this binding process. There are two general views about this problem. The first assumes that PTI constitutes an intra-hemispheric process and that both hemispheres are equally efficient in its temporal limitation. The alternative view postulates that timing provides specific temporal constraints for human speech and that mechanisms for temporal analysis, including those of PTI, are really the basis for left-hemispheric processing. Consequently, one can predict that the hemisphere specialized for speech would also be the most effective hemisphere for the binding process.

To test hemispheric asymmetry in temporal binding, the method of dichotic presentation was employed. This method attributes ear advantages to the notion that contralateral pathways are stronger than ipsilateral ones because the contralateral pathway activity blocks or inhibits the ipsilateral pathway activity. Accordingly, metronome beats were presented either to both ears simultaneously or to the left or right ear separately; the other ear was masked by white noise. Although a general tendency to shorter MIIL for higher frequencies was observed (for detailed discussion, see above), the clear result was that there were no hemispheric differences in MIIL (Figs. 3 and 4).

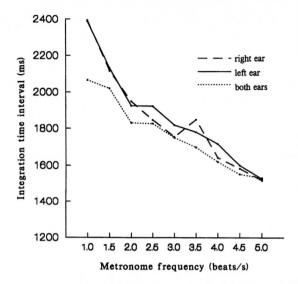

Fig. 3. Mean integration interval length for metronome beats presented to both ears and to the left or right ear separately (verbal response).

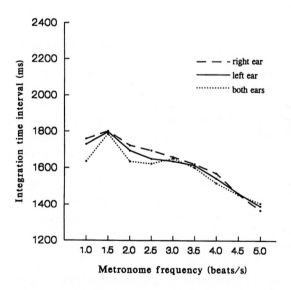

Fig. 4. Mean integration interval length for metronome beats presented to both ears and to the left or right ear separately (nonverbal response).

In other words, MIIL for beats presented to the left hemisphere was similar to that found in the right hemisphere. This was true independent of the type of subjects' responses (verbal versus nonverbal) in the experiment. Based on this result, one might argue that PTI constitutes a fundamental supra-modal process of great complexity. Relevant to this point is the well-documented observation that this mechanism is involved in the processing of both visual and auditory, verbal and nonverbal information and also in the performance of motor action (e.g., Pöppel, this volume). It can be concluded that both hemispheres do not differ in temporal processing in this time domain.

4   Individual Differences in Temporal Integration

There is considerable evidence concerning large individual differences among human beings in their cognitive abilities, such as intelligence quotient, rate of information processing, memory span, etc. Accordingly, the purpose of the present section is to discuss whether such individual variations can also be related to differences in temporal information processing.

4.1   Temporal Integration Across the Life Span

A later-life impairment of several mental functions (e.g., memory, intelligence, verbal fluency, capacity for new learning, spatial ability, reasoning, and problem-solving) is often discussed in the recent literature and well documented by clinical and experimental investigations (e.g., Huff 1990, Huppert 1991). Such declines probably constitute a part of the normal aging process. Based on these observations, an important question is: are there also age-related changes in the length of the period during which temporal information can be integrated in the nervous system?

The experiment using the metronome paradigm (binaural stimulus presentation and verbal responses) was performed with younger (mean age about 35 years) and older adults (mean age about 60 years). Subjects were divided into age groups based on results from longitudinal studies (e.g., Schaie 1983) indicating a pervasive decline in cognitive functions beyond an age of 50 years and on results of the KAI-test ("Kurztest für allgemeine Basisgrößen der Informationsverarbeitung") employed in the present study. This psychometric test is commonly used for assessing cognitive functions such as memory span, rate of information processing, and intelligence quotient. The results of the KAI-test proved that older people included in the present study performed worse than younger ones. Moreover, it was found (Szelag et al.1996) that the mean time interval during which the older subjects were able to integrate temporal information into one perceptual unit (about 1.8 s) was longer than for younger individuals (about 1.4 s).

It should be stressed that MIIL in both age groups depended on the presented frequency. In general (as discussed above) it was found that the higher the metronome frequency, the shorter the time interval in which subjects were able to integrate the temporal information into a perceptual unit.

The results of the present study suggest that this hypothetical integration process slows down in elder adults. This relationship proved to be valid regardless of the frequency of presented beats, i.e., independent of the information content of the presented stimuli. Similarly, a prolonged reproduction of temporal intervals was found in other investigations on mentally ill individuals as compared with the matched normal group. For example, Fraisse (1984) suggested that schizophrenics overestimate a one-second duration. Moreover, in well-studied populations of learning-impaired individuals, including dyslexics (Tallal et al. 1985, Shapiro et al. 1990), integration times and stimulus persistence measures were significantly slowed down. The atypical temporal processing discovered in our experiment may be a possible neurologic correlate of a decline in perceptual and intellectual functioning associated with normal chronological aging.

Several hypotheses were formulated about the neuropsychological basis of the normal aging process (see Hellige (1993) for a review). One hypothesis is that aging involves greater loss of right-hemispheric than left-hemispheric functions and, therefore, the right hemisphere ages more rapidly (e.g., Schaie and Schaie 1977). Another possibility is that the two hemispheres decline to the same extent with advancing age (e.g., Goldstein and Shelly 1981, Mittenberg et al. 1989).

To sum up, the lack of hemispheric asymmetry in MIIL as reported above as well as the specific processing disorder found in the present study, do not support "the right-hemisphere aging hypothesis" and indicates a deterioration in aging individuals of certain functions which typically cannot only be ascribed to the right hemisphere. By analogy, Goldstein and Shelly (1981) found that performance in subtests from the Halstead-Reitan neuropsychological battery associated with each hemisphere decreased linearly across the age range tested, and there was no subtest-by-age interaction. Similarly, there is no evidence for reduced performance in tests associated with right- rather than left-hemispheric functions, as reported by Mittenberg et al. (1989) or shown in studies investigating perceptual asymmetries using dichotic listening (Ellis and Oscar-Bermann 1984) or the visual-half-field method (Obler et al. 1984). It can be assumed that age-related changes in basic measures of integration periods might reflect a more general intra-hemispheric functional impairment.

4.2   Gender Differences in Temporal Integration

In humans, there are relationships between biological sex and mental cognitive abilities. In tests of general intelligence, gender differences consistently have been observed with respect to specific abilities (see Hellige (1993) for a review). Males tend to perform tasks that are spatial in nature, including block design, maze performance, mental rotation, mechanical skills better than females. Conversely, women tend to outperform men on some tests that require the use of language, such as verbal production, speed of articulation, and grammar. Recent studies suggest that sex differences in verbal and spatial abilities may be related to different cortical organization in males and females.

The crucial question to be considered here is whether there exist gender differences in the processing of temporal information in the domain of PTI. With this idea in mind, we reanalysed MIIL in males and females observed in the experiment with the metronome (see above). As can be seen in Fig. 5, for verbal responses MIIL in males was less dependent on the presented metronome frequency than MIIL in females. Referring to three potential strategies responsible for the integration task (see Fig. 1), it can be stated that males performed the task according to constant time, i.e., independently of the information content of presented stimuli, whereas women relied mostly on counting. These relations were especially pronounced for low metronome frequencies. It can be assumed that the two different strategies applied in the integration task in males and females may correspond to gender differences in specific abilities mentioned above. Better verbal skills in women may increase their preference for serial ordering or sequential labelling. Thus, they prefer to apply the verbal strategy (counting) in the present task. In contrast, the male use of nonverbal, global strategies based mostly on constant time might be interpreted as another specific aspect of spatial abilities. According to Linn and Peterson (1985), these abilities involve multiple processes which generally

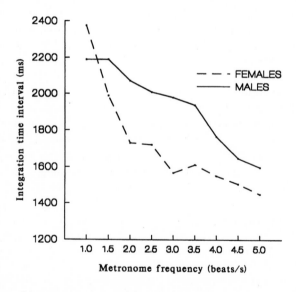

Fig. 5. Measured integration interval length for various metronome frequencies in male and female subjects (verbal response).

refer to the skill in transforming nonlinguistic, symbolic information. Alternatively, it may be postulated that men are more "field-independent" than women. In recent psychometric studies (Linn and Peterson 1985), men tend to outperform women on perception tests required to determine spatial orientation with respect to the orientation of their own bodies in spite of distracting information (Rod and Frame Test) or localization of simple figures within complex figures (Embedded Figure Test). More skilled male performance on these tests can be interpreted as higher "field-independence". By analogy, it seems that male performance might be less influenced by the information content of presented stimuli in the present experiment.

4.3   Cognitive Abilities and Temporal Integration

Another question is whether individual differences in cognitive abilities affect information processing in the considered temporal domain. In our study, the level of such abilities was assessed by the KAI-test given to each subject and scored according to the standard instruction. Next, a correlational analysis was undertaken to examine the relation between MIIL for a subset of metronome frequencies and individual cognitive abilities assessed by this test. (It should be noted that there were no systematic differences in the level of such abilities between men and women).

A negative Spearman correlation was found in individuals showing high and low levels of such abilities depending on the frequency of presented beats (Szelag et al. 1996). In particular, MIIL for the lowest frequency (1 beat per s) correlated negatively with those cognitive abilities, whereas MIIL for the two highest frequencies (4.5 and 5 beats per s) correlated positively with such abilities. In other words, the difference in MIIL observed for the lowest and the highest frequency appeared to be significantly smaller in people showing a high level of cognitive abilities than the difference observed in individuals with a low level of these abilities. We can conclude, therefore, that individuals with a high cognitive ability level probably integrate sensory information mostly using constant time, whereas those with a low level of cognitive functions rely mostly on counting, i.e., they mentally always connect the same number of beats (see Fig. 1). It can also be assumed that individuals characterized by a high level of cognitive ability can process sensory information relatively quickly: they display shorter reaction times, larger memory span, higher IQ, and demonstrate a higher level of behavioral activity. Consequently, they can add more successive events to form a perceptual unit when the information content of presented stimuli is high. Thus, individual differences in MIIL can provide interesting information about individual differences in the cognitive domain.

130   Elzbieta Szelag

5   Concluding Remarks

In this rewiew, an objective picture of the accumulated empirical findings on
the neuropsychological basis of PTI has been presented. We conclude that
PTI may constitute a general supra-modal and intra-hemispheric process of
great complexity. It appears that the sensory systems can hold information
up to the temporal limit of approximately 3 s. Moreover, some observations
suggest that the information can be integrated also in a shorter time interval.
The extent of the hypothetical integration mechanism was found to depend
significantly on several factors: the information content of the presented stimuli, age- and gender-related changes, and individual differences in cognitive
abilities.

Acknowledgments. This research was supported by the State Committee
for Scientific Research, grant 4P05E 09609, and the Alexander von Humboldt
Foundation.

References

Axel R. (1925): Estimation of time. *Arch. Psychol.* **12**, 5–77.

Efron R. (1990): *The Decline and Fall of Hemispheric Specialization* (Lawrence
Erlbaum, Hillsdale).

Ellis R.J. and Oscar-Berman M. (1984): Effects of aging and alcoholism on recognition of dichhaptically presented stimuli. *Int. Neuropsychol. Soc. Bull.* **13**, 14.

Fraisse P. (1984): Perception and estimation of time. *Ann. Rev. Psychol.* **35**, 1–36.

Goldstein G. and Schelly C. (1981): Does the right hemisphere age more rapidly
than the left? *J. Clin. Neuropsychol.* **3**, 65–78.

Hellige J. (1993): *Hemispheric Asymmetry: What's Right and What's Left* (Harvard
University Press, Cambridge).

Huff F.J. (1990): Language in normal aging and age-related neurological diseases.
In *Handbook of Neuropsychology, Vol. 4*, ed. by S. Corkin (Elsevier, Amsterdam),
169–196.

Huppert F. (1991): Age-related changes in memory: learning and remembering new
information. In *Handbook of Neuropsychology, Vol. 5* ed. by J.G. Geffman and
F. Boller (Elsevier, Amsterdam), 123–147.

Linn M. and Peterson A. (1985): Emergence and characterization of sex differences
in spatial ability. *Child Dev.* **56**, 1479–1498.

Mittenberg W.H., Seidenberg M., O'Leary D.S., and DiGiulio D.V. (1989): Changes
in cerebral functioning associated with normal aging. *J. Clin. Exp. Neuropsychol.*
11, 918–932.

Nichelli P. (1993): The neuropsychology of human temporal information processing.
In *Handbook of Neuropsychology, Vol. 8* (Elsevier, Amsterdam), 337–369.

Obler L.K., Woodward S., and Albert M.S. (1984): Changes in cerebral lateraliza-
tion with aging? *Neuropsychologia* **22**, 235–240.

Pöppel E. (1978): Time perception. In *Handbook of Sensory Physiology, Vol. 8:
Perception* (Springer, Berlin), 713–729.

Pöppel E. (1988): *Mindworks: Time and Conscious Experience* (Harcourt Brace Jovanovich, Boston).

Pöppel E. (1994): Temporal mechanisms in perception. *Int. J. Neurobiol.* **37**, 185–202.

Pöppel E., Chen L., Glünder H., Mitzdorf U., Ruhnau E., Schill K., and von Steinbüchel N. (1991): Temporal and spatial constraints for mental modelling. In *Frontiers in Knowledge-Based Computing*, ed. by V.P. Bhatkar and K. Rege (Narosa Publishing House, New Delhi), 57–68.

Ruhnau E. (1995): Time gestalt and the observer. In *Conscious Experience*, ed. by T. Metzinger (Imprint Academic, Thorverton), 165–184.

Schaie K.W. (1983) The Seattle longitudinal study: a 21-year exploration of psychometric intelligence in adulthood. In *Longitudinal Studies of Adult Psychological Development*, ed. by K.W. Schaie (Guilford, New York), 64–135.

Schaie K.W. and Schaie J.P. (1977): Clinical assessment of aging. In *Handbook of the Psychology of Aging*, ed. by J.E. Birren and K.W. Schaie (Van Nostrand-Reinhold, New York), 692–723.

Shapiro K.L., Ogden N., and Lind-Blad F. (1990): Temporal processing in dyslexia. *J. Learn. Disord.* **23**, 99–107.

Szelag E., von Steinbüchel N., Reiser M., Gilles de Langen E., and Pöppel E. (1996): Temporal constraints in processing of nonverbal rhythmic patterns. *Acta Neurobiol. Exp.* **56**, 215–225.

Tallal P., Stark R., and Mellitis D. (1985): The relationship between auditory analysis and receptive language development: evidence from studies of developmental language disorder. *Neuropsychologia* **23**, 527–534.

Neurophysiological Relevance of Time

Andreas K. Engel[1], Pieter R. Roelfsema[1,2], Peter König[3], and Wolf Singer[1]

[1] Max-Planck-Institut für Hirnforschung, Deutschordenstrasse 46,
D–60528 Frankfurt, Germany
[2] The Netherlands Ophthalmic Research Institute and Department of Medical
Physics, University of Amsterdam, P.O. Box 12141, NL–1100 AC Amsterdam,
The Netherlands
[3] The Neurosciences Institute, 10640 John Jay Hopkins Drive, San Diego, CA
92121, USA

> This surprising tendency for attributes such as form, color, and move-
> ment to be handled by separate structures in the brain immediately
> raises the question of how all the information is finally assembled, say
> for perceiving a bouncing red ball. It obviously must be assembled
> somewhere, if only at the motor nerves that subserve the action of
> catching. Where it's assembled, and how, we have no idea.
> (David H. Hubel, 1988)

1 Introduction

What does it mean, in neural terms, to see, to move, to communicate, or
to be conscious of the world around us? How does our nervous system enable us to behave in such complex ways, and to do so effortlessly minute for
minute of our daily lives? – Certainly, these issues are among those which
motivate neurobiological research and which continue to entangle scientists
into the fascinating endavour of studying animal and human brains. We are
still far from being able to answer these questions. Obviously, during the
past decades, neuroscientists have been tremendously successful in elucidat-
ing the constituents of neuronal networks at the cellular and molecular level.
Myriads of experimental studies have been concerned with the morphological
and physiological features of neurons, have inquired into their mechanisms of
growth, maintainance and death, of impulse generation and signal exchange,
have charted the ingredients of their membranes and cell bodies, - with the
final goal of tracing the functional properties of neurons down to the level
of individual molecules. Yet, despite the victorious career of this analytic
approach we are still lacking an appropriate understanding of the brain's integrative functions: How do all the known components interact as a system,
how can they develop synergy and be integrated into a functional whole? How
do networks of neurons aquire those emergent functional properties that become evident in unified perception and behaviour? It is only recently that
neurobiologists have seriously begun to address these enigmatic issues. In

part, this is due to the appearance of a new research paradigm, which has overturned many of the traditional assumptions that used to pervade neurobiological thinking: the connectionist approach to the workings of brain and mind (Rumelhart et al. 1986; McClelland et al. 1986).

It is part of this connectionist approach to believe that any complex behaviour as well as most cognitive functions are based on the activity of large neuronal networks, rather than on small highly specialized patches of brain tissue that can, for a proper understanding of their function, be treated in isolation. As suggested by the large body of evidence available today, even low-level perceptual and memory tasks or the execution of simple movements require the activation of numerous brain regions and a massively parallel processing of information. Consequently, we have to assume that neural representations, i.e., those activity patterns in which the brain processes and stores information, are of a highly distributed nature. Understanding how large neuronal populations are coordinated for the formation of such representational states is one of the central goals of current neurobiological research: How can responses of individual neurons in the brain be integrated into organized patterns of activity which are functionally effective? Many researchers agree that resolving this issue, which is commonly addressed as the "binding problem", would constitute a major step towards understanding integrative processes in the nervous system (Crick 1984; Sejnowski 1986; Damasio 1990, Treisman 1996). In the present contribution, we shall propose a specific hypothesis concerning the problem of integration in distributed neuronal networks. This hypothesis predicts that time may be the key to solving the binding problem (Abeles 1982; von der Malsburg 1981, 1995; Engel et al. 1992; Singer and Gray 1995; Pöppel, this volume).

2   The Binding Problem:
How Is Visual Information Integrated?

In what follows, we will focus on one specific example for the problem of neural integration, namely, the problem of feature binding and perceptual integration in vision. Visual perception clearly provides a paradigmatic case for studying the binding problem. Psychophysical studies have shown that the analysis of visual scenes involves at least two crucial processing steps (Treisman 1986, 1996). In a first step, specific features of objects such as, for instance, colour, motion or the orientation of contours, are detected by the visual system in a local and parallel manner. Subsequently, these perceptual components are integrated and bound into organized units to provide the basis for a coherent representation of the respective objects. As already shown by the studies of the Gestalt psychologists, this process of feature binding follows certain "Gestalt criteria" that include, for instance, the proximity or similarity of features (Köhler 1930).

Of course, the binding problem as characterized in the psychophysical domain has its correlate at the physiological level. From a neurobiological point of view, a mechanism for integration is required for several reasons. First, neurons in visual centers typically have spatially restricted receptive fields which – at least at early stages of processing – are small compared to the size of typical objects. Therefore, neuronal responses need to be bound within retinotopically mapped visual areas to represent the coherence of features across different parts of the visual field. Second, potential mechanisms for the integration of neural responses must act over considerable distances to achieve binding across different visual areas. It is now commonly agreed upon that different classes of object features are processed in distinct cortical areas serving as "feature maps" (Felleman and van Essen 1991; Bullier and Nowak 1995). In the monkey visual system, more than 30 visual areas have been characterized which can be grouped into different processing streams. In the cat, about 20 visual areas have been distinguished that differ considerably in the response properties of their neurons. Thus, neural activity must be integrated across the borders of individual areas to establish coherent representations of complex objects. Third, a versatile and highly flexible binding mechanism is required because, usually, multiple objects are present in the visual field all of which activate neurons in a large number of cortical areas. Thus, any mechanism supposed to solve the binding problem must be able to selectively "tag" feature-selective neurons that code for the same object and to demarcate their responses from those of neurons activated by other objects in order to avoid the illusory conjunction of features (von der Malsburg 1981).

More than a decade ago it has been suggested that the binding problem might be solved by a temporal integration mechanism (von der Malsburg 1981). According to this proposal, distributed cortical neurons could be bound into coherent assemblies by synchronization of their discharges with a precision in the range of few milliseconds (Fig. 1). This "temporal binding model" predicts that perceptual coherence is represented by the correlated firing of feature-sensitive neurons responding to the same visual object. In contrast, the discharges of cells responding to unrelated stimuli should not show such temporal correlation. The synchrony among neural discharges would provide an elegant way of "tagging" responses as being functionally related and of selecting these for further joint processing because synchrony can easily be detected by coincidence-sensitive neurons in other brain areas (Abeles 1982; Singer and Gray 1995; König et al. 1996). The absence of synchrony between different assemblies could then be exploited to achieve figure-ground segregation and segmentation of the visual scene (von der Malsburg 1981, 1995).

This temporal binding mechanism seems attractive because it avoids pitfalls of more classical models. In a seminal paper, Barlow (1972) suggested that complex objects could be represented by the activity of very few or even

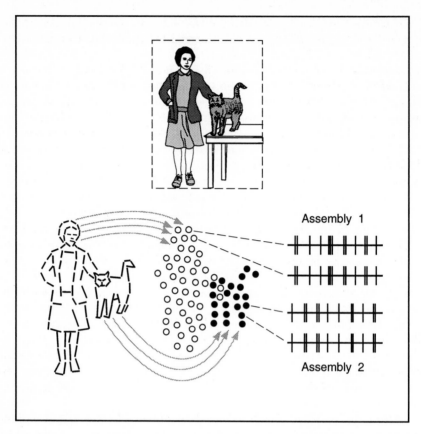

Fig. 1. Feature binding by synchronization. The temporal binding model assumes that objects are represented in the visual cortex by assemblies of synchronously firing neurons. In this example, the lady and her cat would each be represented by one such assembly (indicated by open and filled symbols, respectively). These assemblies comprise neurons which detect specific features of visual objects (such as, for instance, the orientation of contour segments) within their receptive fields (lower left). The relationship between the features can then be encoded by the temporal correlation among these neurons (lower right). The model assumes that neurons which are part of the same assembly fire in synchrony whereas no consistent temporal relation is found between cells belonging to different object representations.

individual neurons. A severe shortcoming of this "single neuron doctrine" is that it leads to a combinatorial explosion with respect to the number of representational symbols needed (Sejnowski 1986). According to this hypothesis, every new object eventually encountered will require the recruitment of a new dedicated cell in the visual cortex. Therefore, the number of neurons required for an adequate representation of a realistic visual environment grows beyond any physiologically plausible estimate. The temporal binding model avoids this serious limitation because the same neurons can, in principle, be recombined into new representational patterns by merely changing the temporal relationships.

Furthermore, the temporal binding model complements and extends the classical notion of object representation by distributed neuronal assemblies (Hebb 1949). The Hebb model implies that only one assembly is activated in a given area which "stands out" from a background of inactive cortical neurons and, thus, becomes salient as a representational unit for other brain areas. However, in natural visual scenes usually multiple objects will be present which, in addition, can be embedded in a complex background. Thus, scene segmentation and figure-ground segregation are likely to require the concurrent representation of multiple sets of coherent inputs. However, coactivation of multiple Hebbian assemblies leads to unresolvable ambiguities. Because all neurons involved raise their average firing rate, specific relationships between subsets of responses cannot be defined in the overall activity pattern. In contrast, a crucial advantage of temporal binding is that several coactivated object representations remain distinguishable (von der Malsburg 1981, 1995; Engel et al. 1992; Singer and Gray 1995). As illustrated schematically in Fig. 1, the temporal relationship between neuronal discharges permits the unambiguous distinction of subsets of functionally related responses. Thus, the overall activity pattern is endowed with significant additional structure which can serve for the fast and reliable selection of visual responses for further processing.

3   Evidence for Temporal Binding in the Visual System

Recent experimental work provides supportive evidence for this concept of a temporal binding mechanism. Numerous studies have shown that neurons in both cortical and subcortical centers can synchronize their discharges with a precision in the millisecond range (for reviews, see Engel et al. 1992; König and Engel 1995; Singer and Gray 1995). This has been demonstrated in particular for the visual system, but similar observations have also been made for the auditory, somatosensory and motor system. Moreover, related synchronization phenomena have been found in cortical association areas. In all these cases, neural synchrony has been revealed using the technique of correlation analysis (Fig. 2). By application of this method to multielectrode data obtained in cats, monkeys and several other species, results have been obtained that agree with the proposal that synchrony might be the "glue" that binds distributed neuronal activity into coherent representations.

In cats, several studies have demonstrated that spatially separate cells within individual visual areas can synchronize their spike discharges (Eckhorn et al. 1988; Gray et al. 1989; Engel et al. 1990; Brosch et al. 1995). Moreover, it has been shown that response synchronization can well extend beyond the borders of a single visual area. Thus, for instance, correlated firing has been

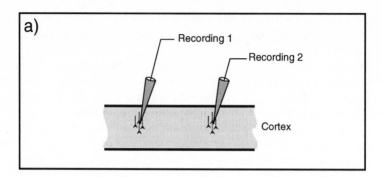

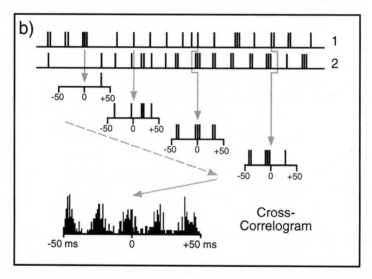

Fig. 2. Operation principle of correlation analysis. (a) To uncover neuronal synchrony, at least two electrodes are used to record from different sites simultaneously. (b) Correlation analysis can be applied to search for coincidences between the discharges recorded at sites 1 and 2. The algorithm counts incidents of precisely synchronized firing as well as the occurrence of discharges which lead or lag in phase relative to a reference spike in the first time series (in this case with phase shifts of up to ±50ms). Successive registration of these events yields the so-called crosscorrelogram. A central peak in the correlogram indicates synchronized neuronal firing. Eventually occurring side maxima reflect an oscillatory temporal modulation of the underlying neuronal responses (cf. Fig. 7).

observed between neurons located in different cerebral hemispheres (Engel et al. 1991a). Figure 3 illustrates one of the measurements where this type of long-range interaction has been studied. In terms of the temporal binding hypothesis, this result is important because interhemispheric synchrony is required to bind the features of objects extending across the midline of the visual field. In addition, temporal correlations in cat visual cortex have been studied for neurons located in different areas of the same hemisphere (Eckhorn et al. 1988; Engel et al. 1991b; Nelson et al. 1992). Of particular interest is the finding that interactions can occur between areas with quite different response properties such as the primary visual cortex (area 17) and the posteromedial area in the lateral suprasylvian sulcus (PMLS), an extrastriate area involved in the processing of object motion (Engel et al. 1991b). As mentioned already, different areas are assumed to process different aspects of visual stimuli (Felleman and van Essen 1991; Bullier and Nowak 1995) and, hence, interareal synchronization could mediate binding across feature maps which presumably is required for the complete representation of visual scenes. Finally, recent evidence shows that synchronous firing is not confined to the cortex but occurs already at the level of subcortical visual structures such as the retina, the lateral geniculate nucleus and the superior colliculus (Neuenschwander and Singer 1996; Brecht and Engel 1996). The majority of these studies on the cat visual system have been carried out in anesthetized preparations. However, the investigation of awake animals has yielded very similar results (Roelfsema et al. 1995).

A key finding in these studies on cat visual cortex is that response synchronization depends critically on the stimulus configuration. It could be demonstrated that spatially separate cells show strong synchronization only if they respond to the same visual object. However, if responding to two independent stimuli, the cells fire in a less correlated manner or even without any fixed temporal relationship. Figure 4 gives one example for this effect that has been observed both for the synchrony within (Gray et al. 1989; Engel et al. 1991c; Freiwald et al. 1995) as well as across visual areas (Engel et al. 1991b). These experiments demonstrate that Gestalt criteria such as continuity or coherent motion, which have psychophysically been shown to determine scene segmentation, are important for the establishment of synchrony among neurons in the visual cortex. As illustrated schematically in Fig. 5, the change in stimulus configuration is likely to induce a rearrangement of spatially distributed assemblies in the visual cortex. These observations provide strong support for the hypothesis that correlated firing could provide a dynamic mechanism which permits binding and response selection in a flexible manner.

Studies performed in other species suggest that the results obtained in the cat may be generalized. Neuronal synchronization with similar characteristics has been observed in the visual system of non-mammalian species as different as pigeon (Neuenschwander et al. 1996) or turtle (Prechtl 1994). Of particular relevance is, of course, the demonstration of comparable synchro-

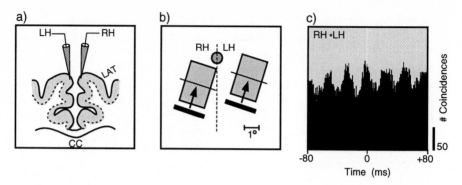

Fig. 3. Interhemispheric interactions in cat visual cortex. (a) In the anaesthetized animal electrodes were placed in the primary visual cortex of left and right hemisphere (LH, RH). LAT, lateral sulcus; CC, corpus callosum. (b) Plot of the receptive fields for the two recording sites. The neurons had identical orientation preferences. The receptive fields were located in the respective contralateral hemifield in the vicinity of the vertical meridian (dashed line). The cells were activated with bar-shaped stimuli moved in the same direction across the receptive fields. The circle represents the center of the visual field. (c) Crosscorrelogram computed for the recorded responses. The central maximum present in the correlogram indicates the coincident neuronal firing. (Modified from Engel et al. 1991a.)

nization phenomena in the visual system of primates. In monkeys, precise synchronization of spatially separate neurons has been demonstrated both within striate and extrastriate cortex (Ts'o and Gilbert 1988; Kreiter and Singer 1992; Livingstone 1996). In addition, correlated firing has been reported to occur between neurons of different visual areas (Frien et al. 1994; Salin and Bullier 1995). Moreover, the stimulus-dependence of neuronal interactions has recently been confirmed in monkeys (Kreiter and Singer 1996). Finally, EEG studies have provided similar evidence for precise synchronization in human visual cortex and suggest that the synchrony is relevant for the perception of coherent stimulus arrangements (Tallon et al. 1995). Taken together, these studies suggest that a temporal binding mechanism might be a general operating principle in the visual system of higher vertebrates.

The findings on synchrony reviewed above raise the question of how temporal synchrony is mediated, in particular between remote neurons located in different areas or even hemispheres (for reviews see Engel et al. 1992; Singer and Gray 1995; Salin and Bullier 1995). There is now evidence to suggest that synchrony is established at the cortical level and not mediated by a common driving input from subcortical structures. In the case of interhemispheric interactions it has been shown directly that cortico-cortical connections account for the observed interactions. If the corpus callosum is sectioned, response synchronization between the hemispheres disappears whereas synchronization within either hemisphere is preserved (Engel et al. 1991a). Further ev-

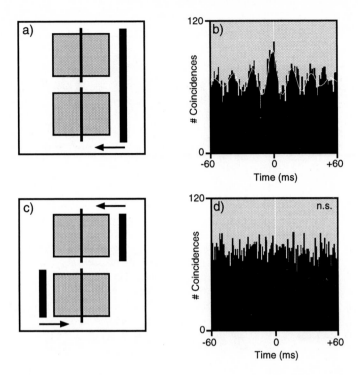

Fig. 4. Stimulus-dependence of long-range synchronization in the visual cortex. Data are from an experiment in which neuronal responses were recorded from two sites separated by 7 mm in the primary visual cortex of an anaesthetized cat. (a,c) Schematic plots of the receptive fields. At both recording sites the cells preferred vertical orientations. The colinear arrangement of the fields allowed the comparison of different stimulus paradigms. The neurons were activated with either a long continuous light bar moving across both fields (a) or two light bars moving in opposite directions (c). (b,d) The respective crosscorrelograms obtained with each stimulus paradigm. Using the long light bar, the two responses were synchronized as indicated by the strong modulation of the correlogram (b). Synchrony totally disappeared with the two incoherently moving stimuli (d). The graph superimposed to the correlogram in (b) represents a damped cosine function that was fitted to the data to assess the strength of the modulation. Abbreviation: n.s., not significant. (Modified from Engel et al. 1992.)

idence for the role of intracortical connections comes from experiments on strabismic cats. In the primary visual cortex, squint induction early in development leads to a breakdown of binocularity (Hubel 1988). The two sets of monocular cells tend to be clustered in columns driven by the left or the right eye. In a recent anatomical study of cats with divergent squint, Löwel and Singer (1992) demonstrated that only territories with the same ocular dominance are linked by tangential intracortical connections. In a subsequent

a)

b)

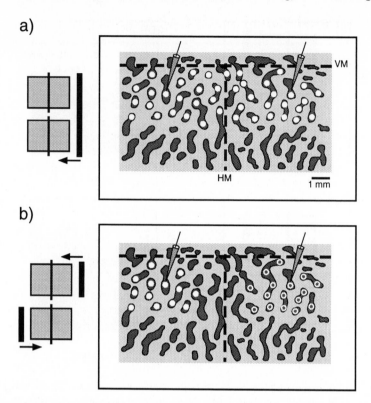

Fig. 5. Stimulus-dependent assembly formation in visual cortex. The experiment illustrated in Fig. 4 is shown in projection onto a map of the primary visual cortex. Dark bands represent territories where neurons prefer vertical stimulus orientations. Electrode symbols demarcate the recording sites. (a) Stimulation with a continuous light bar yields one large assembly of neurons (white dots). In this case, all cells responding to the stimulus are firing in synchrony. (b) If the neurons are activated with two stimuli moving in opposite direction, two spatially separate assemblies emerge (open and filled symbols). Each assembly is composed of synchronously active neurons, but there is no consistent phase relationship between the two assemblies. Abbreviations: HM, representation of the horizontal meridian; VM, representation of the vertical meridian. (Modified from Engel et al. 1992.)

correlation study it was shown that in these animals response synchronization occurs also preferentially between neurons driven by the same eye (König et al. 1993). Evidently, this correlation of functional interactions and connection topology provides further support for the notion that synchrony is achieved by coupling at the cortical level. Recent modelling studies provide additional support for this conclusion (Finkel and Edelman 1989; Sporns et al. 1989; König and Schillen 1991).

4   Functional Relevance of Synchronization

Although the temporal binding model offers an attractive conceptual scheme for understanding the integration of distributed neuronal responses, definitive evidence that the brain actually uses synchronization in exactly this way has not yet been obtained. Nonetheless, the study on cats with divergent squint provides the first hint that neuronal synchronization is indeed functionally relevant and related to the animal's perception and behaviour (König et al. 1993). Typically, humans and animals with a divergent strabismus develop a pattern of alternating fixation with the two eyes. Usually, monocular vision is undisturbed in these subjects, but they show a striking inability of combining information into a single percept that arrives simultaneously through the two eyes. As mentioned above, this deficit is accompanied by a loss of synchronization between neuronal populations with different ocular dominance (König et al. 1993). This correspondence between a functional deficit and loss of neuronal interaction clearly argues for the importance of correlated neuronal firing in normal visual perception.

Further evidence for the functional relevance of synchrony among cortical neurons comes from a recent correlation study of cats with convergent squint (Roelfsema et al. 1994a). Unlike subjects with a divergent strabismus, convergent squinters often use only one eye for fixation. The non-fixating eye then develops a syndrome of perceptual deficits called strabismic amblyopia. Symptoms of strabismic amblyopia include a reduced acuity of the affected eye, temporal instability and spatial distortions of the visual image, and the so-called crowding phenomenon: discrimination of details deteriorates further if other contours are nearby. Clearly, at least some of these deficits indicate a reduced capacity of integrating visual information and an impairment of the mechanisms responsible for feature binding. However, it turned out to be difficult to identify the neural correlate of strabismic amblyopia. Cortical neurons driven by the amblyopic eye do not show a loss of responsiveness and have essentially normal response properties. The results of the correlation study by Roelfsema et al. (1994a) indicate that the perceptual deficits of subjects with strabismic amblyopia may be due to a disturbance of intracortical interactions. As illustrated in Fig. 6, clear differences can be observed in the synchronization behaviour of cells driven by the normal and the amblyopic eye, respectively. Responses to single moving bar stimuli recorded from neurons dominated by the amblyopic eye showed much weaker correlation than responses of neurons driven by the normal eye. In addition, synchronization between neurons dominated by different eyes was virtually absent, confirming the results that had been obtained in cats with divergent squint (König et al. 1993). However, in terms of average firing rates the responses were very similar for neurons driven by the normal and amblyopic eye. These results indicate that strabismic amblyopia is accompanied by an impairment of intracortical interactions. Indirectly, this result corroborates the idea that

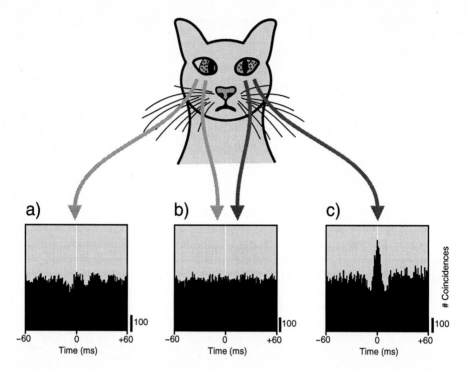

Fig. 6. Neuronal synchronization in the primary visual cortex of cats with strabismic ambylopia. The lower panel shows examples of crosscorrelograms between cells driven by the normal eye, by the amblyopic eye and between cells dominated by different eyes. Temporal correlation is strong if both recording sites are driven by the normal eye (c). Synchronization is, on average, much weaker between cells dominated by the amblyopic eye (a) and is in most cases negligible if the recording sites receive their input from different eyes (b). (Modified from Roelfsema et al. 1994a.)

neuronal synchrony is indeed employed for feature integration and, hence, an important variable for cortical information processing.

5   Potential Significance of Oscillatory Firing Patterns

A striking observation made in many of the correlation studies described above is that the emergence of synchronized states in the cortex is frequently associated with oscillatory firing patterns of the respective neurons (Eckhorn et al. 1988; Gray et al. 1989; Engel et al. 1990, 1991a, 1991b, 1991c; Kreiter and Singer 1992; König et al. 1993; Frien et al. 1994; Roelfsema et al. 1994a; Brosch et al. 1995; König et al. 1995; Freiwald et al. 1995; Neuenschwander et al. 1996; Neuenschwander and Singer 1996; Kreiter and Singer 1996; Brecht and Engel 1996; Livingstone 1996). These temporal patterns, which induce

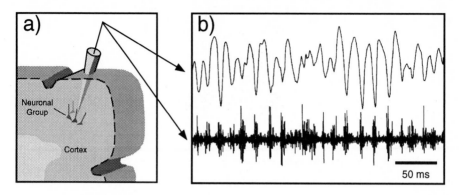

Fig. 7. Example of an oscillatory response in cat primary visual cortex. (a) In this experiment, the activity of a small group of neurons was recorded by means of a low-impedance electrode. (b) By filtering in different bandpasses, a local field potential (top, positivity upward) and multiunit spike activity (bottom) can be extracted from the raw electrode signal. As a light bar is passed through the cells' receptive field a clear oscillatory response is observed in the field potential, indicating that the neurons of the recorded cluster have engaged in a coherent and rhythmic firing pattern. Note the variability of both amplitude and frequency. The multiunit recording shows directly that different neurons (as indicated by spikes of different sizes) synchronize their discharges into a burst-and-pause firing pattern. Note that the spike bursts are in phase with the peak negativity of the field potential. (Modified from Roelfsema et al. 1994b.)

troughs and additional peaks in correlograms (Figs. 3,4), were first described in the visual system by Gray and Singer (1989). They observed that, with appropriate stimulation, neighbouring neurons tend to engage in grouped discharges which recur at frequencies in the gamma range between 30 and 80 Hz (Fig. 7).

In the preceding sections, we have argued that synchronous firing may serve for the dynamic binding and selection of distributed neuronal responses. This, then, raises the question of what role the oscillatory firing patterns might play as part of such a temporal binding mechanism. One possible answer is suggested by simulation studies that have investigated the interactions of coupled neuronal elements (König and Schillen 1991). The results of these studies indicate that an oscillatory modulation of neuronal responses may have crucial advantages for the establishment of synchrony between widely remote sites in the cortex (Fig. 8) and, thus, may be instrumental as a "carrier signal" for temporal binding in the brain (Engel et al. 1992, König et al. 1995). As described above, both the experiments on interhemispheric interactions and on squinting cats provide evidence that response synchronization is mediated by connections at the cortical level and not by synchronously driving subcortical input (Engel et al. 1991a; König et al. 1993). Interest-

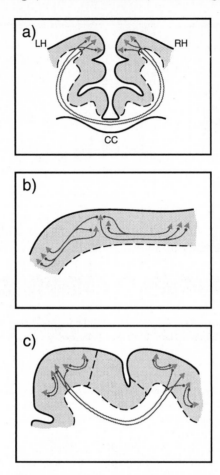

Fig. 8. Potential advantages of an oscillatory temporal structure in neuronal responses. (a) Synchronization of neuronal discharges without phase lag despite considerable conduction delays in long-range cortico-cortical connections. This constraint occurs e.g. in the case of interhemispheric interactions (cf. Fig. 3). (b) Synchronization of cell groups which are not directly coupled via intermediate neuronal groups. (c) Synchronization despite anisotropic delays which may occur due to large variation in the length or myelination of coupling connections. Abbreviations: LH, left hemisphere; RH, right hemisphere; CC, corpus callosum. (Modified from Roelfsema et al. 1994b.)

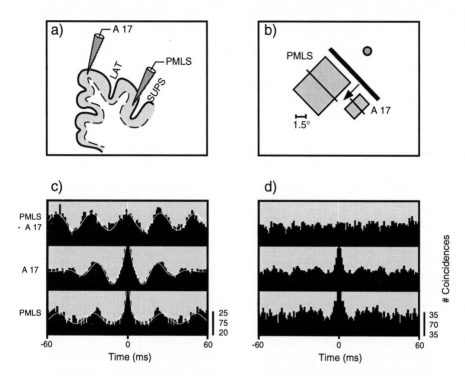

Fig. 9. Long-range synchronization is accompanied by oscillatory firing patterns. (a) Data are from an experiment in which one of the recording sites was located in the primary visual cortex (A 17), the other in extrastriate area PMLS. LAT, lateral sulcus; SUPS, suprasylvian sulcus. (b) The cells at the two recording sites had nonoverlapping receptive fields but their response properties were sufficiently similar to allow for coactivation with a single moving stimulus. In this case, a large number of trials were recorded using the same stimulus. Figures (c) and (d) show data from two representative response epochs. (c) Example of a trial where both cell groups showed narrow-banded oscillations, as indicated by the presence of multiple peaks and troughs in the autocorrelograms (middle and bottom). This is reflected in a strong modulation of the corresponding crosscorrelogram (top). (d) Example of a sweep where no temporal correlation was evident in the crosscorrelation function (top). The two autocorrelograms (middle and bottom) showed only a center peak without significant satellite peaks. As in Fig. 4, the continuous line superimposed on the correlograms represents a damped cosine function fitted to the data. (Modified from Roelfsema et al. 1994b.)

ingly, interareal and interhemispheric interactions have consistently been observed to occur with near-zero phase lag (Engel et al. 1991a, 1991b; Nelson et al. 1992; Frien et al. 1994; Roelfsema et al. 1995; cf. Fig. 3), although the underlying connections are known to exhibit considerable transmission delays. As suggested by the simulation studies, the establishment of synchrony without phase lag may be facilitated under these conditions if the respective neurons show oscillatory firing patterns (König and Schillen 1991). These models demonstrate that due to the recurrent temporal structure of such patterns reciprocally coupled neurons can entrain each other and improve synchrony within a few oscillatory bursts. Further advantages of oscillatory activity suggested by these models are that oscillating neurons can be synchronized via polysynaptic linkages without adding up of small phase-lags and that synchrony in such networks is robust despite considerable variation of the conduction delays (Fig. 8).

If the predictions suggested by these modelling studies hold true, cortical long-range synchronization should be closely correlated with the occurrence of oscillatory activity. However, such a relation would not necessarily hold for synchrony between closely spaced cells, since these tend to be strongly coupled without major delays in the respective connections. Recently, physiological evidence has been obtained in support of this hypothesis (König et al. 1995). Indeed, the physiological data demonstrate a close relationship between long-range synchronization and oscillatory firing patterns. If synchronization occurs over large distances within striate cortex, between areas or across the hemispheres it is almost always accompanied by oscillatory response patterns (Fig. 9). In contrast, short-range interactions can occur both with and without oscillatory modulation of the correlograms. These data support the hypothesis that oscillatory firing patterns may facilitate the establishment of synchrony between widely separate neuronal populations in the brain (Engel et al. 1992; König et al. 1995). If so, these discharge patterns may actually be a prerequisite for the binding of distributed neurons into assemblies and for the buildup of coherent representational states. Clearly, however, the role of oscillatory firing patterns needs to be substantiated by further experiments. Based on the available data, the possibility cannot be ruled out that synchrony is achieved by mechanisms which do not strictly require a band-limited temporal structure of neuronal responses.

6   Visual Information Processing: Implications of Temporal Binding

The key assumption of the temporal binding model is that time constitutes an important variable for neural information processing. The evidence reviewed above suggests that synchrony is not a mere epiphenomenon of cortical circuitry as assumed in classical cross-correlation analysis but, rather, reflects a dynamic functional coupling which is causally relevant in the system. By making time available as an additional coding dimension, the binding mechanism suggested by von der Malsburg allows to express specific relationships and, thus, to endow neuronal activity patterns with a rich internal structure. In the current debate about the implementation of representational states in connectionist models (Rumelhart et al. 1986) it has been argued that a combinatorial or "syntactic" structure of internal states is required to encode complex objects and to represent conjunctions of facts or events in the network (Fodor and Pylyshyn 1988). This problem can be settled by temporal binding because appropriate synchronization can establish specific bindings among the constituents of neural representations. This is not possible if average firing rates or activation levels serve as the only coding dimension (von der Malsburg 1981, 1995; Fodor and Pylyshyn 1988).

As described earlier in this chapter, sensory information processing is assumed to be massively parallel and distributed (Rumelhart et al. 1986; McClelland et al. 1986; Felleman and van Essen 1991; Bullier and Nowak 1995). For a number of reasons, the temporal binding model agrees well with this view on the architecture of sensory systems: Temporal binding permits the integration of sensory information without anatomical convergence of different processing streams (Felleman und van Essen 1991) to a single visual "master area". Furthermore, synchronization can implement parallel interactions between visual neurons and cells located in other cortical or subcortical systems. Of course, the notion of parallel processing is also in accord with the idea that representations are not instantiated by localized symbols ("grandmother neurons"), but by temporally defined assemblies extending across cortical areas.

However, several implications of the temporal binding model seem in conflict with still prevailing intuitions about sensory processing. For instance, the hypothesis advocated here implies a firmly holistic view on the architecture of the visual system. The representational code implemented by synchrony is strictly relational: the activity of individual neurons – considered in isolation – has a rather limited causal efficacy, because the functional significance of a neuron depends on the context set by the other members of the assembly. In this sense, processing has "gestalt-quality". This holistic stance, which converges with classical gestalt psychological notions (Köhler 1930), is incompatible with the atomistic framework that emphasizes the single neuron as the relevant level of description (Barlow 1972) – a view still pervading much of neurobiological thinking.

Moreover, the temporal binding model conflicts with the now-classical assumption that the visual cortex is characterized by a serial-hierarchical structure comprising independent processing pathways (Felleman und van Essen 1991). With respect to the cat visual system, the findings on interareal interactions reviewed above suggest that neurons in different processing pathways are by no means independently active but, rather, can be tied into the same assembly (Eckhorn et al. 1988; Engel et al. 1991b; Nelson et al. 1992). For the primate visual system we are still lacking the experimental proof of synchrony between areas that are part of different processing pathways. However, the temporal binding model predicts that also in primates synchrony should be observed across processing streams (in accordance with known anatomical cross-talk) which would serve for the binding of different classes of features. Interestingly, temporal correlations have recently been found between monkey primary (V1) and secondary (V2) visual cortex (Frien et al. 1994; Salin and Bullier 1995). This observation contradicts the standard assumption that these two areas represent different levels in the presumed hierarchy of visual areas (Felleman and van Essen 1991). As in the case of intra-areal interactions, synchrony between these two areas occurs, on average, with zero phase lag. Therefore, this finding suggests a parallel activation of V1 and V2 neurons rather than a serial-hierarchical transfer of information. This conclusion is supported by a number of additional anatomical and physiological findings (for a review, see Bullier and Nowak 1995).

Finally, it seems worth emphasizing that the temporal binding model is compatible with the idea that "top-down" influences play an important role for scene segmentation and figure-ground segregation (Treisman 1986) – which does not necessarily hold for more classical models of visual segmentation such as, for instance, the theoretical approach by Marr (1982). In a recent study (Munk et al. 1996) it could be shown that stimulus-induced synchrony in visual cortex is not fully determined by the peripheral stimulus and by the binding criteria implemented in the architecture of the respective visual area. Rather, the experiments demonstrate a profound influence of central modulatory systems which participate in the control of arousal and attention. Similarly, recent measurements in awake behaving cats show that temporal correlation between cortical cell populations critically depends on the behavioural situation the animal is engaged in (Roelfsema et al. 1995).

7 Synchrony as a General Mechanism for Integration and Response Selection

In this chapter, we have reviewed experimental evidence suggesting the existence of a temporal binding mechanism. The available data support the hypothesis that correlated firing may be functionally relevant in the brain for the binding of distributed neurons into coherently active assemblies and the dynamic selection of their responses for joint processing. Moreover, the results suggest that oscillatory firing patterns with frequencies in the gamma range may play a crucial role as carrier signals for the establishment of synchrony. So far, we have restricted our discussion to evidence obtained in the visual system. However, it seems likely that binding problems similar to those observed in vision are faced in other functional systems as well. The problem of visual feature integration just exemplifies a much more general problem of integration that always occurs in neuronal networks operating on the basis of "coarse coding" and distributed representation (Sejnowski 1986; Damasio 1990). Since information processing in other sensory modalities and in the motor system is also highly parallel, the needs to organize and bind distributed responses are similar to the visual system. Furthermore, information must be flexibly coordinated both across sensory modalities as well as between sensory and motor processes in order to allow for adaptive behaviour of the organism. The hypothesis pursued here predicts, therefore, that temporal binding mechanisms should exist in other cortical systems and, moreover, that synchrony should occur between different neural systems.

Indeed, several recent findings suggest that the temporal binding model can be generalized to other functional neural systems. Studies in non-visual sensory modalities and in the motor system indicate that synchrony and oscillatory activity may actually be quite ubiquitous in the nervous system. Synchronization with prevalence of the gamma frequency range is well known to occur in the olfactory bulb and entorhinal cortex of various species, where these phenomena have been related to the integration of odor information (Freeman 1988). In the auditory cortex, synchronized gamma oscillations have been described by several groups. In humans, these phenomena have been observed with EEG and MEG (magneto-encephalographic) techniques (Galambos et al. 1981; Pantev et al. 1991; Madler et al. 1991). In addition, temporal correlations in auditory cortex have been observed in animal experiments (Eggermont 1992; deCharms and Merzenich 1996). In the somatosensory system, interactions in this frequency range have recently been described in the awake monkey (Murthy and Fetz 1992). Finally, synchronized oscillatory firing has been observed in the hippocampus (Bragin et al. 1995).

Similar evidence is available for the motor system where neural synchronization in the gamma frequency range has been discovered in monkeys (Murthy and Fetz 1992) and, more recently, also in humans (Kristeva-Feige et al. 1993). It seems interesting to relate these findings to the dynamics of so-called central pattern generators, i.e., networks of coupled oscillators found in the spinal cord of vertebrates and in the nervous system of invertebrates which are involved in the execution of basic motor programs (Grillner et al. 1991). Although the frequency of rhythmic activity is far below the gamma range in these systems, the principles of pattern generation appear similar. The same sets of neurons can be used to generate a large diversity of different patterns by changing the coupling strength among oscillatory modules and by modulating oscillation frequencies.

Remarkably, there is evidence to suggest that synchrony may also play a role for sensorimotor integration. In monkeys, synchronization between sensory and motor cortical areas has been reported (Murthy and Fetz 1992; Bressler et al. 1993). Similar results have been obtained in a recent study on awake behaving cats that were trained to a visuomotor coordination task

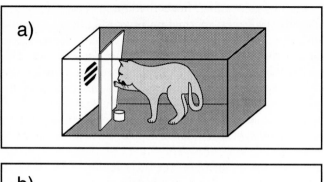

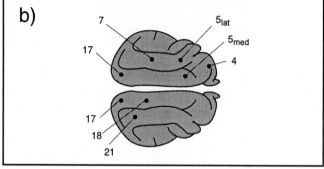

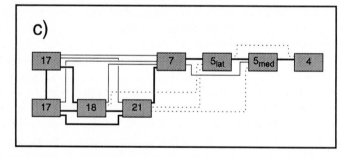

Fig. 10. Synchronization between visual, parietal and motor cortex in awake be-having cats. (a) The cats were situated unrestrained in a testing box and had to watch a screen through a transparent door. At the beginning of each trial, a grating was projected onto the screen. The cat had to respond by pressing the door with the forepaw and had to hold it until the grating was rotated. Upon change of the visual stimulus, the animal had to release the door. After correct trials, a reward was presented in a food well at the bottom of the box. (b) Location of the recording sites. Electrodes were implanted in areas 17, 18 and 21 of the visual cortex, in areas 5 and 7 of parietal cortex and in area 4 of the motor cortex where the forepaw is represented that the cat used for pressing the door. (c) Synchrony between these areas during the epoch where the cat was watching the grating and waiting for its rotation. Thick lines indicate strong correlation, thin and hatched lines show weak, but still significant interactions. Areas have been placed according to their position in the processing stream that links the visual cortex to the motor cortex. The diagram shows that precise synchrony is a global cortical phenomenon and is not restricted to the visual cortex. (From Roelfsema et al. 1995.)

(Fig. 10; Roelfsema et al. 1995). In these animals, local field potentials were recorded with electrodes chronically implanted in various areas of the visual, parietal and motor cortex. The results of this study show that synchronization of neural responses does not only occur within the visual system, but also between visual and parietal areas as well as between parietal and motor cortex (Fig. 10). Moreover, the experiments demonstrate that the temporal coherence between the areas recorded changes dramatically in different behavioural situations. Both findings suggest that synchrony may indeed be relevant for visuomotor coordination and may serve for the linkage of sensory and motor aspects of behaviour. The specificity of such interactions might allow, for instance, the selective channeling of sensory information to different motor programs which are concurrently executed. We assume that further studies of the synchronization between sensory cortices and the motor system may lead to valuable insights into how temporally coded sensory information might be used for suitable behavioural adjustments.

In conclusion, comparable synchronization phenomena are found in a large number of different functional systems. Thus, it seems justified to generalize the results obtained in the visual cortex and to suggest that temporal binding may be of general relevance for neural information processing. Importantly, these studies provide increasing evidence that temporal binding, in particular at frequencies in the gamma band, may also occur in the human brain. Yet, it must be kept in mind that for most of the functional systems considered above the available data demonstrate merely the presence of correlated neural activity. To prove that these temporal correlations do indeed serve for binding and response selection, additional data must be provided showing that the synchrony can be influenced by changes stimulus configurations and by alterations of the behavioural context. At this point, the studies of the visual system seem to provide the most convincing evidence for the existence of temporal binding mechanisms. Yet, even here further studies are required to directly demonstrate the causal relevance of temporal structure for perception.

References

Abeles M. (1982): *Local Cortical Circuits. An Electrophysiological Study* (Springer, Berlin).

Barlow H.B. (1972): Single units and sensation: a neuron doctrine for perceptual psychology? *Perception* **1**, 371–394.

Bragin A., Jandó G., Nádasdy Z., Hetke J., Wise K., and Buzsáki G. (1995): Gamma (40-100Hz) oscillation in the hippocampus of the behaving rat. *J. Neurosci.* **15**, 47–60.

Brecht M. and Engel A.K. (1996): Cortico-tectal interactions in the cat visual system. In *Artificial Neural Networks - ICANN 96*, ed. by C. von der Malsburg, W. von Seelen, J.C. Vorbrüggen, and B. Sendhoff (Springer, Berlin), 395–399.

Bressler S.L., Coppola R., and Nakamura R. (1993): Episodic multiregional cortical coherence at multiple frequencies during visual task performance. *Nature* **366**, 153–156.

Brosch M., Bauer R., and Eckhorn R. (1995): Synchronous high-frequency oscillations in cat area 18. *Eur. J. Neurosci.* **7**, 86–95.

Bullier J. and Nowak L.G. (1995): Parallel versus serial processing: new vistas on the distributed organization of the visual system. *Curr. Opin. Neurobiol.* **5**, 497–503.

deCharms R.C. and Merzenich M.M. (1996): Primary cortical representation of sounds by the coordination of action-potential timing. *Nature* **381**, 610–613.

Crick F. (1984): Function of the thalamic reticular complex: the searchlight hypothesis. *Proc. Natl. Acad. Sci. USA* **81**, 4586–4590.

Damasio A.R. (1990): Synchronous activation in multiple cortical regions: a mechanism for recall. *Semin. Neurosci.* **2**, 287–296.

Eckhorn R., Bauer R., Jordan W., Brosch M., Kruse W., Munk M., and Reitboeck H.J. (1988): Coherent oscillations: a mechanism for feature linking in the visual cortex? *Biol. Cybern.* **60**, 121–130.

Eggermont J.J. (1992): Neural interaction in cat primary auditory cortex. Dependence on recording depth, electrode separation, and age. *J. Neurophysiol.* **68**, 1216–1228.

Engel A.K., König P., Gray C.M., and Singer W. (1990): Stimulus-dependent neuronal oscillations in cat visual cortex: inter-columnar interaction as determined by crosscorrelation analysis. *Eur. J. Neurosci.* **2**, 588–606.

Engel A.K., König P., Kreiter A.K., and Singer W. (1991a): Interhemispheric synchronization of oscillatory neuronal responses in cat visual cortex. *Science* **252**, 1177–1179.

Engel A.K., Kreiter A.K., König P., and Singer W. (1991b): Synchronization of oscillatory neuronal responses between striate and extrastriate visual cortical areas of the cat. *Proc. Natl. Acad. Sci. USA* **88**, 6048–6052.

Engel A.K., König P., and Singer W. (1991c): Direct physiological evidence for scene segmentation by temporal coding. *Proc. Natl. Acad. Sci. USA* **88**, 9136–9140.

Engel A.K., König P., Kreiter A.K., Schillen T.B., and Singer W. (1992): Temporal coding in the visual cortex: new vistas on integration in the nervous system. *Trends Neurosci.* **15**, 218–226.

Felleman D.J. and van Essen D.C. (1991): Distributed hierarchical processing in the primate cerebral cortex. *Cerebral Cortex* **1**, 1–47.

Finkel L.H. and Edelman G.M. (1989): Integration of distributed cortical systems by reentry: a computer simulation of interactive functionally segregated visual areas. *J. Neurosci.* **9**, 3188–3208.

Fodor J.A. and Pylyshyn Z.W. (1988): Connectionism and cognitive architecture: a critical analysis. *Cognition* **28**, 3–71.

Freeman W.J. (1988): Nonlinear neural dynamics in olfaction as a model for cognition. In *Dynamics of Sensory and Cognitive Processing by the Brain*, ed. by E. Basar (Springer, Berlin), 19–29.

Freiwald W.A., Kreiter A.K., and Singer W. (1995): Stimulus dependent intercolumnar synchronization of single unit responses in cat area 17. *Neuroreport* **6**, 2348–2352.

Frien A., Eckhorn R., Bauer R., Woelbern T., and Kehr H. (1994): Stimulus-specific fast oscillations at zero phase between visual areas V1 and V2 of awake monkey. *Neuroreport* **5**, 2273–2277.

Galambos R., Makeig S., and Talmachoff P.J. (1981): A 40-Hz auditory potential recorded from the human scalp. *Proc. Natl. Acad. Sci. USA* **78**, 2643–2647.

Gray C.M. and Singer W. (1989): Stimulus-specific neuronal oscillations in orientation columns of cat visual cortex. *Proc. Natl. Acad. Sci. USA* **86**, 1698–1702.

Gray C.M., König P., Engel A.K., and Singer W. (1989): Oscillatory responses in cat visual cortex exhibit inter-columnar synchronization which reflects global stimulus properties. *Nature* **338**, 334–337.

Grillner S., Wallén P., Brodin L., and Lansner A. (1991): Neuronal network generating locomotor behavior in lamprey: circuitry, transmitters, membrane properties, and simulation. *Annu. Rev. Neurosci.* **14**, 169–199.

Hebb D.O. (1949): *The Organization of Behavior* (Wiley, New York).

Hubel D.H. (1988): *Eye, Brain, and Vision* (Freeman, New York).

Köhler W. (1930): *Gestalt Psychology* (Bell and Sons, London).

König P. and Schillen T.B. (1991): Stimulus-dependent assembly formation of oscillatory responses: I. Synchronization. *Neural Comput.* **3**, 155–166.

König P. and Engel A.K. (1995): Correlated firing in sensory-motor systems. *Curr. Opin. Neurobiol.* **5**, 511–519.

König P., Engel A.K., Löwel, S., and Singer W. (1993): Squint affects synchronization of oscillatory responses in cat visual cortex. *Eur. J. Neurosci.* **5**, 501–508.

König P., Engel A.K., and Singer W. (1995): The relation between oscillatory activity and long-range synchronization in cat visual cortex. *Proc. Natl. Acad. Sci. USA* **92**, 290–294.

König P., Engel A.K., and Singer W. (1996): Integrator or coincidence detector? The role of the cortical neuron revisited. *Trends Neurosci.* **19**, 130–137.

Kreiter A.K. and Singer W. (1992): Oscillatory neuronal responses in the visual cortex of the awake macaque monkey. *Eur. J. Neurosci.* **4**, 369–375.

Kreiter A.K. and Singer W. (1996): Stimulus-dependent synchronization of neuronal responses in the visual cortex of awake macaque monkey. *J. Neurosci.* **16**, 2381–2396.

Kristeva-Feige R., Feige B., Makeig S., Ross B., and Elbert T. (1993): Oscillatory brain activity during a motor task. *Neuroreport* **4**, 1291–1294.

Livingstone M.S. (1996): Oscillatory firing and interneuronal correlations in squirrel monkey striate cortex. *J. Neurophysiol.* **75**, 2467–2485.

Löwel S. and Singer W. (1992): Selection of intrinsic horizontal connections in the visual cortex by correlated neuronal activity. *Science* **255**, 209–212.

Madler C., Keller I., Schwender D., and Pöppel E. (1991): Sensory information processing during general anaesthesia: effect of isoflurane on auditory evoked neuronal oscillations. *Brit. J. Anaesth.* **66**, 81–87.

Marr D. (1982): *Vision* (Freeman, San Francisco).

McClelland J.L., Rumelhart D.E., and the PDP Research Group (1986): *Parallel Distributed Processing Vol. 2: Psychological and Biological Models* (MIT Press, Cambridge).

Munk M.H.J., Roelfsema P.R., König P., Engel A.K., and Singer W. (1996): Role of reticular activation in the modulation of intracortical synchronization. *Science* **272**, 271–274.

Murthy V.N. and Fetz E.E. (1992): Coherent 25- to 35-Hz oscillations in the sensorimotor cortex of awake behaving monkeys. *Proc. Natl. Acad. Sci. USA* **89**, 5670–5674.

Nelson J.I., Salin P.A., Munk M.H.J., Arzi M., and Bullier J. (1992): Spatial and temporal coherence in cortico-cortical connections: a cross-correlation study in areas 17 and 18 in the cat. *Vis. Neurosci.* **9**, 21–37.

Neuenschwander S. and Singer W. (1996): Long-range synchronization of oscillatory light responses in the cat retina and lateral geniculate nucleus. *Nature* **379**, 728–733.

Neuenschwander S., Engel A.K., König P., Singer W., and Varela F.J. (1996): Synchronization of neuronal responses in the optic tectum of awake pigeons. *Vis. Neurosci.* **13**, 575–584.

Pantev C., Makeig S., Hoke M., Galambos R., Hampson S., and Gallen C. (1991): Human auditory evoked gamma-band magnetic fields. *Proc. Natl. Acad. Sci. USA* **88**, 8996–9000.

Prechtl J.C. (1994): Visual motion induces synchronous oscillations in turtle visual cortex. *Proc. Natl. Acad. Sci. USA* **91**, 12467–12471.

Roelfsema P.R., König P., Engel A.K., Sireteanu R., and Singer W. (1994a): Reduced synchronization in the visual cortex of cats with strabismic amblyopia. *Eur. J. Neurosci.* **6**, 1645–1655.

Roelfsema P.R., Engel A.K., König P., and Singer W. (1994b): Oscillations and synchrony in the visual cortex: evidence for their functional relevance. In *Oscillatory Event-Related Brain Dynamics*, ed. by C. Pantev, T. Elbert, and B. Lütkenhöner (Plenum Press, New York), 99–114.

Roelfsema P.R., Engel A.K., König, P., and Singer W. (1995): Synchronization between transcortical field potentials of the visual, parietal and motor cortex in the awake cat. *Society for Neuroscience Abstracts* **21**, 517.

Rumelhart D.E., McClelland J.L., and the PDP Research Group (1986): *Parallel Distributed Processing Vol. 1: Foundations* (MIT Press, Cambridge).

Salin P.-A. and Bullier J. (1995): Corticocortical connections in the visual system: structure and function. *Physiol. Rev.* **75**, 107–154.

Sejnowski T.R. (1986): Open questions about computation in cerebral cortex. In *Parallel Distributed Processing Vol. 2*, ed. by J.L. McClelland and D.E. Rumelhart (MIT Press, Cambridge), 372–389.

Singer W. and Gray C.M. (1995): Visual feature integration and the temporal correlation hypothesis. *Annu. Rev. Neurosci.* **18**, 555–586.

Sporns O., Gally J.A., Reeke G.N.Jr., and Edelman G.M. (1989): Reentrant signaling among simulated neuronal groups leads to coherency in their oscillatory activity. *Proc. Natl. Acad. Sci. USA* **86**, 7265–7269.

Tallon C., Bertrand O., Bouchet P., and Pernier J. (1995): Gamma-range activity evoked by coherent visual stimuli in humans. *Eur. J. Neurosci.* **7**, 1285–1291.

Treisman A. (1986): Properties, parts and objects. In *Handbook of Perception and Human Performance*, ed. by K. Boff, L. Kaufman, and I. Thomas (Wiley, New York), 35.1–35.70.

Treisman A. (1996): The binding problem. *Curr. Opin. Neurobiol.* **6**, 171–178.

Ts'o D.Y. and Gilbert C.D. (1988): The organization of chromatic and spatial interactions in the primate striate cortex. *J. Neurosci.* **8**, 1712–1727.

von der Malsburg C. (1981): The Correlation Theory of Brain Function. Internal Report 81-2, Max-Planck-Institute for Biophysical Chemistry, Göttingen. Reprinted in *Models of Neural Networks II*, ed. by E. Domany, J.L. van Hemmen, and K. Schulten (Springer, Berlin 1994), 95–119.

von der Malsburg C. (1995): Binding in models of perception and brain function. *Curr. Opin. Neurobiol.* **5**, 520–526.

Sensations of Temporality: Models and Metaphors from Acoustic Perception

Manfred Euler

Fachbereich Physik, Universität GH Paderborn, Warburger Strasse 100, D–33098 Paderborn, Germany

> The objective world simply is, it does not happen.
> (Hermann Weyl, 1949)

1 The Timeless Present in Physics Versus the Presence of Time in Perception

Attempts to grasp the nature of time are amongst the deepest and most puzzling challenges to the human mind. Our immediate perception of the passage of time is connected with consciousness, an equally enigmatic phenomenon. Everyone will agree to experience time as a continuous stream. However, a closer look via introspection breaks up the continuum and reveals discrete elements showing up as thoughts or mental events. Thus, in our conscious experience of time continuity and discreteness are delicately interwoven. We are endowed with a feeling of nowness binding together events from external or internal perceptions which occur within a certain window of space and time. The immediate conscious presence appears to possess a temporal extension, which allows to access voluntarily a selection of specific events, somehow laid out within an extended present. In a simple picture, we can compare the experience of events in the flow of time with viewing pearls lined up on a string in continuous motion with past events "dropping successively away, and the incomings of the future making up the loss" (James 1950).

Physics abstracts from subjective experience. Decoupling physical time from subjective time is one of the central starting points of modern physics. This is done by identifying time with the reading of clocks. Measuring time means counting periodical processes. The discrete counts are interpolated according to the continuum hypothesis, and finally time is mapped to the continuum of real numbers. This concept of time as a continuous mathematical parameter appears closely related to subjective time, because it is an intuitive generalization of our bodily experience of rhythmic processes, like the heartbeat, which lead to counting. Apparently, however, it expels the subject of experience from the mathematical formulation of spacetime processes. There is a clash between the static spacetime view of relativity theory and our experiences of time, expressed by the introductory quotation. Weyl adds the following explanation (Weyl 1949): "Only in the gaze of my consciousness, crawling upward along the lifeline of my body, does a section of this world come to life as a fleeting image in space, which continuously changes in time."

Figure 1 visualizes the puzzle. In physics, the "now" has no extension. It is a point between the passive light cone of the past and the active light cone of the future. Obviously, such a representation of the physical present leaves no space for our subjective experience of time. We have become strangers in the abstract symbolic world of mathematical physics, which our brains have created. Conscious experience, which immerses us in the strange flow of time, is the primary reality to us, but it has the status of an illusion in physical theories. Is it possible to reconcile our perceptions of time with physics? Is the continuum hypothesis adequate to describe subjective time?

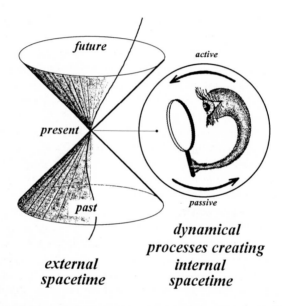

Fig. 1. Recurrent dynamical processes with a broad spectrum of cycle times make up the conscious observer and bring about the sensation of temporality.

2   The Clockwork Metaphor and the Conscious Observer: Views from Within the Mechanism

The conscious observer in Fig. 1 is depicted as an open dynamical system. He or she receives input from and acts on the physical world, which the observer is part of. Subjective time is considered as an emergent internal variable, resulting from complex dynamical processes of our brain. The feeling of nowness, although biologically highly important, still remains an illusion. From the point of view of dynamical systems theory, the observer is an open, self-sustaining system with many levels of feedback and self-reference. Such a view quite naturally leads to the hypothesis that some puzzles of consciously experienced time can be resolved by investigating information dynamics in recurrent networks, where cyclic information flows and time delays on many levels and timescales occur.

As a full theory of conscious perception is beyond reach, we are going to consider simple dynamical processes of the clockwork type, which are complex enough to provide some insight into the perception of temporality. Such a mapping to simple systems is possible because clocks can be regarded as prototypes of systems, which implement a cyclic flow of information. In a provokingly reductionistic metaphor, one could state that we perceive time, because we are made up of clocks. Clockwork-like properties with a more or less pronounced regular oscillatory behavior are ubiquitous in biology and emerge on various timescales and levels of the nervous system. In spite of their mechanistic taste, systems of the coupled clockwork type possess a surprising power in modeling the active biological information processing substrate on various levels. Speaking of clocks, physicists tend to think of ideal Hamiltonian systems like rotating pointers or harmonic oscillators, isolated as far as possible from the disturbing influences of their surroundings; but clocks on biological substrates are of a completely different type. Their operation is based on irreversibility and openness. Due to the inherent nonlinearities, their rhythms are flexible and mutable. They can be modulated, allowing them to carry and to process information. Contrary to our intuition, random noise is an integral part of their function.

What are relevant structure principles, underlying the internal clockwork dynamics and connecting it with processes and events of the external world? How far does the metaphor of the perceived present encoded in dynamical processes lead? These questions will be addressed with a strong emphasis on hearing, because in this sensory channel the temporal structure of signals plays a decisive role. The auditory sense comprises the broadest span of perceptible timescales, ranging from tens of microseconds in binaural hearing roughly up to the range of seconds in the perception of musical rhythms or in speech. With respect to working out the puzzles of temporal effects, it is crucial to have a two level access to auditory perception: an objective one from the outside, analyzing its function by physical and physiological methods, and a subjective one from the inside mediated via introspection and psychophysical methods.

The present psychophysical approach will demonstrate links between temporal phenomena in auditory pattern perception and the behavior of dynamical systems of the coupled clock type. On the lower end, at the sensory periphery, these systems are surprisingly effective, e.g. providing high performance of the sensory organs. On the upper end, even certain voluntary high level perception processes, accessible via guided introspection, fit into the dynamical patterns of this class of models. Thus, the psychophysical approach can provide highly interesting unifying views from within. In spite of its power to establish connections between low level material and high level mental dynamics, it does not resolve the paradox of how mental properties emerge from a material substrate. However, as we can put ourselves into the position of both internal and external observers, we are able to appreciate the analogies between material and mental dynamics (Euler 1994).

3   Hearing as Seeing: Reconstructing Events in Auditory Spacetime from Two Continuous Signal Streams

Figure 2 shows an array of discrete blocks, which appear unconnected, carrying no information at first sight. This puzzle is an attempt to visualize the essence of the present discussion: How do we get the illusion of a continuously flowing sensation, although the conscious mind apparently puts together discrete chunks of information? The chunks are represented by seemingly unrelated patterns. Viewed in an isolated way, they do not make much sense. Nevertheless, the integrating action of our mind is able to connect the chunks and to embed them into a meaningful context. The impatient reader may immediately turn to the solution of the puzzle in Fig. 6, which answers metaphorically the question of how the mind links events and creates time.

The patient reader is asked to switch from seeing to hearing, and to consider the opposite process: How do we construct sensory events from continuous flows of signals? Our two ears receive two streams of sound pressure fluctuations, which represent a superposition of signals from a variety of sources. The extraction of discrete events from these two continua is the central issue of auditory perception by which acoustic messages are decoded and sound sources are located in space and time.

That part of hearing which refers to the analysis of signals is rather easily accessible to investigation, and a formulation of the relevant dynamical prin-

Fig. 2. A collection of seemingly unrelated chunks of information. How is it possible to bind them together in a meaningful way? For a solution of the puzzle see Fig. 6.

ciples is possible. But analysis is only one aspect of the perception process. It has to be complemented by a synthesis of the objects of analysis to make up the perceived stream of events. While analysis can be described in fully mechanistic terms, the synthesis problem is highly puzzling with many holistic and "mind-like" facets. Nevertheless it is tempting to relate and transfer the dynamical processes found in analysis to the synthesis side of the perception process. This switching may appear daring, but it is not completely misplaced, assuming a generic logic of self-organization in physics or biology. Although nature abounds with phenomena, it is very parsimonious with the underlying structure principles. The distinction commonly made between perceptions, sensations, and cognitive processes does not appear fundamental in this respect.

First, in order to discuss adequately temporal effects in auditory perception, we consider how acoustic signals are represented internally, because auditory temporality is interwoven with spatial effects in highly subtle ways. As we cannot develop a full model of hearing, we shall give a general pictorial presentation in terms of auditory scene analysis (Bregman 1990), an approach describing auditory function in terms of biological requirements. Although humans are guided primarily by vision, we encounter many situations, in which it is highly beneficial to have an additional acoustic model of the surrounding world, conveying relevant events through a second, independently working channel. Apart from speech communication, the main function of hearing is to tell where sound sources, if stationary, are receding or approaching.

Think of the task of monitoring the surface of a lake. By a suitable analysis of the wave patterns it is possible to detect and classify the underlying events. For instance, the transient disturbance of an impacting stone differs from the patterns of moving boats. Additionally, it is possible to determine the position of the sources by an evaluation of the wave field, a task which is known as the inverse problem. Nature's solution to the acoustic inverse problem is based on correlating signals from two sufficiently distant places, the two ears. Although this is only an approximate procedure, it proves highly effective.

In the lake surveillance metaphor, binaural hearing can be compared with analyzing and correlating surface waves from two places, e.g., two canals. In simple situations with only one well-defined disturbance, measuring the time delay between the arrival of wavefronts in the canals suffices to detect the direction of the incoming signals. In more complex situations, with several sources involved, the individual signals have to be separated, before a correlation can be performed. In principle, the analysis can be done by specially shaping the canals. For instance, it is possible to create a waveguide with dispersive properties by a suitable variation of canal width coupled with depth. Such a dissipative waveguide is similar to a Fourier analyser, transforming the incoming signals to spatiotemporal patterns along the canal. After such a dispersive preprocessing, the incoming sequential signal streams are sampled in parallel according to their frequency components. This allows a more sophisticated correlation, comparing the vibration patterns at corresponding places along the canals. The spatially dispersive spectrum analysis separates signal components from superpositions, their temporal correlation makes it possible to locate individual events. As a result, the lake metaphor shows the necessity of spatiotemporal processing in order to reconstruct events from two wave streams.

This duality of spatial and temporal representation of sound signals is essential to acoustic perception. A fast mechanical spatiotemporal prepro- cessing of sound signals takes place along the basilar membrane of the inner ear, which acts as a dispersive waveguide, performing frequency analysis via travelling waves and mapping frequencies to strongly localized regions of max- imum vibration, with an approximately logarithmic dependence of frequency on distance along the basilar membrane. In the subsequent neural processing, the mechanical excitation pattern is sampled in parallel by sensory neurons (inner haircells), coded into neural signals, and transferred towards the brain. The neural projections of the auditory pathways preserve the tonotopic or- ganization. Even at the top layer of processsing, frequencies are continuously mapped to neural space in the auditory cortex. In a way, this mode of audi- tory processing converts time (frequency) to space.

In addition to the spatial mode, there is a second channel of processing, representing the temporal structure of acoustic signals. This is the so-called periodicity principle, by which the acoustic nerve transmits the low frequency temporal stucture of acoustic stimuli in a quasi-analog way, similar to a bunch of telephone lines. The periodicity principle is based on modulating the timing of neural spikes. The ability to encode information neurally via the timing of spikes or the phase of neural cycles is an important addition to the canonical textbook principles, which are restricted to neural information encoding via spike rate. Metaphorically speaking, auditory neurons not only have to tell how much they are excited. It is equally important for them to tell when they are excited. The phase of oscillators carries information, and the possibility to modulate the timing of internal oscillators by external stimuli has important consequences for the present question of perceiving time. The temporal chan- nel must be fast enough to allow real-time processing of neural periodicity up to 1 kHz, which is roughly the limiting frequency of the phenomenon of binaural beats (Oster 1973) and of frequency-following potentials (Euler and Kießling 1981). Fast real-time correlation is essential to binaural processing. There must be specialized networks to process binaural time delays down to the 10^{-5} sec range, a precision easily achieved by humans in directional hearing (Keidel 1975).

In analogy to the above lake surveillance metaphor, neural representations of the acoustic world are based both on spacelike and timelike representations. As a simple model of the internal arena of inspecting acoustic events we may think of a two-dimensional manifold with one dimension connected with the basilar membrane position. The second dimension represents the flow of time. Here one may think of the spreading of an excitation pattern in a time delay network. Combining both dimensions results in a hypothetical two-dimensional spatiotemporal internal representation of acoustical patterns quite similar to the symbolic notation of music, where the vertical coordinate is equivalent to the logarithm of frequency and the sequence of events in time is arranged along the horizontal coordinate. We shall call this platform of internal auditory pattern analysis "auditory spacetime" for short.

Metaphorically speaking, hearing can be regarded as seeing spacetime patterns. In vision, two-dimensional retinal patterns are processed. In hearing, one dimension is spacelike (cochlear coordinate) and the other is timelike (neural periodicity). Obviously, a "frozen" window of auditory spacetime is accessible to the conscious internal observer, similar to a static two-dimensional picture. The analogy to vision, though it appears daring at first sight, will turn out to be quite productive, for instance in explaining the parallels between certain auditory and visual illusions.

4 Noisy Clockworks and Their Unreasonable Effectiveness in Biological Information Dynamics

The clockwork metaphor applies to different levels of auditory processing, both mechanical and neural. The basic perception paradigm can be considered as driving internal clocks (self-sustained oscillators) by external signals. However, in this biological clockwork world, noise is essential and possesses many constructive facets. A variety of effects occurs, when a clockwork-like process is subject to external perturbations. Huygens, the inventor of modern clocks, already discovered the phenomenon of phase locking. He was fascinated by the fact that two pendulum clocks, attached to the same wall, tend to oscillate in synchrony after some time and to maintain their common rhythm, although their coupling is very weak (Huygens 1893). This perfect phase locking is a very dramatic way of how a periodic signal can be "perceived" by a dynamical system. More generally speaking, perceptions are encoded by modulating the timing or the phase of internal oscillations. This scheme describes mechanical vibrations at the sensory periphery and for neural periodicity as well. It even provides clues for a generic dynamic background of various effects of perceiving time.

Activity and autonomy on the lowest level of information processing in the receptor organ come as a surprise. The inner ear is not a passive sound receiving system like a microphone. It is a mechanically active system, which pumps metabolic energy into basilar membrane vibrations by nonlinear feedback processes. This active function is essential to bring about the extreme performance of hearing in terms of sensitivity, frequency discrimination, and dynamic range. As a side effect, the sound receiver "ear" becomes a sound generator. Weak signals are sent out spontaneously or as active echos of incoming sounds. These otoacoustic emissions can be considered as the fingerprints of autonomy, of the ears' internal active processes. In most healthy ears a few rather stable narrowband emissions occur with an equivalent sound pressure in the hearing threshold range (Zurek 1985).

The clockwork character of the ears' internal gears can be demonstrated by the Huygens effect of synchronisation. The spontaneous internal modes are attracted by sound signals with suitable frequencies. As a result of this pulling over, the internal oscillation is phase locked, amplifying the external signal and suppressing the noise. Outside the locking range, additional interaction effects like combination products occur, which are highly relevant to acoustic perception. For a more detailed description see Euler (1996a).

The driven clockwork model also provides important insights into the periodicity principle, by which the neural system continuously represents low frequency acoustic signals, although the underlying information carriers are discrete neural spikes. The basic limitations are set by the sampling theorem. As an individual nerve cell can fire at a maximum of 500 spikes per second, a limiting frequency of the order of less than 250 Hz is to be expected theoretically. However, a system of nerve fibers is able to transmit higher frequencies. Basically, this is made possible by the redundancy of processing spike trains from several fibers in parallel. Spike sequences from a group of neighbouring neurons, driven by the same signal, are superposed synaptically (Spekreijse and Oosting 1970). By parallel processing and by the integrating action of synapses a continuous signal is reconstructed from a superposition of several coherent spike sequences.

Again, phase locking is important to generate coherence in the firing sequence of neurons. Neural spike trains usually come as a Poisson sequence (Keidel 1989). Information is encoded by modulating the timing of spikes. This is the noisy version of the Huygens phaselocking effect. In contrast to the synchronization of ideal clocks, phase locking is not tight because of the inherent noise. There is a considerable phase jitter, degrading the temporal coherence of the spike sequence, which becomes more pronounced if the driving signal increases.

The phaselocking of ideal clocks works practically in an "all or nothing"-way, with a sharp phase transition from the unlocked to the locked state. Noise makes these sharp borders fuzzy, facilitating a smooth transition from discreteness to continuum. In moving from one neural processing level to the next, synapses have to be passed, where a change in representation occurs. Discrete events (neural spikes) are summed up and transformed to continuous signals (postsynaptic potentials). The noisy timing of spike sequences has a smoothing effect on the superposition and the integration of spike trains. As a result, the extremely nonlinear behavior of the neural system on the level of individual spikes is essentially linearized by synaptic action. Information is encoded continuously on the next level of processing, building up on a substrate that works essentially nonlinear and discrete. Such a switch between continuous and discrete representations takes place whenever synapses are passed. It is repeated at each processing stage.

Beyond this smoothing and linearizing action of noise in the transition from discreteness to continuum, there are even more unexpected and farther reaching beneficial effects. We have to rethink the role of noise in biological information dynamics completely. In the case of driving a neuron by a periodic signal, an addition of noise at the input will even result in a better signal to noise ratio at the output (Longtin et al. 1991). This surprising phenomenon of improving the signal to noise ratio by an addition of noise is called stochastic resonance. Although the effect is pretty dramatic and generic to a variety of nonlinear systems, it has been discovered only recently (see Wiesenfeld and Moss (1995) for a review).

Stochastic resonance in inner ear mechanics provides another instructive example of the constructive use of noise in an active biological substrate, not degrading but even increasing its perceptiveness to low power sound signals close to threshold. Obviously, the inner ear uses thermal noise to tune the cochlear amplifier and to increase threshold sensitivity. A mechanical threshold amplitude in the subatomic range is achieved far below the minimum amplitude required for a mechanical gating of molecular switches in haircells (Euler 1996b).

In his influential book *What is Life?*, Schrödinger gave a vivid description of biological clocks. He regarded their suffering from "thermal fits" highly important to their function (Schrödinger 1944). His emphasis on the relevance of Brownian motion for the function of biological clocks has proved prophetical, as the surprising phenomenon of stochastic resonance shows. By restricting our views to the behavior of linear systems we are used to consider noise nocuous, e.g., degrading the function of technical clocks. We have to come to terms with the counterintuitive effectiveness of noise and "blind" chance. In nonlinear dynamics, we have just begun to understand its constructive function, a feature that evolution has exploited in many creative ways since the beginning of biological time on our planet.

To describe the above effects of timing and of information processing by noisy clocks mathematically, we consider a prototype clock, the van der Pol oscillator (van der Pol 1926). Basically, this is a harmonic oscillator where, instead of a frictional resistance proportional to velocity, the amplitude x feeds back on the coefficient of friction in a nonlinear way according to $\gamma(x^2 - 1)$. A negative friction produces an exponentially increasing oscillation at small amplitudes, which will reach saturation as soon as the in- and outward energy flows are in balance. For small values of the nonlinearity parameter γ the system will exhibit limit cycle oscillations with the natural frequency ω_o of the passive oscillator (Hayashi 1964). If the system is driven by a periodical external force, a beating effect resulting from internal and external periodicity is expected. As the frequency ω of the driving force approaches the the eigenfrequency of the oscillator, the beating disappears all of a sudden. As soon as the detuning $(\omega - \omega_o)$ is below a certain threshold, the system responds only at the driving frequency ω. This is the Huygens effect of phaselocking, also called frequency entrainment or synchronization.

The periodically driven van der Pol oscillator is governed by the equation (Hanggi and Riseborough 1983):

$$\ddot{x} + \gamma(x^2 - 1)\dot{x} + \omega_o^2 x = \gamma E \sin(\omega t). \tag{1}$$

Assuming that the solution is of the form $x(t) = a(t)\cos(\omega t + \phi(t))$, the equation can be reduced to two coupled order parameter equations for the amplitude $a(t)$ and the phase $\phi(t)$ (mathematical details in Hanggi and Riseborough (1983)). For the present discussion of timing effects in clockwork-like systems, we confine ourselves to a discussion of the phase. For small detuning, the first order differential equation for the phase reads:

$$\frac{d\phi}{dt} = (\omega - \omega_o) - \left(\frac{\gamma E}{2a\omega}\right)\cos\phi. \tag{2}$$

The evolution of the phase is governed by a potential V, which looks like a tilted washboard (Fig. 3):

$$V(\phi) = -(\omega - \omega_o)\phi - \left(\frac{\gamma E}{2a\omega}\right)\sin\phi. \tag{3}$$

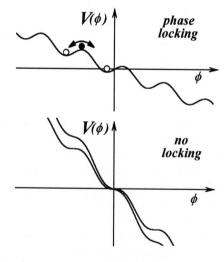

Fig. 3. Tilted washboard potential, underlying the evolution of the phase of the driven van der Pol oscillator. A synchronization of the oscillator is possible if detuning is sufficiently low (upper diagram). At an increasing detuning, the phase of the oscillator changes from the locked state to a continuously changing phase (lower diagram).

The washboard potential diagram gives a visual solution of the phase locking problem. The dynamics of the phase can be symbolically linked with the heavily damped motion of a material particle in a potential landscape with a phase rate proportional to the potential gradient:

$$\frac{d\phi}{dt} = -\frac{\partial V(\phi)}{\partial \phi}. \tag{4}$$

There is a maximum detuning $(\omega - \omega_o)$, for which a phase locked solution exists. The particle and, correspondingly, the phase are trapped in one of the many minima of the washboard potential. For an increasing detuning the potential is increasingly tilted. At a certain critical detuning, the minima disappear and the particle starts a downhill motion, comparable to skiing on a humpy track. This corresponds to an effectively free running phase, which includes phase modulation effects. Additionally, a more realistic consideration of biological clocks has to consider noise. The addition of noise to the periodic forcing contributes to the unlocking before the critical slope is reached and smoothes the sharp transition between the unlocked and the locked state (Stratonovich 1967).

The simple model of driven active oscillations allows us to give a rather coherent view of the highly effective signal processing on the level of receptor organs in the inner ear. Moreover, the same dynamical principles that make sensory systems sensitive are equally important for representing temporal effects internally. The phase of oscillators carries information, and the effects of modulating this timing provide an adequate conceptual framework for discussing various effects of perceiving time.

5 Perceiving Tone Sequences and the Internal Creation of Temporal Modi

Kant distinguished between three temporal modi, which represent categories of empirical knowledge: perseverance (Beharrlichkeit), simultaneity, and succession (Kant 1993). In modern terms, perseverance corresponds to the concept of invariance or, in perception, to object identity. We will now consider acoustic perception experiments showing the emergence and the breakdown of these modi and relate their internal generation to the dynamics of coupled clockworks.

The smooth and continuous flow of perceived time reveals a grained structure if we consider the perception of rapid tone sequences. As soon as the time delay of successive tones drops below a certain critical time, their percepts fuse. There is a rather sharp, discontinuous transition between simultaneity of internal events and succession. The existence of a temporal threshold makes one think of internal time being somehow "quantized". The coupling of two clocks can already be interpreted as a simple dynamical model of perceiving such a minimum time. The Huygens effect of phase locking operates within a certain frequency window. Depending on coupling strength, the oscillations of two different clocks merge and appear as one single phenomenon, if their frequency difference $\Delta f = (\omega_2 - \omega_1)/2\pi$ is below a critical threshold $(\Delta f)_{th}$. Synchronization in the frequency domain can be interpreted as fusion in the temporal domain, which occurs if the time delay Δt between two events is sufficiently small. $(\Delta f)_{th}$ is transformed to $(\Delta t)_{th}$ according to $(\Delta t) = (\Delta f)/f^2$. Thus the driving of clocks and the merging of active oscillations represent dynamical prototypes for the the fusion of events on active substrates. Beyond these temporal coupling effects we also have to consider spatial effects, as the discussion of the acoustic spacetime concept has shown. In order to demonstrate the close connection between spatial and temporal effects in internal representations we are going to discuss the perception of tone sequences.

It is a common naive expectation that one is able to follow the temporal structure in a sequence of tones and to reproduce their beat correctly. This is suggested by the ability to repeat a simple musical beat, e.g., by finger tapping. However, a correct timing will work only if simple stimulus sequences are presented, like a succession of clicks, for instance. If the patterns are more complex like tone sequences, the perceived rhythmical patterns will depend both on time and on the frequency interval between the tones. This was first explored comprehensively by van Noorden (1975). Depending on the frequency intervals (ratio of frequencies), we tend to group a sequence of alternating tones in a variety of ways, although the physical timing is quite regular. Fig. 4 gives a simplified description of the relevant percepts. The ordinate is the time delay between successive tones, and the abscissa is the logarithm of the frequency interval, which corresponds roughly to basilar membrane distance. If we focus on tone pairs, percepts related to temporal order are simultaneity, succcession, and pitch motion. If we put the tones in a broader context of tone sequences, fission, coherence, and fusion are the main percepts.

First we consider the perception of tone pairs. The diagram shows a critical time of more than 40 ms, which is necessary to perceive the temporal order of tonal events correctly. Below a time interval of 40 ms, there are two percepts of tonal succession, which depend on the frequency interval. If the frequency interval is large enough we perceive two individual tones. However, it is impossible to tell which tone comes first. Low-high tone sequences sound the same as high-low. Both tones appear to be simultaneous. If the frequency interval is sufficiently small, temporality is encoded in a different percept. Instead of two different tones we perceive one continuous event with a gliding frequency. The two neighbouring tones are linked via pitch motion. The separation between the percepts of "simultaneity" and "pitch motion" is marked by the dotted line in Fig. 4.

Above 40–50 ms it is possible to perceive the temporal sequence of two tones correctly. However, the situation is complicated by the fact that we can

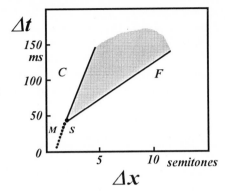

Fig. 4. Schematic and simplified diagram, showing how sequences of alternating tones are perceived (based on data by van Noorden (1975)). The frequency interval (ratio of frequencies) of the sine tones is the abscissa, the ordinate is the time interval. For sufficiently high time intervals, the temporal order of the successive tones is judged correctly. The percepts of connecting tones are tonal coherence (C) or fission (F). For time intervals below 40 ms the percepts are pitch motion (M) or simultaneity (S).

voluntarily focus on different percepts of tonal succession, which coexist in a certain parameter range. Instead of tone pairs, tone sequences are required for this experiment. The sequence of two sufficiently separated alternating high and low tones can be heard as one stream of tones belonging together or as two separate high and low pitch streams. Either we connect the tones to form one coherent melody line, moving up and down, or we perceive two separate lines. The first case is called temporal coherence, the latter fission (van Noorden 1975). In case of fission, we have the voluntary option of symmetry breaking: We can direct our attention to either sequence, perceiving the other stream as a background.

Considering the coexistence of and the transitions between the percepts shown in Fig. 4, we take a fixed time interval of tones (say 100 ms) and vary their frequency interval. Starting from high frequency intervals, we perceive two separate tone sequences (fission). As soon as the frequency interval drops beneath the left demarcation line of the shaded area, fission changes to temporal coherence. Moving the opposite way, we start with perceiving temporal coherence and switch to fission at the phase border to the right. As the critical parameters depend on direction, there is a typical hysteresis behavior. What we perceive, depends on history. In between the two forced phase transitions, i.e. in the whole shaded area, there is an option to switch voluntarily between fusion and fission. This time-ordered phase disappears below 40 ms, where we have no chance to detect the true temporal course of tonal events or to direct our attention voluntarily to a certain percept. Either we perceive fission with two separate tones, which seem simultaneous, or the two tones fuse to one single event with gliding frequency (pitch motion).

The critical zone of 40–50 ms can be regarded as a threshold for creating the Kantian modi of time. Beyond that time interval, a time-ordered phase exists. Obviously, bringing about perseverance or object identity requires a certain amount of time, below which events fuse. Temporal order coevolves with the creation of separate internal events. Below 40 ms, two tonal events have an indefinite temporal relationship, because a time delay in the usual sense of the word does not yet exist internally. If one thinks of time as emerging dynamically, this threshold can be considered roughly as the cycle time of dynamical processes, which establish events coevolving with the temporal order concept (Pöppel, this volume).

In perception, motion appears more basic than temporality, because it is implemented on lower levels and shorter timescales, prior to the emergence of internal time. This comes pretty close to the Aristotelian view that motion is primordial and time a derived concept! From a purely kinematical point of view, a limited velocity of internal processing is a highly plausible hypothesis in explaining the interlocking of internal time and space, which adds a relativistic touch to the interpretation of Fig. 4. Kinematics in physical spacetime must take into account the speed of light as the absolute limit of information exchange. As a result, temporal successions and causal connections are defined only if the events lie within the respective light cones (cf. Fig. 1). If event B lies in the future light cone of event A then A precedes B. Speaking of the temporal sequence of the events A and B makes no sense, if B is outside A's past or future light cones. In such a case of spacelike separation of events, no causal connection is possible.

There is a feature common to all the percepts in Fig. 4. The time delay, which is required to establish a connection between tonal events, increases with the frequency interval. Accordingly, due to tonotopic organization, it also increases with the distance of internal representation. This applies to the fast process of pitch motion as well as to tonal coherence in slower tone sequences. In both percepts, a connection of events is not established immediately, but requires a minimum time interval which increases with distance. This points to a connecting process, involving a real motion, e.g., of an internal neural excitation pattern. Its velocity sets the limiting "neural light cone" for a causal connection of events. Depending on the respective neural processes, there are different causal cones, e.g., for linking events by pitch motion or by tonal coherence. Events outside the respective light cones cannot be linked by the respective percepts, so that they are perceived as unrelated. In the case of fission, each sequence has to be accessed separately by the conscious observer; the rest appears as a more or less diffuse background.

The metaphor of the Cartesian theater is well known from vision: The notorious internal observer watches the movie, which the optical nerves project to some internal screen. Auditory pattern perception shows that the Cartesian movie is viewed in some kind of "relativistic" spacetime. The relativistic effects result from a limiting velocity of neural processing, setting speed limits to move the focus of attention. In addition to spatial motions, the internal observer is also able to shift her or his internal focus temporally to some limited extent. At rapid repetition rates of tone sequences, the naive concept of temporal order, as laid down in musical notation, can break down. It is possible to rearrange the temporal course of acoustical events. We shall not discuss these effects in more detail here and refer to the literature (van Noorden 1975).

6   The Continuity Illusion as a Metaphor of Consciousness: Tagging and Untagging of Acoustic Events

As the discussion on auditory scene analysis has shown, hearing is organized in such a way that the continuous stream of signals is interpreted as a succession of discrete individual events, which are connected with sound sources of the external world. In order to create object-like internal representations of acoustic events, the timing of signals plays a decisive role. Their onset is an essential clue, establishing individuality similar to putting a label to the internal excitation pattern. It appears that the first wavefront of a sound signal triggers some kind of a "tagging" mechanism, which contributes to create object identity of the acoustic event or its perseverance in the Kantian sense.

This can be shown by a highly perplexing auditory demonstration, the Franssen effect (Franssen 1960), which was discovered in the early days of stereophony. Two separate loudspeakers are supplied with the same sine tone, but with a different modulating envelope. In the left speaker, the tone is abruptly switched on and decays gradually to zero amplitude. At the end of the demonstration, the amplitude is smoothly turned on before it is switched off suddenly. The right channel is supplied with the complement of that envelope (i.e., the tone is softly turned on and off with a long period of constant amplitude). During the whole demonstration of 10 sec or more, the audience localizes the tone in the left speaker, although most of the time this source does not emit any sound at all. The sharp transients in switching the left channel on and off attract attention completely. It is in fact very difficult to convince the audience, that most of the time they focused on a phantom.

The spatiotemporal tagging process not only puts labels to auditory events but also attracts an attention mechanism. This is especially important for binaural hearing, which builds up a spatial acoustic representation of the world, allowing acoustic events to be correlated with optic events. The Franssen effect is related to what is called the "law of the first wavefront" in room acoustics, putting emphasis on direct sound. The underlying, still widely unknown neural mechanisms are important for processing signals in complicated sound fields, e.g., in echoic rooms. Additionally, this mechanism may help in focusing on single sound sources as in the famous cocktail-party effect. In the clockwork metaphor, temporal tagging can be regarded as a synchronization process triggered by the onset of acoustic events.

There is a related puzzling auditory illusion, which allows to untag individual tones, so that a continuous stream is heard, although only a discrete sequence of signals is presented in reality. In the literature, this effect is described as pulsation threshold or as continuity illusion (Thurlow 1957, Thurlow and Elfner 1959). Figure 5 shows the corresponding stimulus patterns. Bursts of tone signals alternate with bursts of disturbing signals, e.g., broad band noise. If the level of both signals is comparable, a sequence of discrete tones alternating with noise bursts is clearly perceived. If the noise level is increased, suddenly a continuous tone with constant loudness is heard, even though objectively the signal is still alternating with the noise.

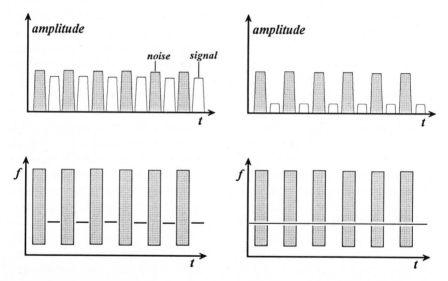

Fig. 5. The continuity effect in auditory perception. In an alternating sequence of noise and sine tones with approximately equal level, the tone is perceived correctly as a repetitive sequence of discrete events (left). If the amplitude of the sine tone is reduced, a transition to a continuous tone sensation occurs (right).

The transition from the discrete sine tone sequence to a continuous signal is illustrated schematically in Fig. 5. In analogy to optical perception, the continuity illusion is usually explained as a high-level interpretative effect of the figure ground type (Bregman 1990). The tone is extrapolated continuously throughout the interruptions, similar to reconstructing an object hidden behind a periodical obstruction, for instance behind a lattice fence. In vision, these interpretative schemes are discussed in terms of Gestalt psychological principles, like the principle of good continuity. In our opinion, in addition to these high level interpretation schemes, low level dynamical effects of the coupled clock type are equally important. Perceiving the pulsating tone as a sequence of discrete events requires the above mechanism of repetitive temporal tagging in synchrony with the signal envelope. It is possible to selectively inactivate the synchronization effect by a sequence of periodic signals with a suitable spectrum and phase. This process can be viewed as a masking phenomenon.

The masking hypothesis is supported by the fact that the spectral characteristics of the continuity inducing sound are typical for other masking phenomena like simultaneous or tone on tone masking. The bursts of broad band noise can be replaced by narrow band noise or sine tones of suitable frequencies. The effect shows the typical asymmetry of simultaneous masking, which is also present in cochlear tuning curves; i.e., signals with frequencies lower than the test tone induce continuity much better than higher frequencies.

It is necessary that the masking effect works backward and forward in time, masking not only the onset of the tone but also its decay, so that discontinuities are smeared out and a continuous sensation is possible. For a detailed discussion of masking see Zwicker and Fastl (1990). In view of causality, forward masking has a strange appeal, because a future signal may act on something, that has already happened in the past. This sounds as paradoxical as the killing of one's grandfather in time travel fictions. We interpret forward masking as a nonlinear dynamical effect in internal spacetime in analogy to the amplitude dependent velocity effects, well-known in soliton physics. An excitation pattern with a higher amplitude travels faster and overruns an earlier low amplitude pattern, giving rise to the internal masking of a signal, which preceded the masker externally. This is no violation of the causality principle, because perception only looks into the physical past. What the internal observer interprets as nowness, are yesterday's papers for an external observer.

If we accept the masking hypothesis, the continuity effect shows a perfect similarity to the phase-locking of clocks. The transition from the locked to the unlocked state of a driven van der Pol oscillator is governed by the washboard potential in Fig. 4. Both a decrease of the driving signal or an increase of noise bring about a transition from the locked state to an effectively free-running oscillation. The phaselocked state corresponds to sticking the sequence of internal events to the periodic external signal bursts. An increase of noise will unstick the phase. The transition to the free-running phase corresponds to the switch from discrete events to continuity due to masking the tagging process.

The continuity effect, interpreted in terms of the clockwork metaphor, completes the unified view of internal creation of the three Kantian temporal modes. Simultaneity or succession as well as perseverance and object-identity are reduced to the same type of dynamical processes as occurring in

Fig. 6. The solution to the puzzle of Fig. 2, metaphorically describing how the mind binds together discrete events and creates time.

driven noisy clockworks. The clockwork metaphor demonstrates that discrete event-like actions and continuously flowing actions are not fundamentally distinct. Both processes can emerge on a sufficiently complex substrate. The discreteness-continuity illusion, interpreted as a transition betweeen phase locking and unlocking, corresponds to a symmetry breaking process in time. It is tempting to describe the actions of our conscious mind and the creation of internal temporality in terms of temporal symmetry breaking transitions. Viewed from the highest level of introspection, the continuity illusion bears a striking resemblance to our experience of mental activity. As mentioned in the introduction, we experience a continuous flow of time interspersed with discrete events, depending on our state of awareness. The objects of awareness are certainly more complex than a sequence of simple tonal events, triggered or untriggered by external signals. Nevertheless, the spontaneous internal triggering of events and their fading away into a diffuse continuum of background activity is experienced in a way very similar to the continuity illusion. Fig. 6 is the optical counterpart to this illusion, and it presents the solution to the puzzle of Fig. 2. It shows the same chunks of information as in Fig. 2, but this time the additional strokes of the word "mind" help to establish a connection between the otherwise unrelated blocks, which are put together to create the word "time". This optical metaphor is meant to depict the integrating action of our mind, connecting the discrete chunks of conscious experience and embed them in the flow of time. This time, the discreteness-continuity puzzle of time is represented symbolically as a figure-ground problem.

7   The Unreasonable Effectiveness of Clocks – A Classical Model for a Nonclassical Mind?

The puzzles of perceiving time are sometimes considered as nonclassical experiences, pointing to the necessity of a quantum description of the brain (Penrose 1989). On the other hand, information processing on the level of neural spikes is classical and requires as much quantum theory as the action of semiconductor devices like NAND-gates or microprocessor chips. In this case, quantum theory is only required to account for the semiconducting properties of the material substrate. As soon as one takes the electrical properties of the chip as granted, the actions of the system on the level of voltages and currents and logical states are fully classical. The present discussion demonstrates that there is much more to "stupid" classical mechanisms than hitherto expected – as, for instance, the counterintuitive cleverness of stochastic resonance in information preocessing reveals. In hearing, the classical approach obviously suffices to describe the micromechanics down to scales in the subnanometer range, where classical concepts definitely become fuzzy. It seems worthwile to exploit the potential of this classical approach before prematurely applying quantum principles as a loophole for the mind. Maybe Occam's razor is an appropriate guiding line: "Don't resort to quantum theory of brain functions unless forced by necessity."

The Huygens effect of phase locking in clocks is an old metaphor of the interaction and the synchrony of mind and matter. Leibniz used it to refute interactionalism and to defend his model of preestablished harmony (Leibniz 1906). It is interesting to note that this historical clockwork model of coupling external "matter-like" and internal "mind-like" oscillations is a highly powerful dynamic prototype of perceiving acoustic signals and creating different modes of internal temporality. Of course, this model is dualistic only with respect to a methodological division between the inside and the outside. Beyond that, it is based on a monistic attitude, assuming that "mind-like" oscillations are governed by the same structure principles as material oscillations. Psychophysical methods provide glimpses from both sides. If it is necessary to include quantum clocks in that picture, only time can tell.

References

Bregman A.S. (1990): *Auditory Scene Analysis* (MIT Press, Cambridge).

Euler M. and Kießling J. (1981): Frequency following potentials in man by lock-in technique. *Electroenceph. Clinical Neurophysiol.* **52**, 400–404.

Euler M. (1994): Sensory perceptions and the endo-exo interface: towards a physics of cognitive processes. In *Inside Versus Outside*, ed. by H. Atmanspacher and G.J. Dalenoort (Springer, Berlin), 309–330.

Euler M. (1996a): Biophysik des Gehörs I: Von der passiven zur aktiven Wahrnehmung. *Biologie in unserer Zeit* **26**(3), 163–172.

Euler M. (1996b): Biophysik des Gehörs II: Die kontraintuitive Effektivität nichtlinearer Dynamik in der biologischen Informationsverarbeitung. *Biologie in unserer Zeit* **26**(5), 304–312.

Franssen N.V. (1960): Some considerations on the mechanism of directional hearing. PhD Thesis, Delft.

Hayashi C. (1964): *Non-Linear Oscillations in Physical Systems* (McGraw-Hill, New York).

Hanggi P. and Riseborough P. (1983): Dynamics of nonlinear dissipative oscillators. *Am. J. Phys.* **51**, 347–352.

Huygens C. (1893): *Oeuvres Complètes, Vol. V* (Martinus Nijhoff, La Haye), 243–244.

James W. (1950): *Principles of Psychology*, reprint of the 1890 edition (Dover, New York).

Kant I. (1993): *Kritik der reinen Vernunft*, hrsg. v. R. Schmidt (Meiner, Hamburg).

Keidel W.D., ed. (1975): *Physiologie des Gehörs* (Thieme, Stuttgart).

Keidel W.D. (1989): *Biokybernetik des Menschen* (Wiss. Buchgesellschaft, Darmstadt).

Leibniz G.W. (1906): *Hauptschriften zur Grundlegung der Philosophie, Vol. II* (Verlag von Felix Meiner, Leipzig).

Longtin A., Bulsara A., and Moss F. (1991): Time-interval sequences in bistable systems and noise-induced transmissions of information by sensory neurons. *Phys. Rev. Lett.* **67**, 656–659.

Oster G. (1973): Auditory beats in the brain. *Scientific American* **229**, 94–102.

Penrose R. (1989): *The Emperor's New Mind* (Oxford University Press, Oxford).

Schrödinger E. (1944): *What is Life?* (Cambridge University Press, Cambridge).

Spekreijse H. and Oosting H. (1970): Linearizing, a method for analyzing and synthesizing nonlinear systems. *Kybernetik* **7**, 22–26.

Stratonovich R.S. (1967): *Topics in the Theory of Random Noise, Vol. II* (Gordon & Breach, New York).

Thurlow W.R. (1957): An auditory figure-ground effect. *Am. J. Psychol.* **70**, 653–654.

Thurlow W.R. and Elfner L.F. (1959): Continuity effects with alternately sounding tones. *J. Acoust. Soc. Am.* **31**, 1337–1339.

van der Pol B. (1926): On relaxation-oscillations. *Phil. Mag.* **2**, 978–992.

van Noorden L. (1975): Temporal coherence in the perception of tone sequences. PhD Thesis, Eindhoven.

Weyl H. (1949): *Philosophy of Mathematics and Natural Sciences* (Princeton University Press, Princeton).

Wiesenfeld K. and Moss F. (1995): Stochastic resonance and the benefits of noise: from ice ages to crayfish and SQUIDS. *Nature* **373**, 33–36.

Zurek P.M. (1985): Acoustic emissions from the ear: a summary of results from humans and animals. *J. Acoust. Soc. Amer.* **78**, 340–344.

Zwicker E. and Fastl H. (1990): *Psychoacoustics* (Springer, Berlin).

Cognitive Aspects of the Representation of Time

Gerhard J. Dalenoort

Department of Psychology, University of Groningen, P.O. Box 72,
NL–9700 AB Groningen, The Netherlands

1 Introduction: The Basis of Our Notions of Time

What is time? In the first place, time is a personal, subjective experience. In philosophical terms, it belongs to the so-called "qualia". The time of our clocks and of history is a construct. It is used as an index to indicate and measure change and speed of change. In physics, it is a construct in the form of a parameter, used to describe the sequential order of states and the speed of transitions from one state to another.

The historical past does not exist as a concrete experience, it only "exists" in our imagination, in our memory. If no brains would be left in the world, any sense of the past would be absent. Neither do sequences of physical states exist; only one state exists at a time. It is only in our minds and memories that we can bring the past to "life", with the help of photographs, texts, diagrams, and formulas, through archeological finds and fossils.

Although these considerations may seem a bit far from reality, they become very concrete if we ask, for example, whether animals have memories that are comparable to ours. In considering on what grounds we may or may not assign a sense of time to animals, we come to realize the uncertainty we have of our notion of time ourselves. May we conclude from the behaviour of dogs and elephants that they remember people they met in the past in the same way we do? It is well known that an elephant who has once met a human who played him a trick or has beaten him (or her), will never forget that person, and the person better beware if he meets that elephant again. And owners of dogs know the excitement when they return from a long journey. There can hardly be doubt that the dog "recognizes" his owner. On the other hand, we are much more doubtful about recognition in the case of lower animals like reptiles and insects. But how much more is a dog's memory than smelling again familiar smells that meant safety, food, caresses?

Can we project our own feelings, coming with the memory of a beloved person, onto a dog in the same way as we project them onto someone else, under the supposition that he/she feels the same as we feel? I think it is reasonable to ascribe a memory of the past to dogs, even if it may be much less elaborate than our own memory of the past. How else could we interpret the behaviour of a dog that feels guilty, for example a short time after we discover that it has snatched a sausage from the table? If we want to be consistent, we must ascribe at least some memory of the past to animals, as we do to other people. (See also the discussion of the experience of time in children and animals by Guyau (1890), p. 9–10).

Our "images" of time are indirect. We infer that some time has passed by comparing two states, one having evolved from the other. No one has demonstrated more clearly the paradoxical character of our conception of change and, related, of time than Zeno. We cannot imagine the difference of two arrows, both at a single instant of time ("now"), one arrow in free fall, the other on its way after having been shot from the bow. From the physical point of view, we know that their speeds are different. But we can only imagine this difference in speed by comparing the positions of the arrows with those at a later instant. We cannot imagine the difference of speeds in the same manner as we can imagine the difference of positions.

It has often been observed that our thinking and imagery are essentially spatial. It has been forcefully argued by Guyau that we derive our notion of time from our notions of space: the movements we make in space, from intended movements, and from the places where we locate events of the past (Guyau 1890, Chap. 3, p. 29). The conclusions drawn from Zeno's paradox of the arrow are completely in line with Guyau's views. In Guyau's view we derive our sense of the future from the ability to make plans. We derive an expected duration of time from the number of steps to be made to reach that position and from the actual number of actions taken. Future does not come upon or over us, but we move towards the future (Guyau 1890, p. 33).

Much of our thinking is in terms of notions derived from spatial relations, and we speak about many things in metaphors of spatial relations. An important aspect of such relations is that we can point to things in space in a way that is entirely different from "pointing" to temporal relations. Spatial relations seem to be more objective, as a consequence of their visual aspect. Sight is in the first place a spatial sense, whereas hearing is temporal. If we want to investigate something in more detail, we attempt to obtain a "closer look".

Some of our temporal illusions, such as a duration that seems to be much shorter than another one, have an analogy with spatial experiences. Just as we estimate a distance to an object longer if many objects are in between, we estimate a duration longer if many events take place in it. We must distinguish between the speed at which time seems to run and our estimation of that duration retrospectively (Guyau 1890, p. 59–108).

There is an asymmetry as to our experience and ability to imagine two different positions at one time and two different instants of time at one position. We can quickly switch between focussing at two alternative positions, such that we see them almost simultaneously. We can look again and again. We cannot go back to an experience in time, we can only replay a record or film. We know the difficulty of comparing pieces of music. We must compare the memory of one piece with the actual experience of the other piece. This asymmetry affects our conception of time. Once we have acquired a sense of time, we can also experience pure temporality, like music. But it seems unlikely that this experience can develop in an individual who has no sense of space. There is an analogy with the impossibility to have vivid visual images like the blue of the sky or the shape of a flower if one never has been able to see. (Of course, the ability to have experiences of space is not only visual – or, even stronger, it cannot be only visual. Our ability to move and our sense of touch are essential for the development of our notions of space.)

The difficulty of comparisons that have to be made at different times is also critical with respect to the senses of taste and smell. It is hard to compare tastes. We cannot taste two things simultaneously. Our memories of tastes are hard to grasp and to be made concrete. The main reason is that no metaphors of space can be used to characterize tastes and smells. This may all be due to the essentially spatial character of our memories. No memories are better and stronger than the ones of places and spatial relations, even to such a degree that many mnemonic techniques are based on spatial relations.

2   General Systems Theory and the Physical Picture of the World

As has been often argued, any scientific construction and, a fortiori, any conceptual system whatsoever, has to start from metaphysical principles, implicitly or explicitly (see, e.g., Ruhnau 1994). For the epistemological aspects of the present discussion I think three metaphysical principles are necessary. I also think they are basic principles of the general theory of systems (GST).[1] They are:

1. For the description of any system a multiplicity or pluriformity of models is needed.
2. Our full understanding of a complex system depends on the degree to which we can construct and describe the correspondences between the models we have available.
3. All properties we assign to things, objects, and systems are products of interactions – in other words: all properties are emergent.

Although these postulates undoubtedly have a metaphysical aspect, they can also be tested to the degree to which they improve our insight into, or even enable the solution of, controversial issues. They can even help us see that some problems are in fact pseudoproblems. I shall now elaborate on these three postulates; they have been described and discussed in detail elsewhere (Dalenoort 1980, 1990).

---

[1] There is no generally accepted basis of GST, there are differences as to what is considered essential or basic.

Our cognitive capacities are restricted. We can have only one view of an object at a time, and we can think of only one thing at a time. This is also expressed as a restriction concerning the number of things that we can attend to at a time. If we are seemingly able to attend to two or more things at a time, like knitting and having a conversation, this is only possible because we have automatized one of the activities. An alternative possibility is that we quickly switch between the activities in such a manner that it seems to us that we see them simultaneously. This illusion is partly due to the way we experience time, e.g., to the fact that there is a threshold for our resolution of experienced time.

In some cases it is obvious that we can only experience one view of an object at a time. For example, we can only see one perspective of the Necker cube at a time, but we can switch between both views (not quite by will, but we may enhance a switch by staring). Claims that we can entertain both perspectives simultaneously at low contrast are perhaps based on the illusion that events or experiences seem to be simultaneous, whereas they occur subsequently within a duration that we experience as a single time. The sequentiality then applies to the peripheral processes in the nervous system, the simultaneity is the property that emerges due to the architecture of central parts of the brain in accordance with the third postulate of general systems theory mentioned above.

Although one may claim that we can entertain two views of an object simultaneously, we can only think of and argue about a complex system from one point of view at a time; in other words, in terms of one type of representation. We cannot think of a gas as a gas and as a collection of molecules at the same time, but we can switch between the two and think of the correspondences between both representations. As a consequence, we need a variety of models of a system for each possible point of view. Because we do not like or accept the idea that the different models of one and the same system are unrelated, we want to understand how the different models are related. This is the second postulate.

The relations between different models of a system can be of various types. There is the reductive correspondence, where we explain high-level concepts in terms of the properties of the subsystems. In order to avoid what I have called a "naive reductionist view" (Dalenoort 1987), we must realize that we assign properties to the subsystems on the basis of their function within a larger whole. For example, our conception of a cogwheel is to a large extent derived from our perception of cogwheels in assemblies of cogwheels, where we see how the movement of one "causes" the movement of the other. Of course, once we have formed our conception of an object like the cogwheel on the basis of our experience with one system, we can use this conception to facilitate or even to enable our understanding of another system. So we may compare the old type of alarm clock and the gearbox of a car. Thus we can say that the way we assign properties to systems and objects is a two-way process: we move from wholes to parts, and from parts to wholes. This insight was – probably – for the first time clearly expressed by Blaise Pascal in about 1670 (Pascal 1962, p. 56):

> "Donc toutes choses étant causées et causantes, aidées et aidantes, médiates et immédiates, et toutes s'entretenant par un lien naturel et insensible qui lie les plus eloignées et les plus différentes, je tiens impossible de connaître les parties sans connaître le tout, non plus de connaître le tout sans connaître particulièrement les parties."
> ("Thus, all things being caused and causing themselves, being supported and supporting themselves, being mediated and mediating themselves, and all maintaining themselves by a natural and invisible bond that ties the most distant and the most different ones, I hold it for impossible to know the parts without knowing the whole, as well as to know the whole without knowing the parts.")

There are other types of correspondence. For example, one and the same process can be described in terms of causes and in terms of goals. In physics, only causal explanations are accepted, although some models may seem to be goal-directed, as the model implied by Fermat's principle.

In the fields of biology and psychology, the problems are more fundamental. Who would deny the "self-evident" feeling we have in making plans to meet at a certain place? We would have a hard time to realize and accept a purely causal picture of such a process that would ignore our feeling of free will and imply that we are only the spectator of a process of which the outcome is determined by physical laws. Our consolation may be that few people think so exclusively materialistic as to assume that the physical laws are the only and ultimate models of nature. These laws have no place for feelings and emotions, for our conscious experience. (Whereas one may argue that free will is an illusion, it would be rather unfruitful to deny feelings and emotions. The fact that within physics there is no place for them leads to the conclusion that any purely physical description of the world is incomplete.) As Descartes noted, our feeling and thinking are the starting points for investigating the world. It would be most irrational to throw them away once we have got some grasp of that world. However, this does not inevitably lead to a dualistic view of the world, as Descartes would have it.

A third important distinction must be made between models of systems in terms of self-organizing processes and models in terms of construction. This is closely related to the distinction between causal and goal-directed representations. One of the central problems of the study of self-organization indeed is to understand how goals emerge from the interactions of a large number of similar elements. Does the fluid in my teacup behave towards a goal in establishing a horizontal surface (in a gravitational field)? Is a plant striving after a goal? At least not consciously, according to our usual conviction. Does our brain, or better: our body as a whole, strive after a goal if the neurons play their complex game of excitations? The problems of interpreting self-organizing systems and processes have been discussed elsewhere in more detail (Dalenoort 1994). It is obvious that our ability to make plans for the future, which Guyau considered as a basis for our notion of time, could not be present if we were not goal-directed systems. Given our usual view of neurons as causal machines, we then have the problem of how our high-level abilities emerge from the causal processes at lower levels. This question can only be stated in terms of the conditions that must obtain at the lower level in order that properties at the higher level can emerge, not in terms of "reducing" the high-level properties to those of lower levels (cf. Pascal as cited above).

The formal models of physical systems, considered as cognitive tools of the physicist, abstract from the way we acquire concepts, especially in our very early childhood. Modern physics has provided us with some further examples of our cognitive restrictions, next to those provided by Zeno: the dualistic nature of elementary particles and relativistic time. We cannot imagine "things" that sometimes behave as particles and sometimes as waves or, even worse, that simultaneously behave in both ways. And we cannot imagine why velocities do not add up in a Galilean fashion. Such examples are a warning to us to be careful in drawing conclusions from our common-sense imagination; it may dramatically fool us.

It is not hard to see how the view that all properties are the products of interactions is compatible with our usual models and equations of physical systems. Take as an example the charge of the electron or its mass. Both occur in the equations of motion of atoms. If we state that the solution of Schrödinger's equation for an atom is the set of Bohr trajectories of the atom, we can easily interpret them as the outcome of the interaction of electron and atomic nucleus, where the charge and mass of the electron are parameters of the trajectories. The fact that we can use the same parameters for the movement of an electron in an arbitrary electro-magnetic field is, of course, of fundamental importance. It allows us to make representations (models) of the world in an economic fashion, as Ernst Mach coined the term. But it does not allow us to abstract to the independent existence of charge and mass of the electron. Both parameters do only acquire meaning in contexts of physical interactions. If we talk of the charge of an electron, we implicitly refer to situations where the electron interacts with another charge or with a field.

3   Some Observations of Psychological Nature and on the Architecture of the Brain

There is a close relation between our notions of causation and time (see, e.g., Watanabe 1975). In this context, it is important to refer to David Hume (1777). Hume had already realized, according to Mackie, that our notion of causation is a mental construct – and he considered this his most important discovery (Mackie 1974, Chap. 1).

In addition, there is a close relationship between our notion of causality and our ability to be conditioned. It is now generally accepted that the physiological basis of learning has to do with the creation of new connections between neurons and the strengthening of existing ones. The only condition for this to happen is the more or less simultaneous excitation of the neurons involved, maybe under the influence of other excitations in the network of which the neurons are a part. (At some point the process must stop, of course, and not all neurons that are simultaneously active must be connected.) No corresponding mechanism has ever been proposed that does not use high-level notions such as the relevance of a stimulus or the attention paid to an event by the organism. Such notions must be the outcome of a process. They are relevant because they are learned. We pay attention to certain events as a result of neural processes, not vice versa.

Our nervous systems may be considered as very good correlators. We are very good at seeing similarities, symmetries, close temporal sequentiality, direct movement (our movement detectors are inborn). This ability to relate events close in time, even the compulsory impression of a causal relation between subsequent events must be the basis of our notion of causality. This is also born out by the experiments of Michotte on our perception of causality (Michotte 1963).

If two events often occur together, we incorporate their relation into our common and general knowledge and we tend to forget the event of observing that relation for the first time. In cognitive psychology, sometimes the distinction is made between episodic memory and semantic memory. The former type of memory concerns events to which we attach a label of time, since the event is unique. The latter type of event is of a more general character. It takes place so often that we forget about the first time, and it obtains the status of a property that we can assign to events and things whenever the occasion arises.

We may then consider time as one of the qualia, emerging from the observation or experience of subsequent events. The third postulate of systems theory implies that all properties are emergent in the sense that they are products of interactions, as argued above. Qualia are subjective properties, and they are special in that sense. From the point of view of general systems theory, we do not distinguish between ordinary properties and emergent properties. (The latter were proposed by A.O. Lovejoy in 1927; see Nagel 1961, p. 374.) We do rather distinguish between properties that we can externalize or point at and properties that we can only refer to in an indirect manner. In this sense, our experience of time is qualia-like; it is subjective.

We see that even "objective" basic physical notions have a cognitive basis. We have a large degree of liberty in choosing our representations of the world. They depend on the questions we want to answer. For some questions we can even use the notion of an absolute time, as Newton did. We do not have to be aware of the relativity of our notions in going about our daily routines or even in doing many down-to-earth scientific experiments. But from the scientific point of view it is important to be aware of the relativity of those notions.

4   Some Notes on Kant, Mach, and Guyau Again

In the section on time in his *Kritik der reinen Vernunft*, Kant (1787) maintains that we must have an a priori notion of time. In his view, such a notion cannot be based on even more primitive notions of succession or simultaneity, since these are only possible if a notion of time is already present in the individual concerned.

There may be some truth in this position, but the question is whether it must be discussed in terms of true and false. Whereas for Newton time was still a notion of something existing outside of ourselves, of which we can have experiences, time had a strongly subjective aspect for Kant: as a "form" of our inner experience ("Form der Anschauung"). For Kant, time was not something that is given by observation of our environment. Newton's position is well known (see, e.g., Mach 1883, Sect. 2-6, p. 217): time is something flowing regularly "outside us".

It is perhaps futile to make a distinction between the notion of time on one hand and the notion of temporality ("before", "after", and "simultaneous") on the other. As Guyau (1890) has quite forcefully argued, our notion of time seems to be derived from our notion of space. If we say that event A takes place before event B, in our imagination we move from the place of A to the place of B, at which we arrive later. It seems impossible to quarrel about the mutual relations of our primitive notions since they are so entangled. It is useless to discuss the relative importance of two aspects of a system that are both vital to its existence. The example of the dictum of LaoTse is exemplary here: "the essence of the vase is the empty space inside". But, of course, it cannot be denied that also the vase itself is vital, and it should be watertight as well for most cases.

Discussions about the relevance of our primitives are often based on the intuitive notions of the discussants concerned, and the general problems with corresponding arguments are well-known. It may be that some day we shall know which neuronal structures of our brain carry the primitives concerned and discover which are the more basic ones in an evolutionary sense. But, in line with Pascal's dictum, we shall not be able to reduce our primitive notions to neuronal processes. The most we can hope to achieve is to find the correspondence between our cognitive notions and neuronal structures. This can increase our understanding of the architecture of the neuronal structures and provide a better understanding of the relations between our cognitive primitives. Even if our notions of space would have evolved before our notions of time, the necessary relating notion of motion cannot be reduced to notions of space alone.

4.1   Ernst Mach (1883)

In many respects, Mach held notions that were just a little bit different from those of his contemporary colleague-physicists. He was one of those physicists whose texts can evoke strong controversies even nowadays. There are few other physicists whose works have been reprinted as often as Mach's works. Mach has contributed a number of important insights that are still important for our present-day discussions; he can still set us thinking.

His observations on space and time, particularly their being constructs of our cognition, are fundamentally new contributions to our epistemology. His work may be considered an essential step towards the theory of relativity, as Einstein wrote to Mach (see Wolters 1988). Mach was probably the first to state explicitly that we cannot measure time directly; we can only compare different processes with respect to their speed of change (Mach 1883, Sect. 2-6, p. 217).

One of the sources of the controversies about Mach's views as such is due to misunderstandings of his ideas. This is partly due to the complexity and depth of the topic; it is at the roots of our epistemological assumptions. And it has turned out to be hard to digest and judge Mach's ideas without being caught in one's own epistemological assumptions – the explicit ones, but even more the implicit ones.

4.2   Jean-Marie Guyau (1890)

Most of what Guyau had to say about the psychology of time, insofar as it seems important for the restricted purpose of this essay, has been stated. He was the first to provide an analysis of our cognition of time, in relation to our notions of space. He considered the former as derived from the latter. As far as his analysis is concerned with psychology, we can add little to it. Of course, modern physics, especially the theories of relativity, have provided fundamental new insights into space and time. In a very general sense, they do only add to Guyau's view that time is a mental construct. No one would deny, however, that it is of fundamental significance.

References

Dalenoort G.J. (1980): Some general considerations on representation of informational systems – and processes. In *Representations des connaissances et raisonnement dans les sciences de l'homme*, ed. by M. Borillo (INRIA, Le Chesnay), 265–281.

Dalenoort G.J. (1987): Is physics reductionistic? In *Current Issues in Theoretical Psychology*, ed. by W.J. Baker, M.E. Hyland, H. van Rappard, and A.W. Staats (North Holland, Amsterdam), 15–21.

Dalenoort G.J. (1990): Towards a general theory of representation. *Psychological Research* **52**, 229–237.

Dalenoort G.J. (1994): The paradox of self-organization. In *The Paradigm of Self-Organization, Vol. 2*, ed. by G.J. Dalenoort (Gordon & Breach, London), 1–14.

Guyau J.-M. (1890): *La génèse de l'idée de temps* (Félix Alcan, Paris). Reprinted in French, and also translated into English, in *Guyau and the Idea of Time*, ed. by J.A. Michon, V. Pouthas, and J.L. Jackson (North Holland, Amsterdam 1988).

Hume D. (1777): *Enquiries Concerning Human Understanding and Concerning the Principles of Morals*. Reprinted and edited by L.A. Selby-Bigge, Clarendon Press, Oxford 1902.

Kant I. (1787): *Kritik der reinen Vernunft* (Riga, 2. Auflage), Kap. I (Transcendentale Elementarlehre), Erster Teil (Die transcendentale Aesthetik), 2. Abschnitt (Von der Zeit).

Mach E. (1883): Die Mechanik in ihrer Entwickelung, historisch-kritisch dargestellt (Brockhaus, Leipzig). English translation: *The Science of Mechanics: A Critical and Historical Account of its Development* (Open Court, La Salle, Ill., 1960).

Mackie J.L. (1974): *The Cement of the Universe, a Study of Causation* (Clarendon Press, Oxford).

Michotte A. (1963): *The Perception of Causality* (Basic Books, New York).

Nagel E. (1961): *The Structure of Science* (Routledge & Kegan Paul, London).

Pascal B. (1962): *Pensées*, ed. by J. Chevalier (Publ. of "Le Livre de Poche", Paris, Nr. 84 (original edition about 1670)

Ruhnau E. (1994): Time - a hidden window to dynamics. In *Inside Versus Outside*, ed. by H. Atmanspacher and G.J. Dalenoort (Springer, Berlin), 291–308.

Watanabe M.S. (1975): Causality and time. In *The Study of Time, Vol. 2*, ed. by J.T. Fraser and N. Lawrence (Springer, Berlin), 267–282.

Wolters G. (1988): Preface to the edition of E. Mach (1883) by Wissenschaftliche Buchgesellschaft, Darmstadt 1988.

Part III

Relativity and Gravity

Concepts of Time in Classical Physics

Jürgen Ehlers

Max-Planck-Institut für Gravitationsphysik (Albert-Einstein-Institut),
Schlaatzweg 1, D-14473 Potsdam, Germany

1 Introduction

My aim in this review is to describe the three main concepts of time which
have been, and still are used in physics, viz. the time concepts of Newtonian
physics, of Einstein's special theory of relativity, and of his general theory of
spacetime and gravitation. In physics, the concept of time serves to describe,
qualitatively and quantitatively, natural processes such as motions of bodies,
propagation of waves and fields, and causal dependences between events. It
is closely, indeed inseparably related to notions of space, and therefore the
following account concerns *spacetime*, with special emphasis on the role of
time as a substructure of the more encompassing and, in (not only) my view,
more fundamental spacetime structure.

The word "classical" in the title is intended to convey that in what follows
I am not concerned with the concept of time in quantum physics, except for
occasional remarks. As a matter of fact, in the existing, elaborated and useful
quantum theories (quantum mechanics, quantum electrodynamics, the standard model of particle physics) as opposed to hoped-for, future theories, time
enters as a classical parameter whose meaning is based on the assumption
that quantum phenomena are embedded in and related to processes which
can be accounted for in classical, non-quantum terms. (Needless to say, the
preceding statement is not to be interpreted as a valuation. Considerations
of the possible modifications of "time" in future theories are not only interesting, but unavoidable to make progress, but they are not my task in this
review.) I shall state the assumptions underlying the mathematical models
of spacetime made in the three theories listed above, indicate motivations
of these assumptions and their ranges of validity, and try to say something
about what these theories contribute to the subject of this volume. ("Technical" mathematical, experimental and historical details will be omitted. A
more detailed account can be found elsewhere (Ehlers 1995).)

My point of view in the sequel is to consider (classical) physics as an
extension of geometry: Natural processes are to be represented as figures
(collections of points, lines, surfaces ...) in spacetime.

2   Spacetime as a 4-Manifold

The world as perceived by our senses and organized by our minds appears as overwhelmingly complex, yet structured: There are "things" distinguishable from their surroundings which move and change otherwise. For a mathematical representation of (parts of) the world it has proven useful to consider the natural processes as consisting of parts, the smallest, pointlike elements of which are called *events*. This idealization, analogous to the idea that space consists of points, is useful for a "phenomenological" description of macroscopic things and processes, but presumably it must be replaced in a quantum theory of spacetime by some other, discontinous, fluctuating microstructure as indicated, e.g. by J.A. Wheeler's "foam" (Wheeler 1968), R. Penrose's "spin–network" (Penrose 1972), or A. Ashtekar's "weaves" (Ashtekar 1994).

In classical physics, events are considered as primitive: "The objects of our perception invariably include places and times in combination. Nobody has ever noticed a place except at a time, or a time except at a place" (Minkowski 1909).

Coordinates serve to fix an event and to distinguish it from other events. Experience shows that for this purpose, four (real) numbers denoted as x^α ($\alpha = 0, 1, 2, 3$) are needed. For events near the earth's surface, e.g., one can take latitude, longitude, height above sea level, and central European time. Events close to each other are assigned nearly the same coordinates. It is not required that "all" events can be related in a one–to–one, continuous manner to quadruples of numbers, it suffices that such a numbering can be done separately for overlapping regions which cover spacetime, the set of all events. Coordinates may be, but need not be lenghts or angles or times; all that is required is that they identify events and respect nearness relations (topology). The (assumed) possibility to coordinatize events in terms of four coordinates as indicated is summarized as "spacetime is a 4-dimensional manifold"; I shall denote it as M.

In the remaining sections I shall consider which *structure* has been assumed for the spacetime continuum in Newtonian and relativistic physics, respectively.

3   Newtonian Spacetime

In everyday life we intuitively consider an event as simultaneous with the instant at which we see it, and because of the size of the speed of light this does not lead to confusion. Moreover, until rather recently it appeared that "good" clocks, if properly set, show the same time, whether at rest or moving relative to each other. In accordance with these experiences men came to view the world as "evolving in time", existing only in the "now" which separates (ontologically) past and future, which exist no more and not yet, respectively. This view forms the basis of the Newtonian concept of spacetime whose structure can be outlined, in modern terms, by the following basic statements (Stein 1967).

(1a) *Simultaneity* is an equivalence relation between events, i.e., if A, B, C are events and A is simultaneous with B and B is simultaneous with C, then C is simultaneous with A and with B.
An equivalence class of simultaneous events represents an *instant* of time. It is convenient to denote the set of all events simultaneous with A as $\bar{A}$. An ordered pair (A, B) of events then represents a process which begins at $\bar{A}$ and ends at $\bar{B}$.
(1b) The instants of time can be related in a one-to-one way to the points of an oriented (real) line such that the distance ratio of any two line segments equals the ratio of the *durations* of the corresponding processes (time intervals) measurable by any good *clock*.
This statement formulates the mathematical idealization of the "flow of time"; it may also be considered as the pre-Einsteinian hypothesis about the functioning of good (metric) clocks.
(2) If the events A, B are simultaneous and also C, D are simultaneous, then there exists a *distance ratio* $(C, D/A, B)$, a real number measurable in terms of a *measuring rod* (or tape). Restricted to one set of simultaneous events, the distance ratios obey the laws of *Euclidean geometry*.
Note that spatial distances between non–simultaneous events A, B are not unambiguously definable since there is no way to decide, without arbitrariness, whether A and B are "equilocal", i.e., happen at the same point of space.

A relation between the temporal and spatial metrics is introduced in terms of the *law of inertia* which also provides a basis for introducing forces:
(3) *Free particles*, i.e. particles not subjected to any influences ("forces"), are represented in M by *world lines* which obey the laws for straight lines of 4-dimensional, affine geometry.
This assumption also implies that there are *parallel* world lines representing free particles having constant (spatial) distance and whose simultaneous relative positions define spatial vectors, which define *constant directions*, as realizable by free gyroscope axes.

The preceding statements (1), (2), (3) about spacetime form the basis of Newtonian dynamics and, more generally, Newtonian physics. In Newtonian spacetime, preferred coordinates (t, x^a) can be introduced such that two events A, B are simultaneous if and only if they have the same time coordinate, $t(A) = t(B)$, and the duration of a process from A to B is given, with respect to a time unit, by $t(B) - t(A)$. Moreover, any world line $(x^a) = const$ represents a possible free particle, and for any instant given by some value t of the time coordinate, the x^a are rectilinear, orthogonal space coordinates. Such coordinates correspond to an *inertial frame of reference*, and the transformations between such coordinate systems form the *Galilei group*.

The role of spacetime structure for the formulation of physical laws may be illustrated by the equation of motion of a test particle moving near a massive body: If the massive body is "at rest" at $x^a = 0$ and has the (active, Keplerian) mass M, the motion $x^a(t)$ of the test particle is determined, except for initial data, by the law

$$\ddot{x}^a(t) = -\frac{Mx^a(t)}{(d(t))^3},$$ (1)

where $d(t)$ is the distance of the event $(t, x^a(t))$ in the history of the test particle from the simultaneous event $(t, 0)$ in the history of the massive body. I am reviewing this well-known law to stress that all four assumptions (1a), (1b), (2) and (3) are needed to formulate and interpret the law. The same holds, mutatis mutandis, for all laws of Newtonian physics and even for quantum mechanics. In spite of relativity, a large part of very succcessful physics is still based on the Newtonian spacetime structure.

A difficulty arises if one tries to incorporate the *propagation of light* into the Newtonian spacetime framework. Suppose, in accordance with experience and with Maxwell's successful electromagnetic theory of light, that the set of events which represents a flash of light emitted in all directions in empty space from an event E, the so-called future *light cone* of E, is determined by E alone, i.e., is independent of the motion of the light source. At a time later than $t(E)$, the light will have reached a closed surface $S(t)$ in space. A necessary condition for light to propagate isotropically, i.e., without a preferred direction, is that $S(t)$ be a sphere. If so, that sphere has a unique central event, $C(t)$. Consequently, the event E then uniquely determines the world line consisting of the centers $C(t)$ for all t later than $t(E)$, and an observer who has this world line is the only one for whom light propagates isotropically. Such observers may be called "optically at rest". Indeed, optical observations (aberration, discovered by Bradley in 1728) seemed to indicate that the set of all possible optically resting observers defines a preferred inertial frame of reference and thus an absolute state of rest, or *equilocality*, in spacetime which could be identified with Newton's absolute space.

If this were true, the Galilean invariance mentioned above would not express a valid symmetry of all laws of nature; it would be broken by the laws of optics and electrodynamics. This expectation which was in essence based on the assumption (1a) of an absolute simultaneity, turned out to be wrong and led to Einstein's first revision of the concept of time in physics.

4   Spacetime According to the Special Theory of Relativity

As is well-known, several attempts to identify experimentally the absolute state of rest, assumed by Newton and apparently required by optics and electrodynamics as indicated above, failed; I only recall the "negative" result of

Michelson–Morley type experiments designed to measure the absolute velocity of the earth. Experiments rather suggest that light in empty space not only propagates independently of the state of motion of the source, as already remarked above, but also isotropically with the same speed c, not only in *one* optically preferred frame of reference, but in frames of reference moving relatively to each other, in contradiction to the Newton–Fresnel–Maxwell expectation which follows from the Newtonian assumptions about the structure of spacetime.

Einstein's resolution of this fundamental difficulty was based on an analysis of the operational meaning of the simultaneity relation between events. Ever since the finiteness of the speed of light had been discovered astronomically by O. Römer (1675) and verified by numerous laboratory experiments, the fundamental assumption (1a) of Sect. 3 about simultaneity could have been questioned but it appears that, apart from remarks by H. Poincaré who recognized the problem around the turn of the century, it was only Einstein who in 1905 made a definite and, as it turned out, successful proposal to solve the problem. From the present point of view, one may describe the main result of Einstein's analysis (the details of which are given in many textbooks (see, e.g., Rindler 1991) as a "negative" (i) and a "positive" (ii) statement, respectively.

(i) A verifiable (testable), oberserver-independent simultaneity relation obeying (1a) does not exist.
(ii) There are infinitely many frames of reference in relation to which light propagates rectilinearly and isotropically with the universal speed c. Free particles move rectilinearly with constant speeds less than c.

Although statement (i) formally contradicts "Newton's" assumption (1a), the latter remains a good approximation as long as it is applied to sufficiently small domains of spacetime and to bodies and reference frames with relative speeds much less than the speed of light. Statement (ii) embodies a *convention* concerning the meaning of synchrony of clocks at rest in the frame considered as well as a testable *hypothesis* about light propagation, as has been explained, e.g., by Weyl (Weyl 1988).

The spacetime structure which incorporates (ii) and which forms the core of the special theory of relativity, is based on the following assumptions, which improve (1a)-(3) of Sect. 3:

(3') Free particles as well as light pulses (photons) are represented in M by world lines which obey the laws for straight lines of 4-dimensional, affine geometry.
(4) The set of all straight lines in M which represent light rays passing through an event A, form a quadratic cone, the *light cone* $C(A)$ of A. Any two light cones $C(A), C(B)$ are related by parallel transport.

An ideal kind of *clock* may be constructed by two free particles with parallel world lines and a light signal moving back and forth between the two particles which provides the ticking of the clock. Using such a measure of time, distances can be defined in terms of travel-times of light-echoes (radar method).

It follows from assumptions (3') and (4) that, similar to Newtonian spacetime, preferred coordinates (t, x^a) can be introduced such that any line $x^a = const$ represents a free particle, the light cone of the event given by $t = 0$, $x^a = 0$ is characterized by

$$c^2 t^2 - (x^1)^2 - (x^2)^2 - (x^3)^2 = 0, \qquad (2)$$

and t coincides with the time registered by a light clock. Moreover, it follows that the radar distances between the particles having fixed spatial coordinates x^a obey the laws of Euclidean geometry and that the x^a are rectilinear orthonormal with respect to that metric. Such coordinates correspond to an inertial frame of reference. Any two coordinate systems of this kind are related by a transformation of the Poincaré group, provided a unit of time is chosen once and for all.

The spacetime structure of special relativity can alternatively be characterized as follows. On spacetime, there exists a *flat Lorentzian* metric

$$ds^2 = \eta_{\alpha\beta} dx^\alpha dx^\beta \qquad (3)$$

which in inertial coordinates (t, x^a) takes the form

$$ds^2 = -c^2 dt^2 + (dx')^2 + (dx^2)^2 + (dx^3)^2. \qquad (4)$$

All of classical physics *except gravity* can be reformulated within this spacetime framework. For low speeds or energies, Newton's laws remain approximately valid, but for speeds comparable to the speed of light and high energies the laws of special relativity differ drastically from the older ones and have been verified in numerous experiments. Also, quantum field theories are based on this spacetime structure, and while their perturbation theories have been quantitatively verified, in some cases to very high accuracy, their mathematical structure suffers from unwanted infinities.

According to special relativity, all good clocks, e.g., atomic clocks, conform with "light clocks" and show *proper time*

$$\tau = \int \sqrt{1 - \frac{v^2}{c^2}}\, dt \qquad (5)$$

even when accelerated relative to an inertial frame.

Time, according to this theory, is the most basic observable; in principle the metric can be determined from time-measurements. In contrast to the older view, *good clocks in arbitrary motion do not indicate the values of one universal scalar function "absolute time" defined on M, but rather they indicate the arc lenghts along their world lines as determined by the metric.* Experimentally this has been established in many cases (time dilatation, twin effect – not "paradox").

The generalization required if gravitation is taken into account (see Sect. 5) does not change this insight, except that the metric has to be changed. There is no indication that biological clocks do not accord with this time structure – after all, organisms consist of atoms and molecules, these obey quantum laws, and the latter include this time structure. Perhaps even more important than the non-integrability, or world line dependence, of clock time is that, due to the absence of an observer-independent simultaneity relation, special relativity does not support the view that "the world evolves in time". (Which time would it be?).

Causal relations are determined by the field of light cones, not by an absolute, universal time function on M. To illustrate this point, it is useful to realize that for a person P on earth, the time span between those events on the sun of which P can have obtained any information and those events on the sun which he can influence by any means – the "causal present of P at the sun" – has a duration of about 15 minutes, and no event on the sun within that time span can, by any known law of physics, be considered as "really simultaneous" with P's present. In this sense, one may well say that *time, in the sense of an all-pervading "now", does not exist*; the 4-dimensional world simply *is*, it does not *evolve*. This conclusion was drawn, and formulated succinctly, by Kurt Gödel (Gödel 1949):

> "It seems that one obtains an unequivocal proof for the view of philosophers who, like Parmenides, Kant, and the modern idealists, deny the objectivity of change and consider change as an illusion or an appearance due to our special mode of perception .... If simultaneity is something relative ..., reality cannot be split up into layers in an objectively determined way. Each observer has his own set of "nows" .... A relative lapse of time, however, if any meaning at all can be given to this phrase, would certainly be something entirely different from the lapse of time in the ordinary sense which means a change in the existing. The concept of existence, however, cannot be relativized without destroying its meaning completely."

In spite of this, special relativity can accomodate the fact that conscious beings experience the world as evolving in time: The events on an individual's world line are linearly ordered in terms of her proper time, and his sensations are correspondingly ordered according to his past light cones from which he receives information in his "course of time". This description can be applied to several individuals and their communications without contradiction. Each individual subjectively experiences his sequence of "nows" (which may have finite resolutions depending on the kind of sensation – this is not my concern here), although there is no unique, intersubjective "present".

To end this section, I should like to point out that *the laws of special relativity do not allow true predictions based on experience.* Whatever somebody can know at any time concerns events in or on his past light cone, and that information does not suffice to predict future events. A person's need and freedom to make decisions, i.e., his subjective feeling of free will cannot, therefore, lead to testable contradictions with the assumption of relativistically causal laws, quite apart from quantum indeterminism.

5 Spacetime According to General Relativity Theory

A satisfactory description of gravitation requires a second modification of the structure assigned to spacetime, as recognized by Einstein during 1907–1915. Since this modification again changes the concept of time in physics, this step also has to be taken into account here. This step, however, does not change the two most important qualitative aspects of the relativistic time-structure discussed in the previous section, viz. the absence of an intrinsic, absolute simultaneity relation and the dependence of local, metric proper time of an object or system on its world line (state of motion); it "only" adds the dependence of proper time on gravity.

The basic idea which led Einstein to modify the spacetime structure of special relativity can be described as follows. In both Newtonian physics and special relativity theory one uses the concept of a *free particle*, and one represents the motions of such particles in spacetime as straight lines, as stated in assumptions (3) and (3'), respectively. This assumption is essential in those theories, since the concept of *force*, and thus *interaction*, is introduced as a measure of the deviation of an actual motion from a "free" or "inertial" motion, represented mathematically by the *curvature* of the world line which, in turn, is a measure as the deviation of that line from a straight line. Free motions are also needed to define, in both theories, inertial frames of reference and the corresponding coordinate systems (t, x^a), and the linearity of the associated transformations between such dynamically preferred coordinate systems derives from the law of inertia. But how, Einstein asked, is one to identify a "free" motion in nature?

Clearly, one has to make sure that the particle in question does not touch other matter, that it is not influenced by electric or magnetic forces or by light shining at it, and the like. While one can take precautions to eliminate such "disturbing" influences, it does not seem possible to shield a particle from gravity. In fact, already Galilei and Newton (and even earlier authors) recognized that for all materials, the *weight* is universally proportional to the *inertial mass*, so that freely falling particles, subject only to inertia and gravity, have the same world lines, given the same initial conditions, and this fact has by now been established experimentally with very high accuracy (about 10^{-10} uncertainty). It therefore appears that there are no completely "free" particles – *all particles "feel" gravity.* This led to the program to modify the spacetime structure such that the role of free particles is taken over by freely falling particles. Freely falling particles, however, exhibit relative accelerations in real gravitational fields. Hence, one needs a spacetime geometry which admits preferred curves which generalize straight lines and which exhibit relative curvature, to represent relative acceleration. In accordance with this idea, one has to give up the existence of exact, global inertial frames and be content with local, "approximately inertial" ones, and this forces one to work with curvilinear coordinates and non-linear transformations between them. This, then, also effects the laws governing proper times.

The ideas just outlined are made mathematically rigorous by generalizing the special-relativistic assumptions on spacetime as follows:

(5) On the spacetime manifold M there exists a Lorentzian metric

$$ds^2 = g_{\alpha\beta}(x^\gamma)dx^\alpha dx^\beta. \tag{6}$$

The associated timelike geodesics represent freely falling particles, null geodesics represent light rays, and the arc lenght along timelike world lines represents proper time.

In contrast to Newtonian and special-relativistic physics, in general relativity the structure of spacetime given by the metric (and fields derived from it) is not given once and for all, but is considered as a physical field similar to the electromagnetic field, subject to dynamical laws which relate the metric to the matter and other fields. It is still possible, given any event A, to introduce coordinates such that, at A, the metric takes the special-relativistic form (4) and the light cone $C(A)$ is given by (2), but the form (2) cannot in general be obtained in a finitely extended region. In consequence, the formula (5) governing clock readings gets modified by terms taking into account gravitation.

Local inertial frames can be identified by several kinds of experiments – mechanical, optical, and even quantum mechanical ones – and it is clear by now that the spacetime structure indicated by a "curved" metric, while preserving the approximate local validity of special relativity theory, indeed accounts for gravitational phenomena.

Quantitatively the relation between the metric field $g_{\alpha\beta}$ and the matter is expressed in terms of *Einstein's gravitational field equation*. So far, all observational tests of consequences of this equation have turned out to confirm the theory, sometimes with very high accuracy (see, e.g., Will 1987). In summary, it can be said that general relativity theory incorporates successfully all of macroscopic, classical physics.

Any physical process is to be represented by a spacetime obeying the field equation and other laws, and observable data are contained in this description as coordinate-independent invariants related to events, world lines, and other "figures" in spacetime. In particular, the predictions of the theory concerning the dependence of proper times on the gravitational field and on the motion of clocks have been well confirmed by measurements. The theory thus further confirms the view that there is no universal time; local systems have their own times, related in terms of the metric. On the other hand, experiences of time can be incorporated in the way indicated in Sect. 4.

Instead of representing a physical process in terms of a 4-dimensional spacetime, one can alternatively and equivalently represent it as a 3-dimensional space whose metric and other physical variables evolve according to a timelike parameter, which in general is not simply related to any clock time. While such a $(3 + 1)$-representation is a useful tool to construct spacetime models from initial data, it is not useful to analyse observations. Moreover, a given spacetime can be represented in infinitely many different ways as an evolving 3-space, and while these "histories of space" may look quite differently, they all agree in their predictive, testable content. Therefore, at least for the classical theory, I consider the intrinsic, unique spacetime description as the primary one. Whether this view can or should be maintained in a future quantum theory of space, time (or spacetime), and gravity, is perhaps not clear and remains to be seen.

References

Ashtekar A. (1994): Overview and outlook. In *Canonical Gravity: From Classical to Quantum. Lecture Notes in Physics* **434**, ed. by J. Ehlers and H. Friedrich (Springer, Heidelberg), 327–367.

Ehlers J. (1995): Spacetime Structures. In *Physik, Philosophie und die Einheit der Wissenschaften*, ed. by L. Krüger and B. Falkenberg (Spektrum Verlag, Heidelberg), 165–177.

Gödel K. (1949): A remark about the relationship between relativity theory and idealistic philosophy. In *Albert Einstein: Philosopher-Scientist*, ed. by P.A. Schilpp (Open Court, La Salle, Ill.), 555–562, here p. 557f.

Minkowski H. (1909): Raum und Zeit. *Physikal. Zeitschrift* **10**, 104–111. English translation: Space and time. In *The Principle of Relativity*, ed. by W. Perrett and G.B. Jeffery (Dover, New York 1952), 75–91; here: p. 76.

Penrose R. (1972): On the nature of quantum geometry. In *Magic Without Magic: John Archibald Wheeler*, ed. by J.R. Klauder (Freeman, San Francisco), 333–354.

Rindler W. (1991): *Introduction to Special Relativity*, (Oxford University Press, Oxford, 2nd edition).

Stein H. (1967): Newtonian space-time. *Texas Quarterly* Vol. X, No. 3, 174–200.

Weyl H. (1988): *Raum, Zeit, Materie* (Springer, Heidelberg, 7th edition), esp. p. 23.

Wheeler J.A. (1968): *Einsteins Vision* (Springer, Heidelberg).

Will C.M. (1987): Experimental gravitation from Newton's Principia to Einstein's general relativity. In *Three Hundred Years of Gravitation*, ed. by S.W. Hawking and W. Israel (Cambridge University Press, Cambridge), 80–127.

Nows Are All We Need

Julian B. Barbour

College Farm, South Newington, Banbury, Oxon, OX15 4JG, England

Abstract. An outline of a completely timeless description of the universe is given. This is based on the notion of the possible relative configurations of the universe. Taken all together, they form the relative configuration space of the universe. A classical history of the universe is a curve in such a configuration space. However, Newtonian mechanics provides clear evidence that a satisfactory account of the dynamical behaviour of subsystems of the universe cannot be given without the introduction of additional kinematic structure, this extra structure constituting a rigid framework of space and time.

The situation is quite different if the entire universe, considered finite, is described. It is shown how the universe as a whole can supply the framework of space and time needed for the description of subsystems. It is also noted that general relativity is a theory that from the beginning dispenses with any external spatiotemporal framework. Thus, in classical physics time is redundant, but since an operationally defined time identical to Newtonian time can be recovered for all subsystems the consequences of this insight are rather minor.

In the context of attempts to construct a quantum theory of the entire universe the consequences of the nonexistence of time are very great indeed. Because there is no external time or space, the wave function of the universe can only depend on the possible relative configurations. There cannot be any dependence of the wave function on time. A static description of the universe is necessarily obtained. A radical solution to the problem of recovering the appearance of time within such a timeless framework is proposed.

1 Kinematic Framework

The kinematic framework used in Newtonian mechanics to describe a system of N point particles of masses m_i is decidedly heterogeneous. The total number of coordinates used is $3N + 1$, of which the $3N$ are the $3 \times N$ Cartesian coordinates of the N particles in three-dimensional Euclidean space, and the extra one dimension is the space of absolute times t. Mathematicians have devised an exceptionally beautiful way of depicting the motion of a system of N particles as if it were the motion of a single particle. This is done by means of the notion of the *configuration space*, the number of dimensions of which is the $3N$ in the tally given above.

The position of a single particle in space is, of course, given by three coordinates (x, y, z). If there is a second particle, its position has to be specified by three more numbers. Formally, this can be done by means of the abstract notion of a six-dimensional configuration space, which is actually just two copies of ordinary space considered together. The six coordinates are then denoted $(x_1, y_1, z_1), (x_2, y_2, z_2)$. For N particles, one uses N copies of three-dimensional space and denotes the $3N$ coordinates by $((x_n, y_n, z_n)$, $n = 1, \ldots, N$.

However, the points of Euclidean space are all identical (and invisible to boot), and this opens up an alternative representation of the same situation that is very interesting. One can, for example, choose an orthogonal coordinate system in ordinary three-dimensional space such that particle 1 is permanently at its origin, and this then has the coordinates $(0, 0, 0)$. One can then choose the x axis in such a way that particle 2 is always on the x axis. Then its coordinates are $(x_2, 0, 0)$. Finally, one can position the z axis in such a way that particle 3 is permanently in the xy plane. Its coordinates are therefore $(x_3, y_3, 0)$. After that the remaining particles will have (in general) three nonvanishing coordinates each. However, in this representation only $3N - 6$ coordinates are used, since the first three particles which have been used to define the coordinate system have only three nonvanishing coordinates instead of nine.

These $3N - 6$ coordinates are the relative coordinates, and they are on a quite different status from the six remaining coordinates of the first description, which in their turn split up into two triplets of coordinates. One triplet serves to specify the overall position of the system in space (say the center of mass of the system), while the remaining three specify the overall orientation of the system.

In Newtonian kinematics, we have, in addition, the completely independent one-dimensional space of absolute times, so that the total representation space for Newtonian dynamics consists of four quite distinct subspaces with dimensions $3N - 6, 3, 3, 1$.

This is a very singular state of affairs. It should also be noted that the $3N - 6$ relative coordinates are on a quite different observational status from the remaining seven coordinates. The fact is that these relative coordinates are more or less directly observable, while all the remaining seven are totally invisible. What evidence there is for their existence in the Newtonian scheme only shows up indirectly, through the relative coordinates, as we shall see shortly.

First, I should like to draw attention to a remarkable *principle of unity* that binds together a relative configuration of N point particles in Euclidean space if $N \geq 5$. This is revealed by the following consideration. Between any N particles there are precisely $N(N-1)/2$ relative separations, which we can denote by r_{ij} $(i, j = 1, \ldots, N; i < j)$. Now we could image someone measuring these separations and thereby obtaining $N(N - 1)/2$ positive numbers, the distances between the points. The numbers could be jumbled up in some random order. However, if they were given to a competent mathematician, a remarkable fact could be established, namely, that $N(N - 1)/2 - (3N - 6)$ algebraic relationships exist between them. This is just a reflection of the fact that $3N - 6$ numbers already specify the complete relative configuration, and for $N \geq 5$ we have $N(N-1)/2 > 3N - 6$. As a consequence, algebraic relationships must exist between the relative separations r_{ij}.

These relationships have a characteristic distinctive form and express the very essence of Euclidean geometry. They knit the relative configuration together in a very special way. One can regard the r_{ij}'s as observable things given to us empirically. A priori there is no reason why such quantities, which can be measured independently, should bear any such relationships to each other. But they do. They have a principle of unity.

Moreover, once we are in possession of this principle of unity, we can, given one such relative configuration, imagine all other sets of $N(N-1)/2$ positive numbers that satisfy the same algebraic relationships. This gives us the set of all relative configurations of N particles, or the *relative configuration space* (RCS) of the universe if we suppose it to consists of just N particles.

Now all we ever see in the world is relative motion of objects. We never see space itself, nor do we see time. It therefore seems to be a reasonable assumption that the universe in its history simply passes through a succession of such relative configurations. *The history of the universe is a continuous curve in its relative configuration space.*

We could imagine taking "snapshots" of each relative configuration in the sequence. Such a snapshot can be supposed to define an *instant of time*. In this view, there is nothing more to time than this succession of instants, each of which are defined concretely by a relative configuration of material points. There is no sense in which the succession is "run through" at a given speed. For there is no clock external to the universe that could be used to measure such speed. In the real universe, we use the motion of objects *within* the universe to measure time, for example, the rotation of the earth. However, if we tried to speed up the motion of the *entire* universe along its curve in its relative configuration space, all motions would be equally affected, and, as beings within the universe, we could not possibly notice any difference.

Thus, on fundamental grounds, it is difficult to see how *absolute*, as opposed to *relative*, speed can have any meaning. The same applies to absolute rotation and translation. We must how see how this intuition matches observational fact.

2   Some Fundamental Facts of Dynamics

One of the most important facts about dynamics, which Newton formulated in absolute space and time, is determinism, which has given rise to the image of the clockwork universe. As Laplace especially emphasized, an omnipotent being who knows Newton's laws and, at a certain instant, the initial positions and initial velocities of all the particles in the universe would be able, from such a given initial state, to calculate all the previous and subsequent motions and positions of all the bodies in the universe.

Now, as Poincaré (1902) noted, a vital ingredient in this scenario is that the initial velocities are specified in absolute space. The reason for this is very instructive and leads us straight to the heart of the distinction between the absolute and relative descriptions of motion. Suppose for the moment that an absolute time exists and consider two particles in a fixed Euclidean space. They will have position vectors $\mathbf{r}_1(t)$ and $\mathbf{r}_2(t)$. Then the separation of the two particles will be

$$r_{12}(t) = \sqrt{(\mathbf{r}_2 - \mathbf{r}_1)(\mathbf{r}_2 - \mathbf{r}_1)}. \tag{1}$$

The rate of change of the separation is

$$\dot{r}_{12} = \frac{dr_{12}}{dt} = \frac{(\dot{\mathbf{r}}_2 - \dot{\mathbf{r}}_1)(\mathbf{r}_2 - \mathbf{r}_1)}{r_{12}}. \tag{2}$$

Any component of $\dot{\mathbf{r}}_2$ or $\dot{\mathbf{r}}_1$ that is orthogonal to $\mathbf{r}_2 - \mathbf{r}_1$ will make no contribution to (2). Thus, it can happen that one or both of $\dot{\mathbf{r}}_2$ or $\dot{\mathbf{r}}_1$ can be nonzero but $\dot{r}$ will vanish. In particular, this is exactly what happens in perfectly circular motion of two mass points around their common center of mass.

The important thing in the specification of the Laplacian scenario is that the $\dot{\mathbf{r}}_i$'s in absolute space are given, not the $\dot{r}_{ij}$'s. One can therefore ask what happens if one does not give

$$\mathbf{r}_i \text{ and } \dot{\mathbf{r}}_i, \quad i = 1, \ldots, N, \tag{3}$$

as in the usual specification but only

$$r_{ij} \text{ and } \dot{r}_{ij}, \quad i < j, i, j = 1, \ldots, N. \tag{4}$$

In fact, it is worth going all the way to a relative specification. This means that we cannot use an external time t to specify rates of change, but can only measure how one r_{ij} varies relative to another r_{ij}. Formally, this can be done by choosing r_{12}, say, as an "internal time" and specifying

$$r_{ij} \text{ and } dr_{ij}/dr_{12}, \quad i < j, j = 2, \ldots, N. \tag{5}$$

Of course, the r_{ij}'s and dr_{ij}/dr_{12} specified here must satisfy the special algebraic relations mentioned in Sect. 1. Alternatively, one could specify freely $3N - 6$ relative coordinates to determine the relative configuration and the derivatives of $3N - 7$ of them with respect to the remaining one of them chosen as an "internal time".

If motion were purely relative, and one wished to set up a relative dynamics of the universe, it would be very natural to require that initial data such as (5) be sufficient to determine the entire history of the universe. In the language adopted in Sect. 1, such initial data would amount to specifying the initial point and initial *direction* of the curve in the relative configuration space that we took to be an exhaustive description of the complete history of the universe.

It is now extremely interesting to see to what extent such data suffice to solve the Laplacian problem of celestial mechanics. We assume that we know that our N point particles interact through Newtonian universal gravity and that we have been told the masses of the particles. Let us consider first the three-body problem. Then as relative coordinates we can take directly the three sides of the triangle r_{12}, r_{23}, r_{13}. If we know the masses, we can immediately calculate the position of the center of mass. We know from the relativity principle of classical dynamics that any motion of the center of mass is completely undetectable in the relative motion. Thus, the three-dimensional subspace associated with overall translatory motion of the system in the $3N-6, 3, 3, 1$ decomposition of the $(3N+1)$-dimensional system of reference appears to be redundant. We certainly do not expect any of the $3N - 6$ (in this case 3) relative coordinates to be redundant, but what about the three dimensions associated with orientations and the single dimension of the absolute times?

At this point, it is appropriate to recall that Newtonian dynamics associates with any isolated system 10 fundamental constants of the motion. These are quantities that can be calculated algebraically from the positions and velocities of the particles at any instant and have values that are unchanged in time. They are conserved quantities. One set of three such constants gives the total momentum of the system, and another set of three gives the position of the center of mass of the system. Both of these relate to the overall motion of the system in absolute space, or in an allowed inertial frame of reference, to use modern terminology. By Galilean relativity, they are completely decoupled from the internal relative motion, as just noted.

The remaining four conserved quantities are the total energy E and the intrinsic angular momentum $\mathbf{M}$ with respect to the center of mass of the system. This last is a vector with three components. For any given instantaneous position of the triangle formed by the three particles, the angular momentum is a vector with origin at the center of mass that points in some direction relative to the plane of the triangle. The total energy is just a number. In general evolution, the direction of $\mathbf{M}$ relative to the plane of the triangle changes. However, one normally expresses this differently, with respect to the center-of-mass inertial frame, not with respect to the triangle formed by the three particles. In the inertial frame, the direction of $\mathbf{M}$ remains fixed (as does its length in both representations), and the plane through the center of mass perpendicular to $\mathbf{M}$ likewise remains fixed. This plane was called the *invariable plane* by Laplace. In the inertial frame, it is the orientation of the triangle formed by the three bodies that changes.

Now the important thing about both E and $\mathbf{M}$ is that neither can be determined from the purely relative data r_{ij}, dr_{ij}/dr_{12} but both have a drastic effect that shows up in the subsequent evolution of the system. This fact was highlighted in a famous study that Lagrange made of the Newtonian three-body problem in 1771, for which he was awarded the prize of the Royal Academy of Sciences of Paris in 1772 (Lagrange 1873). Lagrange still used an absolute time, but he found three equations that contain only r_{12}, r_{23}, r_{13} and their time derivatives. These equations have a very remarkable structure. They contain E and $|\mathbf{M}| = M$ as constant parameters and otherwise contain the time derivatives of the r_{ij} to various orders: two of the equations contain second time derivatives, and one contains *third* time derivatives.

If one were to replace the time derivatives by derivatives with respect to r_{12}, one would then obtain two equations that contain third derivatives of r_{23} and r_{13} with respect to r_{12}. Then to solve the initial-value problem with purely relative initial data it would be necessary to specify

$$r_{12}, r_{23}, r_{13}, \frac{dr_{23}}{dr_{12}}, \frac{dr_{13}}{dr_{12}}, \frac{d^2 r_{23}}{dr_{12}^2}, \frac{d^2 r_{13}}{dr_{12}^2}, E, M. \tag{6}$$

This need to specify two second derivatives and also E and M is precisely the fact that justifies Newton's use of absolute space and time. The need for knowledge of M and the two second derivatives arises because the change of the orientation in absolute space cannot be deduced from the r_{ij}'s and dr_{ij}/dr_{12} (or dr_{ij}/dt), but the change of orientation at the initial instant does show up in the subsequent evolution of the purely relative r_{ij}'s. It is necessary to specify E because the total energy contains the velocities with respect to the absolute time.

This state of affairs shows up even more remarkably in the N-body problem with $N > 3$. It follows from a famous paper by Jacobi (1842) (entitled "On the elimination of the nodes in the three-body problem"), with generalization to the N-body problem, that, if attempted in relative form, solution of the N-body problem of celestial mechanics requires specification of E, M, $3N - 6$ relative coordinates, $3N - 7$ relative rates of change of $3N - 7$ of the relative coordinates with respect to the one remaining relative coordinate, and any *two* second derivatives of the relative coordinates with respect to the same "internal time" relative coordinate. If N is very large, say half a million as in a globular cluster, it seems very strange that just two second derivatives out of so many are needed.

Also remarkable is the status of the two constants E and M, since they influence the dynamics radically. If the system is to remain bounded for a long period, E must be negative. The existence of a nonvanishing M (and with it an invariable plane of definite orientation) is shown dramatically in systems such as Saturn with its marvellous system of rings and by all spiral galaxies.

All these facts from dynamics and observational astronomy show clearly that relative configurations by themselves cannot possibly provide a satisfactory framework for describing subsystems of the universe. However, as Dziobek (1888) noted, there are some very interesting special cases of the three-body problem. In the special case that the plane of the triangle of the three particles remains parallel to the invariable plane, the third derivatives vanish from Lagrange's relative equations, though M and E still remain. This gives one special case. Further special cases arise if $M = 0$ and if both $M = 0$ and $E = 0$. Each of these special cases gives rise to qualitatively different systems of equations that describe the relative motion. The same happens in the general N-body problem.

For our purposes, the most significant fact is that if $E = 0$ and $M = 0$ all evidence for the existence of absolute space and time vanishes from the equations that describe the purely relative motion in the N-body problem. In particular, the initial-value problem can be solved given just a minimal set of $3N - 6$ relative coordinates and the relative rates of change of $3N - 7$ of them with respect to the one chosen as the "internal time". This fact was independently rediscovered by Zanstra (1924).

Now it is well known that Mach (1883) wished to formulate mechanics in purely relative terms but never succeeded in doing so. In fact, attempting to implement Mach's idea in a particular manner, Bertotti and I [6] found a very simple geodesic variational principle in a relative configuration space that leads to precisely the same relative motions of the bodies in the N-body problem as follows from Newtonian theory in the special case $E = 0$, $M = 0$.

It is worth describing the scheme very briefly, since it highlights how relative configurations, which I have identified with instants of time (or "nows"), are quite sufficient to set up dynamics. I shall describe it for the three-body case, since the relative configurations are then triangles, which are very easy to visualize. Each such triangle represents a point of the relative configuration space (RCS). Is it possible to set up a variational principle that distinguishes curves in the RCS using only structure within the RCS and nothing else?

We can do this by defining a "distance" between any two neighbouring points of the RCS, that is, between any two triangles that differ infinitesimally. This "distance" must depend on the two triangles and nothing else, since the "nows" (the triangles) are to suffice to do all of dynamics. We assume masses m_1, m_2, m_3 at the vertices 1, 2, 3 of each triangle and that we can distinguish the masses (and hence the vertices of the triangle). Let us place one triangle on top of the other in some arbitrary position but in such a way that the corresponding vertices are close to each other. We can do this because we have assumed that the triangles differ very little. Then the vertices at which m_1 is situated will be separated by some distance dx_1, and similarly there will be dx_2 and dx_3 for m_2 and m_3. We can now form the trial quantity

$$D_{\text{trial}} = \sqrt{\sum_{i<j} \frac{m_i m_j}{r_{ij}} \sum_i m_i dx_i^2}, \qquad (7)$$

where the potential $\sum_{i<j} m_i m_j / r_{ij}$ has been included to give rise to a nontrivial dynamics and also to reproduce Newtonian gravity. It is obvious that, because it contains the dx_i (and our placing was arbitrary), the trial quantity (7) is not determined by the two triangles alone. However, we can now move

the one triangle around on top of the other into all possible positions and find the unique configuration for which (7) is minimized. This is then a measure of the purely intrinsic difference between the two relative mass distributions and can be used as a measure of the infinitesimal intrinsic difference, or "distance", between the two corresponding points. The procedure generalizes in an obvious manner to configurations of N particles.

But once we have such an infinitesimal distance between any two points of the RCS we have a metric and can determine geodesics. The corresponding geodesic curves are candidates for our histories of the universe. Bertotti and I (Barbour and Bertotti 1982) showed that the evolutions of the relative separations corresponding to such geodesic histories are identical to the evolutions of the relative separations in Newtonian celestial mechanics with $E = 0, M = 0$ that Dziobek identified as interesting special cases.

It should be born in mind that if N is reasonably large there are still an astronomically large number of motions of the complete system for which $E = 0, M = 0$. However, there can very readily exist more or less isolated systems within such a universe that have nonvanishing E and M. It is only necessary that the E and M values for all the subsystems add up to vanishing values (vectorially for the M's, of course.) This therefore shows how the complete universe can be described in a timeless and frameless manner and simultaneously provide a frame of reference with respect to which its subsystems exhibit standard Newtonian behaviour with nonvanishing E and M.

Bertotti and I (Barbour and Bertotti 1982) also showed that if general relativity is considered as a dynamical system it is very closely analogous to the system of equations that describe the Newtonian N-body problem with $E = 0$ and $M = 0$. This is why we consider general relativity to be very Machian. For details, the reader is also referred to Barbour (1994).

To complete this discussion of classical dynamics, I should show how the conventional notion of time can be recovered from this timeless picture (Barbour and Bertotti 1982, Barbour 1994). Given any curve in the RCS, one can always introduce on its points an arbitrary label λ (a real number) that increases monotonically along one direction of the curve. Then instead of describing the changes of the relative separations by means of an "internal time" one can describe them with respect to λ. The equations of motion then contain an awkward factor having the form

$$\sqrt{\frac{\sum_{i<j} m_i m_j / r_{ij}}{\sum_i m_i (dx_i/d\lambda)^2}},\tag{8}$$

in which the dx_i are the ones determined by the "best-matching" procedure described above for the minimization of (7). An obvious device to simplify the equations of motion is to choose the λ in (8) in such a way that always

$$\sum_{i<j} m_i m_j / r_{ij} = \sum_i m_i (dx_i/d\lambda)^2.\tag{9}$$

The corresponding λ has the property that it casts the equations of motion into a particularly simple form. In fact, in a suitable frame of reference, the equations then become identical to Newton's equations of motion and the specially chosen λ appears in the role of the absolute time t. Thus, we have recovered Newtonian time from a timeless situation. In fact, rewritten in the form

$$dt = \sqrt{\frac{\sum_i m_i dx_i^2}{\sum_{i<j} m_i m_j / r_{ij}}}, \qquad (10)$$

the condition (9) becomes the *definition of time*, and its infinitesimal advance dt is given as an explicit weighted average of all the intrinsic displacements dx_i of the objects in the universe. Time is truly nothing but a measure of intrinsic change.

This concludes the story of time in classical physics. It does not need to be postulated as a part of an external kinematic frame. It can be recovered purely relationally from a timeless geodesic principle in the relative configuration space of the universe. Now we must consider the quantum implications of this picture.

3   The Quantum Implication: A Static Wave Function of the Universe

There are several ways in which one can see that the attempt to quantize Einstein's general theory of relativity may well lead to startling consequences. As noted earlier, general relativity has a dynamical structure very similar to that of the system of equations which describe the solutions of Newtonian dynamics for the special case $E = 0, M = 0$. This system of equations can, in its turn, be derived from a geodesic principle in a timeless relative configuration space.

In the classical theory, we have only changed the conventional notion of time in a relatively minor way. The fact is that we still assume a *unique* history of the universe. The universe passes through a unique succession of relative configurations. All that we have done is deny the existence of an external clock. The existence of a unique succession of states with each state described by a unique configuration of bodies in Euclidean space is extremely characteristic of classical physics.

But now we must consider what Schrödinger did to physics. It seems to me that all other revolutions in physics are as nothing compared with his transition to wave mechanics. The basic step he took was to replace the notion of a unique curve of a dynamical system in its configuration space by a wave function that, in principle, takes values on all possible configurations in the same configuration space. However, several factors tended to disguise quite what a revolutionary step this was. First, Schrödinger, like all quantum physicists, initially only applied his wave mechanics to microscopic systems.

The classical macroscopic world was left untouched. Second, the configurations on which Schrödinger defined his wave functions were configurations of particles in an external inertial frame of reference, and they also depended on the absolute time t.

Symbolically, we can write a Schrödinger wave function of an N-body system in the form

$$\Psi_s = \Psi_s(3N - 6 \text{ rel. coordinates}, 3 \text{ cms coordinates}, 3 \text{ orientations}, t). \quad (11)$$

This corresponds exactly to the $3N-6, 3, 3, 1$ decomposition of the Newtonian kinematic framework discussed earlier. However, in our discussion of classical physics we have seen how the nonrelational coordinates – the 3, 3, 1 part of the decomposition – can become redundant in the description of the entire universe.

This therefore suggests that if we attempt to set up a quantum theory of the entire universe – a quantum cosmology – the "wave function of the universe" will simply have the form

$$\Psi_{\rm QC} = \Psi_{\rm QC}(3N - 6 \text{ relative coordinates}), \quad (12)$$

where QC stands for quantum cosmology.

For a conventional physicist, the most disconcerting thing about (12), as compared with (11), is that it contains no place for a dependence on anything that one could call time. All we have is a static wave function, given once and for all. This is, in fact, exactly the picture that DeWitt (1967) obtained 30 years ago as a result of his attempt to construct a canonical quantization of general relativity for the case of a closed (finite) universe. Such a result is the inescapable consequence of the relational structure of generally relativistic dynamics, as I showed in detail in Barbour (1994).

It is a matter of some surprise to me how little stir was generated by DeWitt's discovery. It is still virtually unknown outside the narrow confines of people working on the quantum theory of gravity, and even within that group very few people worked on the problem until more than 15 years had passed. In part, this was because many people still believed it would be possible to restore a time dependence to the general relativistic analogue of (12) by identifying one of the relative coordinates as a distinguished "internal time" like the one considered in the discussion of Lagrange's work. However, this has always appeared to be very arbitrary – why choose one such coordinate rather than another? If, as seems the only natural choice (Barbour 1994), one defines time operationally by (10), this forces one to adopt a static wave function of the form (12).

During the last decade, a definite line of research has developed that has tried to make sense of a wave function of the form (12). The basic idea is to assume that, at least in some parts of the configuration space, the wave function of the universe is in the WKB regime. (This is the quantum mechanical analogue of geometrical optics.) In these circumstances, there seems to be a hope that one can recover a notion of time and the familiar time-dependent Schrödinger equation as emergent approximations from a timeless wave function of the form (12). This work, which I feel is important and promising, is described in this volume by Kiefer, who has himself made original contributions to this field. I refer the reader to his paper.

In the final section, I want to discuss a general idea that can provide a unifying viewpoint for the work reviewed by Kiefer and simultaneously yield an interpretation of quantum mechanics.

4 Time Capsules and the Many-Instants Interpretation of Quantum Mechanics

Let us consider in very general terms what sense we might attempt to make of a wave function like (12). What it actually states is that the wave function of the universe associates with each complete relative configuration of the universe a definite complex number Ψ. This complex number is given once and for all. (In fact, it is an open question whether Ψ should be real or complex, as I discuss in Barbour (1993). Here I assume Ψ is complex.) Following a basic Born-type interpretation, we could postulate that $\Psi^*\Psi$, where Ψ^* is the complex conjugate of Ψ, gives the "probability" of the corresponding complete configuration of the universe. However, this probability cannot be understood in the usual sense that if we make an observation at a definite time $\Psi^*\Psi$ gives the probability to find the system in the corresponding configuration. For we have nothing over in addition to the configuration to tell us the time. Indeed, in accordance with the classical notions developed earlier, the complete configuration is itself an instant of time. We seem to be at an impasse. However, it is also true that we ourselves cannot "bring about" instants of time. We just experience them. Let us therefore suppose that $\Psi^*\Psi$ is a probability for the given configuration to be "experienced". What meaning can we give to this?

In any experienced instant, we may suppose that the atoms in our brain are in some definite configuration, which in the usual view of things is constantly changing by virtue of neuronal activity. I suppose most people would assume that the psychologically experienced "now", or rather some extended specious present (see the contribution of Pöppel to this volume), is correlated with the state of neuronal activity in the given instant. In this view, consciousness will be correlated with the overall brain configuration and the manner in which it is changing.

However, it could be that the experienced "now" is actually correlated solely with the instantaneous configuration. For example, it is clear that our long-term memory, which provides us with our entire knowledge of ourselves and our history up to the present, must somehow be "hard-wired" into the configuration of our neurons. In this respect, our brain seems to be very like a computer memory. Therefore, it seems to me possible that in any experienced instant we are simultaneously aware of much information that is all coded into an instantaneous configuration, as in a computer memory. Experiencing an instant is thus like reading and comprehending a computer memory all at once.

I do not want to insist on this picture too much. I am certainly not an expert in brain research. I introduce the idea mainly to illustrate the notion of what I call a time capsule. Whatever may be the connection between brain configuration and consciousness, it is certainly a fact that brain configurations are always very special. They must be, since otherwise they could not code all our long-term memory. Such configurations are examples of what I call time capsules.

To be more precise, I define a time capsule as any static configuration that seems to contain many mutually consistent records of processes that took place in time in accordance with definite laws. A beautiful example of a time capsule is provided by any geological section (for example, the face of the Grand Canyon). In fact, it is from painstaking examination of such sections, which are all effectively static, that almost all our knowledge of the history of the earth has been built up. Moreover, it is a remarkable fact that virtually all matter that we can study in the universe seems to be a time capsule in the sense defined above. All objects encode to greater or lesser degree information about what we call their history. This is even true of gas in stars, the element abundances of which are interpreted with impressive success by astrophysicists in terms of primordial nucleosynthesis and "cooking" in stars and supernovae.

Of course, the usual interpretation of these "time capsules" is that they record processes which actually occurred. However, if we can understand how the very special configurations that constitute time capsules could arise in a completely timeless fashion, history in the usual sense would become redundant. Bearing in mind that a time capsule is merely a very special configuration, let us now consider the absolutely standard account of atomic and molecular structure that is provided by quantum mechanics.

This structure is explained by solving the *time-independent* Schrödinger equation for the corresponding system,

$$H\Psi = E\Psi, \tag{13}$$

where H is the Hamiltonian.

The physically allowed solutions are required to satisfy a condition like square integrability. For bound systems at least, this selects a very special set of solutions, the negative-energy eigenstates with only discrete values of E.

The obtained physically allowed solutions of (13) simple give Ψ and, with it, the probability $\Psi^* \Psi$ for all the possible complete configurations of the system. Certain characteristically structured configurations are found to be much more probable than others. This is what makes it possible to say that molecules have structure.

Now, written in symbolic form, the equation that DeWitt found for the wave function of the universe (the so-called Wheeler-DeWitt equation) has the form

$$H\Psi = 0, \qquad (14)$$

i.e., it is like a time-independent Schrödinger equation for energy eigenvalue zero. (I should mention that the Wheeler-DeWitt equation is also often likened to a Klein-Gordon equation with one of the coordinates playing the role of an "internal time".) The picture that this suggests to me is that of the universe as a huge "molecule" in a stationary eigenstate of an equation like the time-independent Schrödinger equation. What we call instants of time are to be identified with any one of the innumerable possible configurations of the universe, which are given different probabilities in allowed solutions of the timeless Wheeler-DeWitt equation.

This picture is not going to get us any further towards understanding the apparent passage of time and our conviction that we have a unique (and common) past without a further important conjecture that I am forced to make. The conjecture is that in allowed solutions of the Wheeler-DeWitt equation the probability $\Psi^*\Psi$ is concentrated to a high degree on configurations that are time capsules. Further, the atoms of our brains in experienced instants of time belong to such time-capsule configurations, and this is why we habitually have the impression that we have a past, move through time, and have just arrived at the present.

In this view, causality must be understood in a manner totally unlike the usual one. According to the standard picture of causality, the universe was "started" in the distant past and has since evolved in accordance with the laws of physics. That, at least, is the picture in classical physics, which is still very widely used in cosmology. The fact that we experience some definite thing now is explained by "creation" of the universe in the distant past and lawful causal evolution since then, a remote consequence of which is our present state.

In the timeless view of things, there is no such thing as creation of the universe in the past. Instead, there is a kind of competition in timelessness between all possible configurations the universe could have. The outcome of the competition determines probabilities for all the configurations. The connection between our experienced "now" and this interpretation of quantum cosmology is that what we see and are aware of in this "now" is simply a presentation of information encoded in the time-capsule configuration of the entire universe in which this personal "now" of ours is embedded.

Causality in the scientific sense of the word operates only to the extent that quantum cosmology determines the timeless rules which accord some of the possible configurations of the universe a high probability but give others a very low one. Why anything is experienced at all remains the greatest mystery, but that is so in any scientific account of the world. However, quantum cosmology in this interpretation is predictive to the extent that it says we shall be in configurations with high probabilities and that such configurations exhibit characteristic features that can be verified empirically. In this respect, my proposal is very like the standard theory of molecular structure, which predicts characteristic structures of molecules.

In this picture, the experiencing of any "now" is a completely self-contained phenomenon. It could be said to correspond to the "realization", or even "creation", of the configuration in which the "now" is embedded. This realization does not depend on whether or not any other configurations are "realized". Our present "now" is not realized or "created" because other "nows" were created in some distant past. The only "creation" we can speak of is realization and experiencing of this "now". All other "nows" remain conjectures.

The reader may well feel that in the above I have merely piled up implausible conjectures in a huge heap. I do not deny that the conjecture that the Wheeler-DeWitt equation spontaneously concentrates its solution(s) on time capsules is a large conjecture. However, many of the other considerations do have a kind of hard inevitability and ring of truth about them. Philosophers have long felt that time should be reduced to change and that position and orientation of the universe as a whole can have no meaning. The notion of the classical history of the universe as a curve in its relative configuration space has the right feel to it and in no way conflicts with the undoubtedly good evidence that Newton presented for the existence of a local effective dynamical frame of reference and absolute time. Indeed, both of these key features can be derived in a beautiful way from a timeless and frameless Machian picture.

Moreover, we know that some features of classical dynamics do survive in quantum mechanics. In fact, Schrödinger retained precisely the configuration space (on which he defined his wavefunction) but also an inertial frame of reference and an absolute time. (Time has always assorted with great difficulty with the true observables of quantum mechanics.) In passing to quantum cosmology, we are merely shedding the most suspect elements of classical mechanics. The residuum is a relative configuration space, and in that context a static wave function defined on it does not seem so outrageous.

Moreover, the evidence that I discuss in Barbour (1994) shows that solutions of the time-independent Schrödinger equation can be concentrated in a remarkable manner on beautiful examples of time capsules. This was demonstrated in a famous paper by Mott (1929), which I discuss in detail in Barbour (1994). Further, the interpretation I give to the Mott solution accords well with the general programme for recovery of standard physics from the timeless Wheeler-DeWitt equation that Kiefer describes in this volume. It is also consistent with the work on decoherence that he describes.

It is not possible in an article of this length to cover all the details of the case for a timeless world. I have tried to give some of the simplest arguments and evidence. I have gone into most detail in the discussion of the N-body problem in celestial mechanics, partly because there are here exact results of great potential significance and partly because they have not hitherto been discussed in the literature. It seems to me that the results are of interest in their own right and ought to be better known.

My main conclusion is expressed in the title: "nows" are all we need to give an account of our experiences. Ruhnau (this volume) argues forcefully that the notion of a flow of time belongs inescapably to the foundations of physics. Quite recently, Shimony (1996) argued against my notion of time capsules along very similar lines. I believe both may be deceived by the rich structure of the universe. As already emphasized, the main conjecture – that the wave function of the universe is concentrated spontaneously on configurations with a very special structure – is a very large one, yet the fact remains that we do find ourselves in an extraordinarily special universe. Time capsules do abound. In the normal interpretation, this is attributed to an initial state of the universe with exceptionally low entropy. This is the great problem of the highly nonequilibrium state of the universe. I suggest that we exploit the known ability of the time-independent Schrödinger equation to give high probability to certain favoured configurations to explain both this fact and the appearance that time exists and flows.

Acknowledgments: I should like to end by thanking Eva Ruhnau and Harald Atmanspacher for organizing a most agreeable workshop in Tegernsee. It was one of the most stimulating time capsules in which I have ever found myself in this universe of so many rich instants. I leave it as an exercise to the reader to work out what "organizing a workshop" means in a timeless quantum universe. If that can be done, we may then understand both time and the universe.

References

Barbour J.B. (1993): Time and complex numbers in canonical quantum gravity. *Phys. Rev. D* **47**, 5422–5430.

Barbour J.B. (1994): The emergence of time and its arrow from timelessness. In *Physical Origins of Time Asymmetry*, ed. by J.J. Halliwell, J. Pérez-Mercader, and W.H. Zurek (Cambridge University Press, Cambridge), 405–414. See also: The timelessness of quantum gravity I. The evidence from the classical theory, II. The appearance of dynamics in static configurations. *Class. Quant. Gravity* **11**, 2853–2873, 2875–2897.

Barbour J.B. and Bertotti B. (1982): Mach's Principle and the structure of dynamical theories. *Proc. R. Soc. London* **A382**, 295–306.

DeWitt B.S. (1967): Quantum theory of gravity I. The canonical theory. *Phys. Rev.* **160**, 1113–1148.

Dziobek O. (1888): *Die mathematischen Theorien der Planeten-Bewegungen* (J.A. Barth, Leipzig).

Jacobi C.G.J. (1842): Sur l'elimination des noeuds dans le problème des trois corps. Compte Rendu, August 8, 1842. Reproduced in: C.G.J. Jacobi, *Mathematische Werke, Vol. 1* (G. Reimer, Berlin, 1846).

Lagrange J.J. (1873): Essai sur le problème des trois corps. In *Oeuvres de Lagrange, Vol. 6* (Gauthier-Villars, Paris), 229–324.

Mach E. (1883): Die Mechanik in ihrer Entwickelung, historisch-kritisch dargestellt (Brockhaus, Leipzig). English translation: *The Science of Mechanics: A Critical and Historical Account of its Development* (Open Court, La Salle, Ill., 1960).

Mott N.F. (1929): The wave mechanics of α-ray tracks. *Proc. R. Soc. London* **A126**, 79–84. Reprinted in: *Quantum Theory and Measurement*, ed. by J.A. Wheeler and W.H. Zurek (Princeton University Press, Princeton, 1983), 129–134.

Poincaré H. (1902): *Science et Hypothese* (Paris). English translation: *Science and Hypothesis* (Walter Scott, London, 1905).

Shimony A. (1996): Physical implications of objective transiency. In *Geometric Issues in the Foundations of Science*, ed. by S. Huggett et al. (Oxford University Press, Oxford), in press.

Zanstra H. (1924): A study of relative motion in connection with classical mechanics. Phys. Rev. **23**, 528–545.

The Quantum Gauge Principle

Dirk Graudenz

Theoretical Physics Division, CERN, CH–1211 Geneva 23, Switzerland

Abstract. We consider the evolution of quantum fields on a classical background space-time, formulated in the language of differential geometry. Time evolution along the worldlines of observers is described by parallel transport operators in an infinite-dimensional vector bundle over the space-time manifold. The time evolution equation and the dynamical equations for the matter fields are invariant under an arbitrary local change of frames along the restriction of the bundle to the world-line of an observer, thus implementing a "quantum gauge principle". We derive dynamical equations for the connection and a complex scalar quantum field based on a gauge field action. In the limit of vanishing curvature of the vector bundle, we recover the standard equation of motion of a scalar field in a curved background space-time.

1 Introduction

The concept of time in quantum field theory is derived from the structure of the underlying space-time manifold. For both flat and curved space-times, in the Heisenberg picture, the state vector of a quantum system is constant[1] and identical for all observers, and the quantum fields fulfil dynamical equations derived by means of a quantization procedure of classical equations of motion based on a classical action functional (Birrell and Davies 1982). This framework is unsatisfactory for two principal reasons:

1) The state vector is an object that describes the knowledge that an observer has about a physical system. The observer, in an idealized case, moves along a worldline $\mathcal{C}$. The state vector should thus be tied to $\mathcal{C}$, and its time evolution should not be related to some globally defined time, but to the *eigentime* of the observer.

2) The observer is free to choose the basis vectors in Hilbert space. In quantum mechanics, a change of basis vectors amounts to a change of the representation, e.g., from the Heisenberg to the Schrödinger picture. In quantum field theory, in particular in perturbation theory, the preferred picture is the interaction picture. The change of pictures is performed by means of unitary operators. If, as proposed in (1), the state vector is tied to a worldline, a change of the basis in the Hilbert space should be allowed to be observer-dependent and arbitrary at any point of the worldline.

---

[1] We do not consider the problem of measurements.

These two requirements amount to what could be called a "local quantum relativity principle": physics is independent of the Hilbert space basis, and the basis may be chosen locally in an arbitrary way. Since a change of the basis is mediated by a unitary transformation, the local quantum relativity principle is equivalent to a local $U(\mathcal{H})$ symmetry, $U(\mathcal{H})$ being the group of unitary operators on the Hilbert space $\mathcal{H}$.

In a recent paper (Graudenz 1994), a formulation of the time evolution of quantum systems in the language of differential geometry has been given. State vectors are elements of an infinite-dimensional vector bundle, and time evolution is given by the parallel transport operator related to a connection in the bundle. In order to keep the paper self-contained, this framework is briefly reviewed in Sect. 2. The goal here is to propose a dynamical principle that yields the connection in the bundle and the field operators at any space-time point. The basic idea is that the geometrical formulation permits the introduction of a local gauge theory, the connection being the "gauge field", and the quantum field operators being linear-operator-valued "matter fields". Such a local gauge theory will be defined in Sect. 3 by means of an action functional. We wish to note that the theory does not have to be quantized, because the dynamical variables appearing in the action are already the components of the quantum operators in an arbitrary frame. It is, however, not yet clear whether in this way canonical commutation relations hold in general.

The theory distinguishes conceptually between the evolution of the state vector, given by the connection in the bundle (for short, called the "quantum connection" in the following), and the dynamical equations of the quantum fields and the quantum connection, derived from the action principle. The time evolution of the state vector, bound to the worldline of a specific observer, and the space-time dependence of the quantum fields are in principle independent, although the generator of the state-vector time evolution is a field that will appear in the equation of motion of the quantum fields. Since the action and the time evolution equation are formulated in a gauge-covariant way, the local quantum relativity principle is fulfilled.

2   The Geometrical Framework

In this section, we briefly review the geometric formulation of quantum theory. For more details, we refer the reader to Graudenz (1994), and – as far as general references are concerned – to Kobayashi and Nomizu (1963) and Wells (1980).

The basic ingredients of the theory are:[2]

(a) A space-time manifold M with metric $g_{\mu\nu}$.

(b) A Hermitian vector bundle $\pi : H \to M$ over M, with the fibres $H_x = \pi^{-1}(x)$ isomorphic to a Hilbert space $\mathcal{H}$. The structure group of the bundle is assumed to be the unitary group $U(\mathcal{H})$ of the Hilbert space, and Hermitian conjugation in a local trivialization is denoted by '$*$'. The metric of the bundle is denoted by G, and Hermitian conjugation with respect to G is denoted by '$\dagger$'.

(c) The bundle $H \to M$ is equipped with a connection D. In a local trivialization, the connection coefficients are denoted by K; they are anti-Hermitian operators

$$K = -K^*. \tag{1}$$

The covariant derivative of a section ψ of the bundle is defined by

$$D\psi = \mathrm{d}\psi - K\psi. \tag{2}$$

The curvature F corresponding to D is $F = D^2$. The covariant derivative of a field A of linear maps of the fibres of $H \to M$ is given by

$$DA = \mathrm{d}A - KA + AK, \tag{3}$$

and similarly the covariant derivative of the metric reads

$$DG = \mathrm{d}G - KG + GK. \tag{4}$$

It is assumed that the connection and the metric are compatible, i.e.,

$$DG = 0. \tag{5}$$

The physical interpretation of these mathematical objects and some assumptions being made are:

(a') The underlying space-time manifold M together with its metric is assumed to be fixed and given by some other theory. It would certainly be desirable that the dynamical laws governing the evolution of the metric

---

[2] In the following, we do not distinguish in the notation between the global, coordinate-independent quantities and their basis-dependent components in, for example, a local trivialization of the vector bundle. What is meant will be clear from the context. Moreover, we do not attempt to fulfil any standards of mathematical rigour; the focus is on the conceptual development. For example, we do not discuss the problems coming from the regularization of operator products. We also sometimes drop technical details. It is, for instance, assumed that local quantities are patched together by a partition of unity, whenever this is required.

field, i.e., the Einstein equations, be incorporated in the formalism developed here. This could be achieved, for example, by adding the Einstein-Hilbert action S_{EH} to the quantum action S_Q to be defined later.[3]

(b′) For an observer B at $x \in M$, we assume that the state vector ψ that B uses as a description of the world is an element of H_x. Observables, such as fields $\Phi(y)$, are assumed to be sections of the bundle $\pi_{\mathcal{L}H} : \mathcal{L}H \to M$, whose fibres consist of linear operators acting on the fibres of $H \to M$. We use the bracket notation for the inner product given by G; for two vectors $\xi, \eta \in H_x$ and a linear operator $A \in \mathcal{L}H_x$, we write $G(\xi, A\eta) = \langle \xi | A | \eta \rangle$. The quantity[4] $\langle \psi | \Phi(x) | \psi \rangle$, for $\psi \in H_x$, is assumed to be the expectation value of the field Φ at x, i.e., the prediction for the average of a measurement by means of a local measurement device carried by B. Predictions by an observer B at x for a measurement of Φ at y can be made by a parallel transport of the state vector along a path joining x and y (see (c)), and taking the expectation value at y. Requiring path independence of the expectation value, i.e., consistency of predictions, leads to a condition

$$[U_\alpha, \Phi(x)] \, \psi = 0 \qquad (6)$$

for state vectors ψ of the physical subspace of $\mathcal{H}_x$, closed loops α attached to x and observables $\Phi(x)$. There is thus a symmetry group (the group of U_α fulfilling (6)) related to the holonomy group of the bundle; for a discussion see Graudenz (1994).[5]

(c′) The quantum connection D can be integrated along curves $\mathcal{C}$ joining x and y to give parallel transport operators $U_\mathcal{C}$ mapping the fibre H_x onto the fibre H_y. The quantum connection D is assumed to govern the evolution of the state vector ψ in the direction of a tangent vector v by means of the equation $D\psi(v) = 0$. For an observer B moving along a worldline $\mathcal{C}(\tau)$, parametrized by B's eigentime τ, the evolution of the

---

[3] Should a genuine quantum theory of gravity be possible, then the theory developed here could certainly no longer be applied, because in the differential geometric formulation we make use of the fact that the manifolds and bundles under consideration are smooth. What would be required in this case would be a geometry of space-time compatible with quantum gravity.

It is not clear a priori whether a quantum theory of gravity can be formulated by quantizing some classical action. Conceptually, space-time is the set of all possible events, and the metric, up to a conformal factor, merely encodes the causal structure. Epistemologically, these notions are much more fundamental and deeper than gauge and matter fields. It is possible that gravity should not be quantized at all.

[4] We do not include the space-time point x in the notation for the metric G.

[5] Invariant operators $\check{A} \in \mathcal{L}H_x$ fulfilling $[U_\alpha, \check{A}] = 0$ for all closed curves α can be constructed from an arbitrary operator $A \in \mathcal{L}H_x$ by means of a "path integral" $\check{A} = \int \mathcal{D}\beta \, U_\beta \, A \, U_\beta^{-1}$ over all closed curves β originating in x, if a left-invariant $(\int \mathcal{D}\beta \, f(\alpha \circ \beta) = \int \mathcal{D}\beta \, f(\beta))$ and normalized $(\int \mathcal{D}\beta = 1)$ measure $\mathcal{D}\beta$ exists.

state vector $\psi(\tau)$ is thus

$$\partial_\tau \psi(\tau) = K_\mu(\mathcal{C}(\tau))\dot{\mathcal{C}}^\mu(\tau)\,\psi(\tau). \tag{7}$$

This equation is nothing but a Schrödinger equation[6] for a path-dependent Hamilton operator[7] or "quantum gauge field" K_μ. The assumption of D and G being compatible means that the transition amplitude $\langle \psi(\tau)|\chi(\tau)\rangle$ of two states ψ and χ is invariant under time evolution.

We now have to discuss the question of where the quantum connection D and the dynamical equation for the quantum matter fields Φ come from. It is desirable to have a common principle for these two objects. We note that the matter fields are related to the vector bundle itself, whereas the connection is naturally related to the principal bundle. We therefore need a means to connect objects related to two different bundles. The following possibilities suggest themselves:

(A) If there is a preferred trivialization of the bundle, i.e., a canonical coordinate system, then in this particular system the quantum connection coefficients K can be defined as a function of the matter fields Φ. An example of this is the translation of the standard formulation of quantum field theory in the Heisenberg picture in Minkowski space to the geometric formulation. The bundle $H \to M$ is nothing but the direct product $M \times \mathcal{H}$ of Minkowski space M and the Hilbert space $\mathcal{H}$. There is a canonical trivialization of the bundle owing to the direct product structure. The quantum connection coefficients K are set to zero in this trivialization; D is thus simply the total differential. Consequently, the state vector is constant and the same for all observers. The metric G is inherited from the metric of $\mathcal{H}$. The dynamical law for the fields $\Phi(x)$ is the Heisenberg equation of motion

$$\partial_\mu \Phi(x) = \mathrm{i}\,[P_\mu, \Phi(x)], \tag{8}$$

where the P_μ are the energy and momentum operators of the theory, being functions of the fields Φ. The crucial step in this construction is the assumption of a trivial bundle $M \times \mathcal{H}$, because with this a preferred trivialization of the bundle comes for free.

(B) A variant of (A) is to single out a specific coordinate system by some physical principle, the prototype being the definition of inertial frames and the application of the equivalence principle in general relativity. Unfortunately, the application of an equivalence principle based on inertial

---

[6] It is possible to include, for example, a one-form $P(\Phi, D\Phi, F)$, polynomial in the fields Φ, the derivatives $D\Phi$ and the curvature F, in the evolution equation, such that $(D - P)\Psi(v) = 0$. This would correspond to an additional interaction term in the Schrödinger equation.

[7] To simplify the notation and suppress factors of the imaginary unit, we require the operator K to be anti-Hermitian, see (1).

frames is not possible in our case, because this would only fix a frame in the tangent bundle of M, but not in the bundle $H \to M$.

(C) Finally, there is the possibility to postulate a dynamical law. This is well suited to the problem at hand, because the connection is essentially a differential operator on the vector bundle. This allows us to define covariant differential equations possibly derived from an action principle.

In this paper, we follow (C) by defining a gauge field action for the special case of a complex scalar field Φ. This and the derivation of the dynamical equations is done in the next section.

3   The Quantum Action and the Dynamical Equations

The action employed to derive dynamical equations for the quantum connection D and a complex scalar quantum field Φ is

$$S_Q = S_K + S_G + S_k + S_m, \tag{9}$$

where

$$S_K = \int dx \sqrt{\sigma g} \, \frac{\alpha}{2} \operatorname{Tr} \left(F_{\mu\nu} F^{\mu\nu} \right) \tag{10}$$

is the action for the quantum connection coefficients,

$$S_G = \int dx \sqrt{\sigma g} \operatorname{Tr} \left(\lambda^\mu D_\mu G \right) \tag{11}$$

is the action implementing the constraint $DG = 0$ by means of a field of linear-operator-valued Lagrange multipliers λ^μ,

$$S_k = \int dx \sqrt{\sigma g} \operatorname{Tr} \left((D_\mu \Phi)^\dagger D^\mu \Phi \right) \tag{12}$$

is the action for the kinetic part of the field Φ, and

$$S_m = \int dx \sqrt{\sigma g} \, \gamma \operatorname{Tr} \left(\Phi^\dagger \Phi \right) \tag{13}$$

is an action reminiscent of a mass term for Φ. Here σ is the sign of the determinant

$$g = \det \left(g_{\mu\nu} \right) \tag{14}$$

of the space-time metric, α and γ are "coupling constants" to be discussed later, and $F_{\mu\nu}$ is the curvature tensor associated with the quantum connection D, defined by

$$F_{\mu\nu} = -\partial_\mu K_\nu + \partial_\nu K_\mu + [K_\mu, K_\nu]. \tag{15}$$

The unusual signs in the first two terms stem from the fact that $F = D^2$, where $D = d - K$ instead of $D = d + K$, as is usually assumed. The trace

'Tr' is the trace operation of linear operators in a local trivialization of the vector bundle.

It can easily be checked that the action S_Q is invariant under a change of basis in the vector bundle. Owing to the Lagrange multipliers λ^ρ, the variables K_ρ, G, Φ, Φ^* and λ^ρ are independent. The action principle $\delta S_Q = 0$ then leads to the equations of motion by varying the fields.[8] In order to achieve a compact notation, we introduce the covariant derivative $\hat{D}_\mu$ for a vector t^μ and for an antisymmetric tensor $t^{\mu\nu}$ by

$$\hat{D}_\mu t^\mu = \frac{1}{\sqrt{\sigma g}} D_\mu \left(\sqrt{\sigma g}\, t^\mu \right) \tag{16}$$

and

$$\hat{D}_\mu t^{\mu\nu} = \frac{1}{\sqrt{\sigma g}} D_\mu \left(\sqrt{\sigma g}\, t^{\mu\nu} \right), \tag{17}$$

respectively. It can be shown that for vanishing torsion these expressions are covariant divergences and transform as a scalar and as a vector, respectively.

(α) The variation of the quantum connection coefficients K_ρ leads to:[9]

$$\frac{\alpha}{2}\, \hat{D}_\mu F^{\mu\rho} + [\lambda^\rho, G] - G \left[\Phi^\dagger, D^\rho \Phi \right] G^{-1} - \left[\Phi, (D^\rho \Phi)^\dagger \right] = 0. \tag{18}$$

In a classical gauge theory, if the vector bundle is finite-dimensional, this is the equation of motion for a gauge field coupled to a matrix-valued complex scalar field in the fundamental representation. In our case, the gauge field is related to the quantum connection D. It should be noted that this equation is different from the one obtained when quantizing, for example, a classical $SU(n)$ gauge field in conjunction with a matrix-valued matter field in the fundamental representation. In this case, for the gauge field, we would have operators $A_{\mu a}$, where a is a colour index in the adjoint representation. The matter field $\Phi^b{}_c$ would come with colour indices b and c in the fundamental representation. In (18), the operators do not carry an explicit colour index, rather the "colour indices" are the indices of the infinite-dimensional matrices, if the equations were written out in a specific Hilbert space basis.

(β) The variation of the metric G results in

$$\left(\hat{D}_\mu \lambda^\mu \right) G + \left[(D_\mu \Phi)^\dagger, D^\mu \Phi \right] + \gamma \left[\Phi^\dagger, \Phi \right] = 0. \tag{19}$$

The solution of this equation yields the Lagrange multiplier λ, eventually to be inserted into the other equations.

---

[8] The variations $i\,\delta K_\rho$, δG, $\delta \Phi$, $\delta \Phi^*$ and $\delta \lambda^\rho$ run through all infinitesimal Hermitian operators δR, so that the condition $\mathrm{Tr}\,(A\,\delta R) = 0$ for all δR leads to $A = 0$.

[9] This is an equation for the covariant derivative $D * F$ of the dual curvature tensor $*F$. The explicit form of $D * F$ is given by (3). There is, of course, also the Bianchi identity $DF = 0$.

(γ) The variation of the complex scalar field Φ leads to

$$\hat{D}_\mu D^\mu \Phi - \gamma \Phi = 0 \qquad (20)$$

and

$$\hat{D}_\mu (D^\mu \Phi)^\dagger - \gamma \Phi^\dagger = 0. \qquad (21)$$

To discuss these equations, let us set the quantum gauge field in the covariant derivative to zero.[10] This can be achieved by the limit $\alpha \to \infty$, for the following reason. Defining $\alpha = 1/a^2$ and $\tilde{K}_\mu = K_\mu/a$ allows the coefficient α to be absorbed into the curvature tensor $\tilde{F}_{\mu\nu}$ of $\tilde{K}_\mu$, where the commutator term in $\tilde{F}_{\mu\nu}$ receives a factor of a. The covariant derivative is $D = \mathrm{d} - a\tilde{K}$. Setting $a = 0$ leads to the desired result. Equation (20) then reduces to

$$(\Box - \gamma)\Phi = 0, \qquad (22)$$

with

$$\Box \Phi = \frac{1}{\sqrt{\sigma g}} \partial_\mu \left(\sqrt{\sigma g}\, g^{\mu\nu} \partial_\nu \Phi \right) \qquad (23)$$

the wave operator on the space-time manifold M. Defining $\gamma = -m^2$, (22) is the Klein-Gordon equation for a scalar quantum field of mass m in a curved background space-time (Birrell and Davies 1982). Moreover, for $a = 0$ the state vector is constant along the worldline of the observer. We are thus able to recover standard quantum field theory in curved background space-times in a certain limit.

(δ) Finally, the variation of the Lagrange multipliers λ^ρ yields the constraint that the metric in the vector bundle be consistent with the quantum connection:

$$D_\rho G = 0. \qquad (24)$$

As can easily be seen, the equations of motion are all explicitly gauge covariant.

4   Discussion

We have proposed dynamical equations for the geometrical formulation of quantum field theory as defined by Graudenz (1994). A classical gauge field action for a complex scalar field in an infinite-dimensional vector bundle[11] gives rise to gauge covariant equations of motion for the quantum connection and for the scalar field. A gauge transformation can be interpreted as a change of frame in the vector bundle, and thus as a space-time-dependent change of the "picture".

---

[10] This corresponds to $F_{\mu\nu} = 0$.

[11] To be more precise, for a complex scalar field in the bundle of linear operators acting on the fibres of a vector bundle.

We wish to point out a similarity of the present theory to the quantum mechanics of a single particle coupled to an electromagnetic field. There, the requirement that the phase of the wave function have no physical meaning motivates the introduction of an Abelian gauge field, which can then be interpreted as the electromagnetic gauge potential. The Schrödinger wave function is in general a section of a complex line bundle. Unobservability of the phase can be rephrased as the independence of physics of the particular choice of basis in the line bundle, admitting arbitrary passive space-time-dependent $U(1)$ transformations. In our case, the situation is slightly different. We are not concerned with the quantum mechanics of a single particle, but with quantum field theory, where, in the geometrical formulation, the space-time dependence relates to the full state vector and not only to the amplitude at a specific space-time point. The quantum relativity principle states that physics be independent of the choice of basis in the Hilbert space, for all possible observers. Instead of the independence of physics of the phase of the Schrödinger wave function, we require that physics be invariant under arbitrary local $U(\mathcal{H})$ transformations. Since the Abelian gauge potential in the quantum mechanics case actually relates to an empirically observable field, it is tempting to speculate whether the quantum gauge connection has some counterpart in physical reality as well. The fact that a dynamical formulation involving the quantum connection coefficients, as done in this paper, is possible, and consequently a resulting set of coupled equations of motion of the quantum connection coefficients and the matter fields can be derived, suggests that quanta of the matter fields can, by quantum fluctuations, be transformed into (hypothetical) quanta of the quantum connection.

The gauge theory structure of the geometrical formulation naturally leads to some additional questions:

- Are there conserved Noether currents and, if so, how should these be interpreted?
- In the present context, "gauge fixing" means the choice of a particular local trivialization of the vector bundle. Locally, the quantum gauge field K_μ can be transformed to zero if and only if the curvature $F_{\mu\nu}$ vanishes. In general, this is not the case. However, for an observer B, it is always possible to choose a specific frame such that $K_\mu = 0$ on the restriction of the bundle to the worldline. This corresponds to a Heisenberg picture for B, since the state vector will be constant. In general, an additional condition $\partial_\mu K_\nu = 0$ along the worldline cannot be achieved. Locally, therefore, the quantum gauge field cannot be transformed away, and this raises the question of its physical significance.

Another set of questions relates to quantum field theory aspects:

- What are suitable initial conditions for the equations of motion?
- How should locality of fields be defined in the geometrical formulation?

- Is it possible to define a vacuum state? Is the vacuum state dependent on the state of motion of the observer (for example, does the Unruh effect lead to a different vacuum state for an accelerated observer)?
- Is it possible mathematically to make sense of the theory; for example, can a perturbative expansion in the coupling constant a of the quantum connection be derived?
- Is there a way to introduce self-interacting scalar fields, fermions and "ordinary" gauge fields in the quantum action?

We close the discussion with a remark concerning a "global Hamiltonian". In order to recover Heisenberg type equations of motion for the quantum gauge field and for the matter fields, we need a "global Hilbert space". The bundle itself can be considered to be such an object. Elements of this space are bundle sections ξ, and a global metric can be defined by

$$\mathcal{G}\left(\xi, \eta\right) = \int \mathrm{d}x \, \sqrt{\sigma g} \, G\left(\xi(x), \eta(x)\right). \tag{25}$$

An interesting problem would be to find a global operator R, mapping sections of the vector bundle into vector-valued one-forms, such that an equation of the type

$$D\varphi = \mathrm{i}\left[R, \varphi\right] \tag{26}$$

holds for all fields φ of the theory, including the quantum gauge field, under the assumption of suitable commutation relations.

Acknowledgments: I am grateful to H. Atmanspacher and E. Ruhnau for inviting me to participate in a very interesting workshop.

References

Birrell N.D. and Davies P.C.W. (1982): *Quantum Field Theory in Curved Space* (Cambridge University Press, Cambridge).

Graudenz D. (1994): On the Space-Time Geometry of Quantum Systems. Preprint CERN–TH.7516/94.

Kobayashi S. and Nomizu K. (1963): *Foundations of Differential Geometry* (Wiley, New York).

Wells R.O. (1980): *Differential Analysis on Complex Manifolds* (Springer, Berlin).

Does Time Exist at the Most Fundamental Level?

Claus Kiefer

Fakultät für Physik, Universität Freiburg, Hermann-Herder-Strasse 3, D–79104 Freiburg, Germany

Abstract. Quantizations of the gravitational field face a conceptual problem in that no time parameter appears at a fundamental level. This constitutes the "problem of time in quantum gravity". After pointing out the crucial role that an external time plays for the standard interpretational framework of quantum theory, various attempts towards a resolution of the problem of time in quantum general relativity are discussed. It is then shown how the notions of an external time and a classical world can be approximately recovered from a fundamentally timeless framework.

1 Introduction

A viable concept of time is an essential ingredient to all known physical theories. Since the structure of these theories has undergone many changes since the advent of Newtonian mechanics, the concept of time has changed, too (Ehlers, this volume). For example, an analysis of the concept of time was Einstein's solution to create the special theory of relativity (Einstein 1982). Only in one respect is the change in the physical framework not followed by a corresponding change in the concept of time: Quantum theory still uses an external, absolute, time parameter – in contrast to the dynamical variables, such as position and momentum, which are represented by operators in some Hilbert space. In the classical theory of gravitation, Einstein's theory of general relativity, space and time obey dynamical field equations. It is thus unimaginable that time can keep its absolute status if gravity and quantum theory are combined into one single framework.

Such a unified theory is not yet available, but all straightforward approaches lead to what is called the *problem of time in quantum gravity*. This arises because the central equations are "timeless" equations, i.e., equations which do not contain any external time parameter and in which other candidates for time variables are not present in an obvious way. Since such a timeless feature is a direct consequence of the fact that an absolute concept of time is absent in the classical theory (although there time *parameters* are still present), the occurrence of this problem is widespread in approaches to quantum gravity and not tied to particular models. One hopes that attempts to solve the problem of time can serve as a guide to the construction of a viable theory of quantum gravity. Thus, the problem of time in quantum gravity may play a role which is analogous to the role that the equivalence principle played in the construction of general relativity.

This contribution is devoted to a discussion of the important question: Does a sensible concept of time exist at the most fundamental level of quantum gravity, or does it only exist if the gravitational field is treated in the semiclassical limit? The organisation of this article is as follows. Section 2 addresses the role that time plays in ordinary quantum mechanics and highlights this role for the interpretational framework of the theory. Section 3 contains a discussion of the problem of time in quantizations of general relativity and points out ways towards its solution. Section 4 then demonstrates how an approximate notion of semiclassical time can be consistently recovered from a fundamentally timeless theory of quantum gravity.

2   Time in Quantum Mechanics

The central dynamical equation in quantum theory is the Schrödinger equation

$$i\hbar\frac{\partial\psi}{\partial t} = \hat{H}\psi. \tag{1}$$

The most important feature for the present discussion is the *absolute* role of the time parameter t appearing on the left-hand side. In contrast to position and momentum, which are contained in the Hamilton operator $\hat{H}$, time is not turned into an operator, but assumed to be given as an external label. In fact, t is part of the classical background structure that is needed in the "Copenhagen interpretation" of quantum mechanics. Connected with this presence of an external time are the following crucial properties of the Schrödinger equation:

− The Schrödinger equation is *deterministic*: The specification of ψ at one instant of time uniquely fixes its evolution.
− The Schrödinger equation is *time-reversal invariant*: The substitutions $t \to -t$ and $\psi \to \psi^*$ do not change the structure of the equation if $\hat{H}^*(-t) = \hat{H}(t)$. The latter condition is also fulfilled in the presence of magnetic fields if the sources are transformed, too (as must be done for consistency).
− The evolution according to the Schrödinger equation is *unitary*: The inner product $\int d^{3N}x\,\psi^*\psi$ is time-independent if $\hat{H}$ is self-adjoint ($3N$ is the dimension of the configuration space). This property is important for the probability interpretation of quantum theory, since the inner product is interpreted as the total probability which must be conserved in time.

In addition to the Schrödinger equation (1), orthodox quantum mechanics contains a second type of dynamics, the infamous "collapse of the wave function", which is assumed to happen in the course of "measurements". In strong contrast to the above features of the Schrödinger equation, it is *not* deterministic, *irr*eversible, and *not* unitary. The occurrence of this second type of dynamics is, however, a contentious issue. If it really happened dynamically (and not only apparently), it would have to be described by novel dynamical equations which are in conflict with the Schrödinger equation (Pearle 1994). Since the "collapse of the wave function" can be interpreted as an *apparent* process only, which results from the irreversible interaction with the environment by *decoherence* (Giulini et al. 1996), this possibility will not be considered here. Moreover, no experimental hints whatsoever are available which point towards a violation of the Schrödinger equation. The occurrence of a fundamental "collapse" also appears to be very artificial in the context of quantum cosmology, where no external measuring agency is available (see below).

Since t is an external parameter, the uncertainty relation between time and energy is *not* analogous to the one between position and momentum (there is no "time operator" canonically conjugate to the Hamilton operator $\hat{H}$). As was already discussed by Pauli (1933a), the presence of such a time operator would be in conflict with the assumed property of $\hat{H}$ that its spectrum be bounded from below. This can be immediately inferred from the canonical commutation relations which demand that the spectrum of the canonical operators must be the whole real axis. The energy-time uncertainty relation, as it is usually formulated, is a relation between the average lifetime of an unstable quantum state and the width of the corresponding spectral line.

It is also not possible to construct a "perfect clock" in the sense that no observable (the "pointer" of a clock) exists which is a function of the canonical variables such that its observed values are monotonically correlated with t. The proof again makes use of the boundedness of the Hamiltonian as well as of the time-reversal invariance of the Schrödinger equation: If it is assumed that the clock can run forward in time, there is a non-vanishing probability that it can run backward (Unruh and Wald 1989). This argument makes use, however, of the standard collapse interpretation. Since the clock is not an isolated system, it cannot obey a Schrödinger equation and must rather be described by a density matrix (see the discussion in Sect. 4). It is, of course, no problem to represent a clock variable by a wave packet with finite width, whose center of mass is monotonically correlated with t.

Note that the appearence of t in (1) is connected with the appearance of the imaginary unit $i = \sqrt{-1}$. This necessarily leads to the occurrence of complex wave functions in quantum mechanics. Complex wave functions, in turn, are crucial for the interpretational framework, since only complex wave functions yield a conserved probability bilinearly in ψ and ψ^* (Pauli 1933b). An understanding of this feature from a fundamental viewpoint is the subject of the final section.

The absolute character of time also holds in the special theory of relativity, where spacetime is assumed to be *given*, i.e., it is not subject to dynamical equations. The four-dimensional metric $\eta_{\mu\nu}$ is abstracted from any motion and allows the existence of perfect test clocks, since there is no dynamical back reaction on the metric (Zeh 1992).

The situation changes drastically if gravity is included. In the general theory of relativity, spacetime obeys dynamical equations. Material clocks now react back onto the geometry. In this sense, the spacetime metric can itself be interpreted as a clock (Zeh 1992). The absolute status of time has thus been irretrievably lost. What, then, happens with the concept of time if the gravitational field is itself subject to quantization? Attempts to answer this question are the subject of the next section.

3   Concepts of Time in Quantum General Relativity

A straightforward approach towards a quantum theory of gravity is the application of standard, "canonical", quantization rules to the classical theory of general relativity. Of course, there exist much more ambitious attempts such as superstring theory which give up the standard picture of quantum field theory from the very beginning. However, the situation there is even less clear than in quantum general relativity. For this reason I restrict myself to the latter here. Although one is still far from having a profound understanding of this approach on a technical level, the conceptual problems are most transparent. It is implicitly assumed that the construction of a sensible theory of quantum gravity is possible without simultaneously unifying all interactions in a single framework (as is done, for example, in superstring theory).

The canonical programme is the following. First, identify the canonical positions and momenta of the theory and demand their non-commutativity by imposing on them the standard commutation relations. Next, represent the canonical variables on a space of states and formulate a dynamics for these states by demanding that they obey a (functional) Schrödinger equation. Then, formulate appropriate boundary conditions for this equation and give a precise interpretational framework. Finally, give observable consequences.

As far as the first step is concerned, it turns out that the canonical position variables are given by the components of the metric on a *three*-dimensional space, while the momenta are related to the "embedding" of this space into the fourth dimension. An immediate consequence is, of course, that a four-dimensional spacetime cannot be attributed any fundamental meaning in canonical quantum gravity.

A central feature of general relativity is the existence of *constraints*, which is an immediate consequence of its invariance properties with respect to arbitrary changes of coordinates – there are thus four constraints per space point. Three of these constraints are relatively easy to deal with, since they express the invariance of the canonical theory with respect to changes of coordinates of the *three*-dimensional space. The interpretation of the fourth constraint, the vanishing of the Hamiltonian, is more subtle, since it is both a constraint and a generator of dynamics. For simplicity, I shall only consider the *Hamil-*

tonian constraint, $H = 0$, in the following. The presence of such a constraint is a typical feature of theories which possess an invariance with respect to arbitrary reparametrisations of a time parameter. One would thus expect that this feature, as well as the consequences discussed below, is a general feature of quantum gravity theories and not tied to quantum general relativity alone.

A standard implementation of the classical Hamiltonian constraint into the quantum theory is to implement an appropriate operator version, the Hamilton operator, onto physically allowed wave functions Ψ:

$$\hat{H}\Psi = 0. \tag{2}$$

The demand for the validity of a Schrödinger equation (1) for Ψ would thus lead to the result that the wave function Ψ does not depend on time at all![1] This is the "problem of time in quantum gravity": There is no obvious candidate for a time variable in (2). Connected with this problem of time is also a problem of Hilbert space, as well as a problem of observables. It is a priori not clear which inner product one must use and what the conservation of probability would mean without external time. As I emphasized in the preceding section, the whole interpretational framework of standard quantum theory depends on these concepts. The concept of an observable, which is usually represented by a self-adjoint operator with respect to the given Hilbert space, then becomes unclear, too. Moreover, since a constraint generates "gauge transformations", observables $\hat{A}$ should commute with the Hamiltonian, $[\hat{A}, \hat{H}] = 0$, and, therefore, be "constants of motion". This seems to contradict the observed change of our world and would thus give rise to a revival of Zenon's apories.

Perhaps the major problem is the interpretation of the wave function itself. This becomes especially crucial in quantum cosmology, the application of quantum gravity to the Universe as a whole. If everything is described in quantum terms, external classical "measuring agencies" which are needed for the orthodox "Copenhagen interpretation" are no longer available. In the next section, it will be explained how the emergence of a classical world from this quantum world can be understood.

It is important to note in the present context that every theory can be cast artificially into a form in which it is invariant with respect to a reparametrisation of time. In Newtonian mechanics, for example, "absolute time t" can be elevated to the status of a dynamical variable. This leads to the constraint

$$\mathcal{H} \equiv H + p_t = 0, \tag{3}$$

where p_t denotes the momentum conjugate to t. Upon quantization, this becomes

$$\hat{\mathcal{H}}\Psi = 0 \quad \Leftrightarrow \quad \left(\hat{H} + \frac{\hbar}{i}\frac{\partial}{\partial t}\right)\Psi = 0. \tag{4}$$

---

[1] For simplicity, I restrict myself here to the case where the three-dimensional space is compact without boundaries. Otherwise the occurrence of boundary terms would change (2).

This, however, is nothing but the Schrödinger equation (1). The constraint (3) has one distinguishing feature: While the ordinary momenta contained in H are quadratic, the one momentum conjugate to t, p_t, occurs *linearly*. This signals that an absolute element, such as Newton's absolute time, has been parametrised. On the other hand, it turns out that the Hamiltonian of general relativity, whose quantum version occurs in (2), contains all momenta *quadratically*. This strongly suggests that all variables have to be treated on an equal footing and that no deparametrisation can be achieved, at least not in an obvious manner. The quadratic occurrence of all momenta would then just signal the timeless character of the fundamental theory. This point has been strongly emphasised by Barbour (1994).

Two main approaches towards a solution of the problem of time in quantum gravity can be distinguished (excellent reviews containing all necessary technical details are Isham (1992) and Kuchař (1992)). The first approach tries to cast the Hamiltonian constraint of general relativity into a form similar to (3) by performing appropriate canonical transformations. In this way, a variable T is found (depending on the original positions and momenta), whose conjugate momentum occurs linearly. Quantization then leads to a Schrödinger equation in a straightforward manner with respect to some "reduced Hamiltonian", and the standard Hilbert space of quantum theory can be employed.

This approach has been proven to be successful in special cases, such as cylindrical gravitational waves or eternal black holes. Its general viability, however, is far from clear. In particular, there are many conceptual and technical problems: The reduced Hamiltonian is explicitly time-dependent, there are many inequivalent choices of times, there is no immediate spacetime interpretation, and others. For the present discussion the perhaps biggest problem is the apparent re-introduction of an *external time* into the fundamental equations.

The alternative approach consists of treating all variables on an equal footing and trying to identify a sensible concept of time *after* quantization. In this case, the central equation (2) is referred to as the *Wheeler-deWitt equation*. The choice for an appropriate Hilbert space structure is then, however, far from obvious. One might in a first attempt try to employ the standard Hilbert space of quantum mechanics and restrict oneself to normalisable wave functions. It is, however, not clear how the corresponding "probabilities" would have to be interpreted in the absence of a distinguished time. The only sensible possibility would be to consider some conditional probabilities. Moreover, one would expect that a Hilbert space structure be only imposed on the space of solutions to (2).

For this purpose, it is useful to have a closer look at the structure of $\hat{H}$ in (2). Its most important property for the present discussion is the indefinite nature of its kinetic term. In fact, the structure is that of a local wave equation (one wave equation at each space point). This indefinite nature is directly connected with the attractivity of the gravitational interaction (Giulini and Kiefer 1994). Since the Wheeler-deWitt equation locally has the structure of a Klein-Gordon equation (although with a complicated potential), a possible solution would be to impose the corresponding Klein-Gordon inner product on the space of solutions. As in relativistic quantum mechanics of single particles, one would then like to project out only the positive frequencies to enable the application of a probability interpretation. Kuchař, however, has shown that this cannot be done (Kuchař 1992). Hence, the Hilbert space problem cannot be solved in this way. It has also been demonstrated that a "mixture" of the above approaches, which consists of demanding the validity of the Ehrenfest equations in quantum gravity, faces severe difficulties (Brotz and Kiefer 1996).

For a preliminary interpretation, it may be sufficient to interpret (2) only as a differential equation without any a priori specification of some Hilbert space. One may ask whether the local structure of a wave function is also present globally, i.e., whether only *one* minus sign remains in the kinetic term after all gauge degrees of freedom have been factored out. It has been shown that, although not true in general, this is true in the vicinity of Friedmann-type cosmologies (Giulini 1995). In the corresponding regions of the configuration space, the Wheeler-deWitt equation is truely hyperbolic. It may, thus, be regarded as a wave equation with respect to some *intrinsic time variable* (Zeh 1992). In spite of its static appearance, (2) describes intrinsic dynamics! In concrete models, the role of the intrinsic time is often played by the volume of the Universe – in this sense the Universe provides its own clock.

This intrinsic structure has important consequences for the interpretation of the wave function, in particular in the case of a Universe which is classically recollapsing (Zeh 1992, Kiefer 1994a, Kiefer and Zeh 1995). For example, wave packets cannot remain narrowly concentrated along a classical trajectory in configuration space, which means that the semiclassical approximation cannot remain valid everywhere. This, in turn, has important consequences for the foundation of irreversibility from quantum cosmology (see the detailed discussion in Kiefer and Zeh (1995)).

It is imaginable that it is sufficient to discuss – and properly understand – a differential equation for Ψ without resorting to a Hilbert space structure. Two important questions in this respect have, however, not yet been answered: What is the fundamental configuration space of quantum gravity? What are the appropriate boundary conditions to be imposed on this equation?

Independent of the exact form of the fundamental theory, a necessary requirement is the recovery of the Schrödinger equation (in its functional form) together with its external time parameter in some appropriate approximation. The following section addresses how this can be derived from the timeless structure of (2).

4   Consistent Recovery of Semiclassical Time

There are basically two steps involved in the derivation of (1) and a classical spacetime structure from quantum gravity. The first consists in the recovery of the Schrödinger equation from a *special* solution to (2). The second step is devoted to an understanding of how a classical world can be recovered from a *generic* solution of (2) in a semiclassical approximation. In this second step the emergence of classical properties by *decoherence* plays a crucial role.

The general method employs an approximation of the Born-Oppenheimer type. This is well known from molecular physics and makes use of the presence of different mass scales. Since the mass of a proton is much bigger than the mass of an electron, the kinetic term for the nucleus of a molecule can be neglected in a first approximation. This leads to the heuristic picture of a "slow" motion for the nuclei, followed by a "fast" motion of the electrons. As Berry (1984) has emphasised, such a kind of approximation is always applicable if a system is divided into subsystems with very different scales.

In the present context, the role of the nuclei is taken over by the gravitational field, while the role of the electrons is played by non-gravitational fields. Formally, this is expressed by the fact that the Planck mass is much bigger than masses of ordinary elementary particles. An expansion with respect to the Planck mass leads, in highest order, to an expression for a solution of (2) of the form (for details see Kiefer (1994b))

$$\Psi[\mathcal{G}, \phi] \approx C[\mathcal{G}] \, e^{iS_0[\mathcal{G}]/\hbar} \, \psi[\mathcal{G}, \phi], \tag{5}$$

where $\mathcal{G}$ denotes the gravitational field on a three-dimensional space, ϕ stands symbolically for non-gravitational fields, and C is a slowly varying prefactor. It is important to note that the functional S_0 obeys the Hamilton-Jacobi equations for the gravitational field alone. Since these equations are fully equivalent to Einstein's field equations, each solution corresponds to a family of classical spacetimes. It is in this sense that a classical background can be recovered approximately. Of course, this scheme stands in a close formal analogy to WKB approximations in quantum mechanics or to the limit of geometrical optics in electrodynamics.

The wave functional ψ in (5) obeys the approximate equation

$$i\hbar \nabla S_0 \cdot \nabla \psi \equiv i\hbar \frac{\partial \psi}{\partial t} \approx H_m \psi, \tag{6}$$

where H_m denotes the Hamiltonian for the non-gravitational fields. The derivative $\partial/\partial t$ is a directional derivative along each of the classical spacetimes which can be viewed as classical "trajectories" in the gravitational configuration space. Hence, the variable t may be interpreted as a classical time parameter. In fact, (6) is nothing but Schrödinger's equation for non-gravitational fields on a classical background spacetime; t controls their dynamics as long as this approximation holds.

In this limit, standard quantum theory – together with its formal framework including the standard Hilbert space of square-integrable wave functions – can be recovered from quantum gravity. It is important to emphasise, however, that all these notions are approximate ones; higher orders of quantum gravitational corrections to the Schrödinger equation (6) spoil, for example, the conservation of the quantum mechanical inner product (Kiefer 1994b).

The derivation of (6) is, however, only the first step in the recovery of standard quantum theory with an external time. It is important to note that this equation only follows from the very special, complex state (5). This is especially peculiar since the fundamental equation (2) is real. Moreover, this equation is linear and, therefore, allows the occurrence of arbitrary superpositions of states such as (5). This would lead to the presence of interference terms which would prevent the derivation of (6). The key notion to understand the non-occurrence of such interference terms is *decoherence*. I therefore present a brief introduction into this topic and refer to the literature for details (Giulini et al. 1996, Zurek 1991).

The basic observation for the understanding of decoherence is provided by the fact that macroscopic systems cannot be considered, not even approximately, as being isolated from their natural environment. In fact, they are strongly quantum-correlated with it. Traditional discussions of the measurement process consider a quantum mechanical system S (described by a basis of states $\{\varphi_n\}$) coupled to an "apparatus" A (described by a basis $\{\Phi_k\}$). Due to the superposition principle, the result of such an interaction is the occurrence of superpositions of macroscopically different states of the apparatus and *not* a single measurement result – in contrast to what is observed. This is the basic issue of the infamous "Schrödinger's cat paradox".

Taking now into account the natural environment $\mathcal{E}$ (described by a basis $\{\mathcal{E}_l\}$) of the apparatus, phase relations between different states of the apparatus become delocalised through correlations with the *huge* number of environmental degrees of freedom (photons, air molecules, . . .). Tracing them out in the total, quantum-entangled state (I consider the simplest case of a correlation)

$$|\Psi\rangle = \sum_n c_n |\varphi_n\rangle \otimes |\Phi_n\rangle \otimes |\mathcal{E}_n\rangle, \tag{7}$$

leads to a reduced density matrix for A of the form

$$\begin{aligned}
\rho_A &= \mathrm{Tr}_{\mathcal{E}} |\Psi\rangle\langle\Psi| \\
&= \sum_{n,m} c_n^* c_m |\varphi_m\rangle \otimes |\Phi_m\rangle\langle\mathcal{E}_n|\mathcal{E}_m\rangle\langle\varphi_n| \otimes \langle\Phi_n| \\
&\approx \sum_n |c_n|^2 |\varphi_n\rangle \otimes |\Phi_n\rangle\langle\Phi_n| \otimes \langle\varphi_n|,
\end{aligned} \tag{8}$$

where the last step follows from the approximate orthogonality of different environmental states (which is what happens in realistic cases). Thus, the density matrix (8) assumes the form of an *approximate* ensemble, and it seems *as if* the system has "collapsed" into one of the states φ_n with a probability $|c_n|^2$.

A simple example for decoherence is shown in Fig. 1. The superposition of two Gaussian states rapidly decoheres due to the interaction with the environment (here taken to be a heat bath for simplicity; Giulini et al. 1996). The off-diagonal elements of the reduced density matrix rapidly become small.

If initially there is no (or almost no) quantum entanglement between $\mathcal{A}$ and $\mathcal{E}$, the local entropy

$$S = -k_B \operatorname{Tr}(\rho_{\mathcal{A}} \ln \rho_{\mathcal{A}}) \tag{9}$$

will increase by this interaction – classical properties emerge in a practically *irreversible* manner, since in realistic cases the environmental degrees of freedom never return to their initial state because of the enormous Poincaré times that are usually involved. For example, dust particles irreversibly acquire classical properties through "continuous measurement" by thermal radiation. This process is even effective in intergalactic space by its interaction with the ubiquitous microwave background radiation (Giulini et al. 1996, Joos and Zeh 1985).

Two important aspects of decoherence are to be emphasised. First, no interferences can be seen locally in a statistical sense. This is immediately recognised from the rapidly decreasing non-diagonal elements of the reduced density matrix. Second, locally there exists a distinguished basis ("pointer basis"; Zurek 1991) which is sufficiently *stable* with respect to the influence of the environment. Hence, the options for a consistent, realistic interpretation are either to stick to a many-worlds interpretation (this seems to be the only reasonable option in quantum cosmology) or to look for deviations from linear quantum theories, which lead to a "collapse" into *one* macroscopic component (Pearle 1994).

Coming back to quantum cosmology, how can one separate the degrees of freedom into a "system" and an "environment"? After all, this separation is not given a priori. However, since our observed Universe seems to be homogeneous and isotropic on large scales, the assumption that the global degrees of freedom, such as the scale factor of the Universe a are the relevant ones seems to be adequate at a first level. In addition, one might wish to consider the homogeneous component of a scalar field ϕ as a relevant degree of freedom. Such fields are assumed to play an important role in inflationary universe models.

While the relevant degrees of freedom are then given by a and ϕ, the irrelevant degrees of freedom are formed by "higher multipoles", which can be found from an expansion of the full three-dimensional metric and the full scalar field into inhomogeneous mode functions. These multipoles describe density perturbations and weak gravitational waves. Tracing them out, one can obtain a description of the loss of interference between different sizes of the Universe and different scalar fields (see Giulini et al. (1996) and Kiefer (1992) for details).

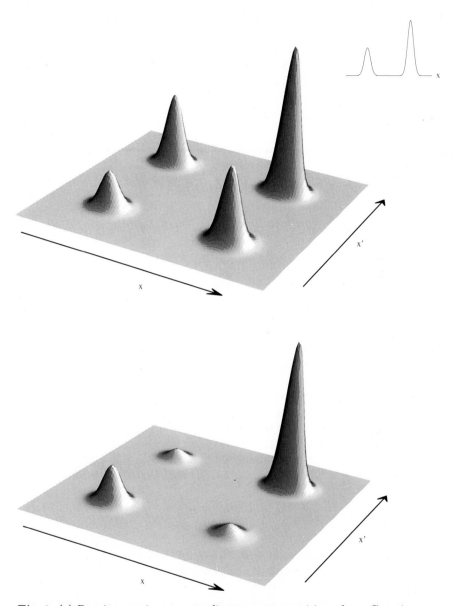

Fig. 1. (a) Density matrix corresponding to a superposition of two Gaussian wave packets. The wave function is shown in the insert. (b) The density matrix after interferences has been partially destroyed by decoherence (from Chap. III by Joos in Giulini et al. (1996)).

Explicit models discuss, for example, a superposition of a state like (5) with its complex conjugate, where only the scale factor a may be considered in $\mathcal{G}$. Integrating out an inhomogeneous matter field as the model for the non-gravitational fields, an explicit calculation yields the following suppression factor $I_\pm$ for the corresponding non-diagonal element of the reduced density matrix (Kiefer 1992),

$$I_\pm \approx \exp\left(-\frac{\pi m H_0^2 a^3}{128}\right), \tag{10}$$

where H_0 denotes the Hubble parameter corresponding to the (semiclassical) background evolution of the scale factor with respect to semiclassical time. If one assumes the values $m \approx 100$ GeV, $H_0 \approx 50$ km/(sec Mpc), and $a \approx H_0^{-1}$ to find an estimate for the present Universe, this yields

$$I_\pm \approx \exp(-10^{43}) \ll 1.$$

Decoherence is, thus, very efficient for large a. Only at the Planck scale and near the turning point of the classical Universe (small a and small H_0, respectively) can this factor become approximately one, and the semiclassical components would interfere.

We note that the decoherence between the $\exp(iS_0)$ and $\exp(-iS_0)$ components can be interpreted as a *symmetry breaking* analogous to the case of chiral molecules (see Chap. IX in Giulini et al. (1996)). There, the Hamiltonian is invariant under space reflections, but the state of the molecules exhibits chirality (see Fig. 2 for illustration). In quantum gravity, the Hamiltonian in (2) is invariant under complex conjugation, while the "actual states" are of the form $\exp(iS_0)$ and, thus, are intrinsically complex. Since this is a prerequisite for the derivation of the Schrödinger equation, one might even say that *time* (the time parameter in the Schrödinger equation) arises from symmetry breaking.

To conclude, although the "problem of time" and the "Hilbert space problem" of quantum gravity have not yet been resolved, one can conceptually understand how an approximate notion of time can emerge in an appropriate semiclassical approximation from a fundamentally timeless framework. In this limit, all the familiar structures, such as Hilbert spaces and the probability interpretation, can be recovered, too. This raises the question of whether these structures are needed at all at the fundamental level of quantum gravity itself. My conjecture is that they are *not* needed. Whether this is true, and what structures substitute them, can only be decided after a mathematically consistent theory of quantum gravity has been constructed.

Acknowledgments: I thank the editors for inviting me to contribute this article. Financial support by the *Wissenschaftliche Gesellschaft in Freiburg im Breisgau* is gratefully acknowledged.

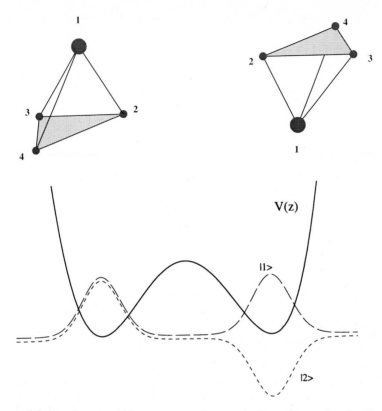

Fig. 2. (a) Simplest possible structure of an optically active molecule. If all four elements (which can be single atoms or groups of atoms) are different, the molecule (left) and its mirror image (right) cannot be transformed into each other by a proper rotation. (b) Schematic picture of the effective potential for the inversion coordinate in a model for a chiral molecule and the two lowest eigenstates. The ground state is symmetrically distributed over the two wells. Only linear combinations of the two lowest states are localised in one well, corresponding to a classical configuration (from Chap. III by Joos in Giulini et al. (1996)). This example from molecular physics stands in close analogy to the recovery of time from quantum gravity.

References

Barbour J.B. (1994): The timelessness of quantum gravity I. The evidence from the classical theory, II. The appearance of dynamics in static configurations. *Class. Quant. Gravity* **11**, 2853–2873, 2875–2897. See also his contribution to this volume.

Berry M.V. (1984): Quantal phase factors accompanying adiabatic changes. *Proc. Roy. Soc. Lond. A* **392**, 45–57.

Brotz T. and Kiefer C. (1996): Ehrenfest's principle and the problem of time in quantum gravity. *Nuclear Physics B* **475**, 339–357.

240   Claus Kiefer

Einstein A. (1982): Lecture delivered in Kyoto 1922. In *Physics Today*, August 1982, 46–47.

Giulini D. (1995): What is the geometry of superspace? *Phys. Rev. D* **51**, 5630–5635.

Giulini D., Joos E., Kiefer C., Kupsch J., Stamatescu I.-O., and Zeh H.D. (1996): *Decoherence and the Appearance of a Classical World in Quantum Theory* (Springer, Berlin).

Giulini D. and Kiefer C. (1994): Wheeler-deWitt metric and the attractivity of gravity. *Phys. Lett. A* **193**, 21–24.

Isham C.J. (1992): Canonical quantum gravity and the problem of time. In *Integrable Systems, Quantum Groups, and Quantum Field Theories*, ed. by L.A. Ibart and M.A. Rodriguez (Kluwer, Amsterdam), 157–288.

Joos E. and Zeh H.D. (1985): The emergence of classical properties through interaction with the environment. *Z. Phys. B* **59**, 223–243.

Kiefer C. (1992): Decoherence in quantum electrodynamics and quantum gravity. *Phys. Rev. D* **46**, 1658–1670.

Kiefer C. (1994a): Quantum cosmology and the emergence of a classical world. In *Philosophy, Mathematics and Modern Physics*, ed. by E. Rudolph and I.-O. Stamatescu (Springer, Berlin), 104–119.

Kiefer C. (1994b): The semiclassical approximation to quantum gravity. In *Canonical Gravity: From Classical to Quantum*, ed. by J. Ehlers and H. Friedrich (Springer, Berlin), 170–212.

Kiefer C. and Zeh H.D. (1995): Arrow of time in a recollapsing quantum universe. *Phys. Rev. D* **51**, 4145–4153.

Kuchař K.V. (1992): Time and interpretations of quantum gravity. In *Proceedings of the 4th Canadian Conference on General Relativity and Relativistic Astrophysics*, ed. by G. Kunstatter, D. Vincent, and J. Williams (World Scientific, Singapore), 211–314.

Pauli W. (1933a): Die allgemeinen Prinzipien der Wellenmechanik. In *Handbuch der Physik, Vol. 24*, ed. by H. Geiger and K. Scheel (Springer, Berlin), 88–272. Reprinted: *Die allgemeinen Prinzipien der Wellenmechanik* (Springer, Berlin 1990), p. 84.

Pauli W. (1933b): Einige die Quantenmechanik betreffenden Erkundigungsfragen. *Z. Phys.* **80**, 573–586.

Pearle P. (1994): Putting wavefunction collapse "in the equations and not just the talk". *Proceedings of the Cornelius Lanczos International Centenary Conference*, ed. by J.D. Brown et al. (SIAM, Philadelphia), 594–596.

Unruh W.G. and Wald R.M. (1989): Time and the interpretation of quantum gravity. *Phys. Rev. D* **40**, 2598–2614.

Zeh H.D. (1992): *The Physical Basis of the Direction of Time* (Springer, Berlin).

Zurek W.H. (1991): Decoherence and the transition from quantum to classical. *Physics Today*, October 1991, 36–44. See also comments on this article by J. Anderson, G.C. Ghirardi *et al.*, N. Gisin, D. Albert and G. Feinberg, P. Holland, V. Ambegaokar, and K.J. Epstein together with Zurek's reply in *Physics Today*, April 1993, 13–15, 81–90.

Part IV

Non-Relativistic Quantum Theory

The Representation of Facts in Physical Theories

Hans Primas

Laboratory of Physical Chemistry, ETH-Zentrum, CH–8092 Zürich, Switzerland

Abstract. The purpose of this contribution is to call attention to a problem which has not received the interest which, in my opinion, it deserves: the problem of representation of facts in physical theories. The crucial point is that, within the framework of fundamental physical theories, the representation of facts requires a breaking of the time-reversal symmetry and nonanticipative measuring instruments. These conditions are satisfied only when the apparatus is described as a system with infinitely many degrees of freedom. In the framework of algebraic quantum theory generalized K-systems can represent facts at least in an asymptotic sense. Such a representation removes the main stumbling block which stands in the way of a fundamental theory of measurement in quantum theory.

1 Introduction

By a *fact* we mean an event in the past which retains its facticity in the future. One crucial precondition for the existence of facts is the possibility to discriminate between the past and the future. This prerequisite is also called "the anisotropy of time" or "the arrow of time". In classical physical theories, we presuppose in addition that any reasonable statement about a fact is either true or false, even if we do not know it. This condition depends on the possibility to *distinguish* different things in our world. Yet facts are not restricted to the macroscopic world where distinguishability seems to be no problem. In quantum theory the basic precondition of distinguishability and separability are not automatically justified. If we postulate the universal validity of quantum theory it is no longer evident that facts in the classical sense exist at all. However, it is not necessary to stick to the classical position. It is sufficient to understand instruments which register facts in the sense of Fock (1957):

> "We call an 'instrument' such an arrangement which on the one hand can be influenced by, and interact with, an atomic object and on the other hand permits a classical description with an accuracy sufficient for the purpose of registering the said influence (consequently, the handling of the instrument so defined does not need further 'means of observation'). It should be noted at once that in this definition of the instrument it is quite immaterial whether the 'instrument' is made by human hands or represents a natural combination of eternal conditions suitable for the observation of the micro-object."

The crucial point is that all what is asked for is "an accuracy sufficient for the purpose of registering". So we have to bear in mind the possibility that in quantum theory facts in the classical sense turn up only in limit of long time. In this case we speak of *asymptotic facts* which manifest themselves after sufficiently long but finite time as *approximate facts*. It may be that the documents which tell us something about facts can be destroyed or changed. Nevertheless, we require that the facts themselves (whether exact or approximate) cannot be changed by any future influences. Since facts never disappear but can in principle always be called back from the past into the present, with every new event *the set of all facts increases with time*.

The only known way to formulate physical laws in a non-phenomenological manner applies to strictly isolated systems. We call a system strictly isolated if all variables which can influence the system can be taken into account in the specification of its initial (or just as well of its final) state (cf. Havas 1965, p. 348). Using the space-time concepts of Newtonian mechanics, the basic principles of Hamiltonian mechanics are in every respect time-symmetric so that all fundamental phenomena are symmetrical with respect to an interchange of past and future. That is, the basic equations of motion are invariant under an involution which exchanges the time parameter t by $-t$.[1] This invariance is called the *time-reversal symmetry*.

In the Hamiltonian formalism for strictly isolated systems time is not a property of the system since for such systems the time coordinate can in principle be eliminated without loss of physical content.[2] *For strictly closed systems time is not an observable.* Accordingly, on the fundamental level causation cannot be defined in terms of time order or by the idea that a cause is ontologically more basic than its effects. All we have at our disposal are time-symmetric *correlations*. On the other side, in everyday life the past is knowable and the future is not. Hence we tacitly presuppose a "principle of retarded causality": no effect can precede its cause. *But at a fundamental level there is no distinction between past and future.* So it makes no sense to speak of cause and effect: *fundamental causality is arrowless* (compare Costa de Beauregard (1987), p. 134).

For example the electromagnetic interaction between two electrons cannot be described by a time-directed notion of causality, e.g., by a retarded or an

---

[1] An involution is an operation whose square is the identity. The involution associated with time-reversal does not only change the direction of time but also of associated quantities like the velocity, the momentum, the angular momentum, the electrical current, and the magnetic field. In quantum mechanics, time-reversal is an antilinear involutive operation which changes any complex number into its complex conjugate. In elementary particle physics, the invariant involution associated with time-reversal $T(T^2 = \mathbb{1})$ also involves the space reflection $P(P^2 = \mathbb{1})$ and the charge conjugation $C(C^2 = \mathbb{1})$ (CPT-theorem).

[2] Using Jacobi's principle of least action, Hamiltonian dynamics can be formulated in a completely geometrical language. Compare, e.g., Synge (1960), sections 82–83, pp. 136–139.

advanced interaction.[3] In a particle theory, a consistent description has to use the unique symmetrized interaction inn which advanced and retarded interactions are combined half and half, thus treating both electrons on an equal footing (compare the example discussed by Feynman and Hibbs (1964), p. 251).

The only systems of interest to experimental science are open. However, there are no fundamental laws for open systems. For example, any dynamical law for an open system contains contextual phenomenological parameters (like relaxation times). In order to discuss open systems from a fundamental viewpoint they must first be combined with all systems with which they interact or are correlated. If we include the whole environment of an open system we can describe the resulting system as a strictly isolated Hamiltonian system by first principles. However, the necessary additional conditions (like initial or boundary conditions) for a description of the open object subsystem are not given by first principles but must be chosen in a way appropriate for the experiments we perform.

In everyday life there is an intrinsic dissimilarity of the past and the future. This historical nature of the world is a precondition of all engineering science. In engineering physics the direction of causation is always assumed to go from past to future. That is, *in order to derive engineering physics from fundamental physics, the time-reversal symmetry of fundamental physics has to be broken.* The first difficulty one encounters in carrying out such a program is technical. Although the phenomenon of spontaneous symmetry breaking is well-understood in modern physical theories, it poses formidable mathematical problems. The second difficulty refers to a conceptually deep problem. If the time-reversal symmetry is broken one gets two representations, one satisfying the generally accepted rules of retarded causality and the other one the strange rules of advanced causality. Advanced causality is a conceptual possibility which is not banned by any fundamental physical law. The usual choice of retarded causality cannot be explained by a statistical mechanical formulation of the "second law" without an a priori postulate imposing an asymmetric evolution toward increasing time.

In this contribution we shall not discuss any reasons for selecting the retarded representation but concentrate on a proper description of open physical systems which can represent facts. This is not a trivial task – not even in classical physics. In Sect. 2 we recapitulate well-known tools from engineering physics which are necessary for the description of facts in classical physics. In Sect. 3 we discuss the additional difficulties which arise in quantum theory. We give a short outline of the problems which are related to the so-called measurement problem. The concepts used in engineering physics for the description of nonanticipative input-output systems are instrumental for the representation of facts by asymptotically disjoint states.

---

[3] Already in 1909, Einstein (1909) argued that there are no fundamental reasons to rule out time-backward advanced solutions.

2   The Representation of Facts in Classical Physics

2.1   Every Laboratory Instrument Is Nonanticipative

Laboratory phenomena are commonly described in terms of cause-and-effect relationships. An input-output system is a mathematical description of an experimental stimulus-response relationship by a dynamical system which, when subjected to the same stimuli, yields the same response as the experimental object. Such an input-output system may be regarded as an abstract operator $\mathcal{R}$ which transforms an input signal $t \mapsto y(t)$ into an output signal $t \mapsto x(t)$ by $x = \mathcal{R}\{y\}$. If this functional relationship is *continuous*, the response operator $\mathcal{R}$ can be represented by a Volterra expansion

$$x(t) = R^{(0)}(t) + \int_{-\infty}^{\infty} ds\, R^{(1)}(t, s)y(s)$$
$$+ \frac{1}{2} \int_{-\infty}^{\infty} ds \int_{-\infty}^{\infty} ds'\, R^{(2)}(t, s, s')y(s)y(s') + ... \qquad (1)$$

If the input and the output are real-valued functions, the integral kernel $R^{(1)} : \mathbb{R}^2 \mapsto \mathbb{R}$ is the linear response function, the kernel $R^{(2)} : \mathbb{R}^3 \mapsto \mathbb{R}$ the quadratic response function, etc. However, not every input-output system of this kind can be realized with a physical instrument in real time. In the world of the experimenter there is a preferred direction of time. Therefore the set of all temporal instants must be ordered such that the past precedes the future. The crucial restriction for a dynamical system representing experimental data in real time is: "No output can occur before the input". More precisely, this condition implies

$$R^{(1)}(t, s) = 0 \text{ for } s > t, \quad R^{(2)}(t, s, s') = 0 \text{ for } s > t \text{ or } s' > t, \quad ..., \qquad (2)$$

so that

$$x(t) = R^{(0)}(t) + \int_{-\infty}^{t} ds\, R^{(1)}(t, s)y(s)$$
$$+ \frac{1}{2} \int_{-\infty}^{t} ds \int_{-\infty}^{t} ds'\, R^{(2)}(t, s, s')y(s)y(s') + ... \qquad (3)$$

In such a system, changes of the output cannot anticipate changes of the input. In the terminology of physicists and engineers such input-output systems are called "causal systems". Since in the philosophical literature the notion of "causality" is used differently, I shall avoid this terminology and speak of *nonanticipative systems*.

    For a given input-output system, the kernels $R^{(0)}, R^{(1)}, R^{(2)}, \dots$ can be determined individually by appropriate experiments. In particular, the linear response function $R^{(1)}$ is well-defined for every nonlinear system. For simplicity we restrict our discussion to the linear response. We consider a continuous, linear nonanticipative system with the linear response function $R^{(1)}$

$$x(t) = R^{(0)}(t) + \int_{-\infty}^{t} R^{(1)}(t, s)\, y(s)\, ds. \tag{4}$$

This input-output map does, however, not characterize the linear system completely. A proper description requires additionally conditions that warrant the controllability and constructibility of the states of the linear system (cf., e.g., Kalman et al. (1969), pp. 324f). Such conditions are automatically fulfilled for systems which are invariant under time-translations. In this case we have $dR^{(0)}(t)/dt = 0$ and $R^{(1)}(t, s) = R(t - s)$. Without restricting the generality, we may neglect the trivial constant $R^{(0)}$ so that a continuous, time-invariant, linear system can be described by

$$x(t) = \int_{-\infty}^{\infty} R(t - s)\, y(s)\, ds. \tag{5}$$

The function $R : \mathbb{R} \mapsto \mathbb{R}$ is called the *linear response* of the system. If the system is nonanticipative, the linear response function fulfills the condition

$$R(t) = 0 \text{ for } t < 0. \tag{6}$$

The Fourier transform of the linear response function is called the *frequency-response function* $H : \mathbb{R} \mapsto \mathbb{C}$

$$H(\lambda) := \int_{-\infty}^{\infty} e^{i\lambda t} R(t)\, dt = \int_{0}^{\infty} e^{i\lambda t} R(t)\, dt, \ \lambda \in \mathbb{R}. \tag{7}$$

In electrical network theory, one defines a so-called *transfer function* as the Fourier-Laplace transform $H^{+} : \mathbb{R} \mapsto \mathbb{C}$ of the linear response function,

$$H^{+}(\lambda + ip) := \int_{0}^{\infty} e^{i\lambda t} e^{-pt} R(t)\, dt, \ \lambda \in \mathbb{R}, p > 0. \tag{8}$$

The transfer function $z \mapsto H^{+}(z)$ is the analytic continuation of the frequency-response function $\lambda \mapsto H(\lambda)$, it is holomorphic in the open upper half-plane $\mathbb{C}^{+} := \{z \in \mathbb{C} \,|\, \mathcal{I}(z) > 0\}$. The frequency-response function is the boundary value of the transfer function:

$$H(\lambda) = \lim_{p \to 0} H^{+}(\lambda + ip), \ \lambda \in \mathbb{R}, p > 0. \tag{9}$$

The analyticity of the transfer function reflects the nonanticipative behavior of a time-invariant linear system.

In network theory, the function $A : \mathbb{R} \mapsto \mathbb{R}^{+}$ with $A(\lambda) := |H(\lambda)|^2$ is referred to as the *amplitude characteristics*. Given a certain amplitude characteristic A, the important engineering question arises whether or not a filter with this amplitude characteristics can be realized by a nonanticipative linear dynamical system. In other words: Does there exist a frequency-response function $\lambda \mapsto H(\lambda)$ such that

$$A(\lambda) := |H(\lambda)|^2 \text{ with } \int_{\infty}^{\infty} e^{-i\lambda t} H(\lambda)\, d\lambda \quad \text{for } t < 0 ? \tag{10}$$

The answer is given by the *Paley-Wiener criterion:*[4]

Necessary and sufficient for a square integrable amplitude function $A : \mathbb{R} \mapsto \mathbb{R}^+$ to be realizable by a *nonanticipative* linear dynamical system with a response function $R : \mathbb{R} \mapsto \mathbb{R}$,

$$R(t) = 0 \text{ for } t < 0, \quad A(\lambda) = \left| \int_0^{\infty} e^{i\lambda t} R(t) dt \right|^2 \tag{11}$$

is the Paley-Wiener criterion

$$\int_{-\infty}^{\infty} \frac{|\ln A(\lambda)|}{1 + \lambda^2}\, d\lambda < \infty. \tag{12}$$

If the Paley-Wiener criterion is satisfied, one can factor the square of the amplitude characteristic A on the real line into a product of two functions which have holomorphic extensions into complex half-planes. Therefore we can write

$$A(\lambda) = H(\lambda)H(\lambda)^*, \quad \lambda \in \mathbb{R}, \tag{13}$$

where $\lambda \mapsto H(\lambda)$ is the boundary value of an analytic function $z \mapsto H^+(z)$, holomorphic in the open upper half-plane $\mathbb{C}^+ := \{z \in \mathbb{C} | \mathcal{I}(z) > 0\}$. Accordingly, $\lambda \mapsto H(\lambda)^*$ is the boundary value of another analytic function $z \mapsto H^-(z)$, holomorphic in the open lower half-plane $\mathbb{C}^- := \{z \in \mathbb{C} | \mathcal{I}(z) < 0\}$,

$$H(\lambda) := \lim_{p \to 0} H^+(\lambda + ip), \quad H(\lambda)^* := \lim_{p \to 0} H^-(\lambda - ip), \quad \lambda \in \mathbb{R}, p > 0. \tag{14}$$

The function $\lambda \mapsto H(\lambda)$ is the frequency-response function of a *past-determined nonanticipative* linear system, while the function $\lambda \mapsto H(\lambda)^*$ corresponds to the frequency-response function of a *future-determined anticipative* linear system.

Since $|H(\lambda)| \in L^2(\mathbb{R})$, the unique extension H^+ is in the Hardy class $H^2(\mathbb{C}^+)$ and therefore admits a unique outer-inner factorization

$$H^+(z) = \Phi(z)\Psi(z), \quad z \in \mathbb{C}^+. \tag{15}$$

---

[4] Paley and Wiener (1934), pp. 16–17. The condition of square integrability in the theorem by Paley and Wiener is not essential. It can be replaced by much weaker conditions, for example by the requirement that the amplitude function is a tempered generalized function. Compare Pfaffelhuber (1971).

The outer function Φ is uniquely given by

$$\Phi(z) = \exp\left(\frac{1}{i\pi}\int_{-\infty}^{\infty}\frac{1+\lambda z}{\lambda - z}\frac{|\ln H(\lambda)|}{1+\lambda^2}\,d\lambda\right), \quad z \in \mathbb{C}^+. \tag{16}$$

There is also a unique representation for an inner function $z \mapsto \Psi(z)(z \in \mathbb{C}^+)$ whose boundary value $\omega \mapsto \Psi(\omega)(\omega \in \mathbb{R})$ represents an all-pass filter with constant amplitude which causes only additional phase delay. If $\Psi(z) = 1$, the filter is said to be of *minimum phase type*. It has the frequency-response function $H(\omega) = \Phi(\omega + i0)(\omega \in \mathbb{R})$.

To summarize: The appropriate tools for the discussion of nonanticipative laboratory instruments are the theory of Hardy spaces, the Paley-Wiener criterion, and the Wiener factorization. In particular, it follows that a nonanticipative linear filter of minimum phase type is uniquely given by its amplitude characteristic.

2.2 Deterministic and Nondeterministic Processes

The laws of any Hamiltonian mechanics are invariant under time-reversal. In particular the dynamics is both forward deterministic and backward deterministic. A present state determines uniquely the future *and* the past states so that exact prediction and exact retrodiction are in principle possible. *Using only mechanical tools it is therefore impossible to distinguish between cause and effect.* This can only be achieved by temporally one-sided processes. Dissipative stochastic processes are examples for one-sided processes. They are backward-deterministic and forward-nondeterministic, they can be retrodicted exactly but predictions are at best probabilistic.

The paradigmatic example for a backward-deterministic and forward-nondeterministic process is the Wiener process, a mathematically rigorous model for idealized Brownian motion. In his model, Norbert Wiener (1923) proved Perrin's conjecture that all paths of an idealized Brownian motion are almost certainly (i.e., with probability one) continuous but nowhere differentiable. That is, in Wiener's idealization, a Brownian path consists entirely of sharp corners.

Wiener's work initiated the mathematical theory of stochastic processes and functional integration. Ten years later, Kolmogorov (1933) laid the foundation for the modern axiomatic treatment of mathematical probability theory in terms of measure theory. However, it would be mistaken to believe that the theory of stochastic processes in the sense of Kolmogorov has superseded Wiener's ideas. In his work on generalized harmonic analysis during 1925–1930, Wiener (1930) based his theory not on equivalence classes of Lebesgue square integrable functions but on *individual* measurable functions $t \mapsto f(t)$ for which the *individual autocorrelation function*

$$t \mapsto A_f(t) := \lim_{T\to\infty}\frac{1}{2T}\int_{-T}^{T} f(\tau)^* f(t+\tau)d\tau \tag{17}$$

exists for all $t \in \mathbb{R}$. Writing A_f as Fourier transform

$$A_f(t) = \int_{-\infty}^{\infty} e^{i\lambda t} dS_f(\lambda), \tag{18}$$

he obtained what is now called the *individual spectral distribution function* S_f.

These relations have their counterparts in Kolmogorov's *ensemble theory* of stationary stochastic processes. A complex-valued stochastic process $\{f(t|\omega)| t \in \mathbb{R}, \omega \in \Omega\}$ in the sense of Kolmogorov is a family of complex-valued random variables $\omega \mapsto f(\cdot|\omega)$ on a common Kolmogorov probability space (Ω, Σ, μ), where Ω is a set, Σ a σ-algebra, and μ a probability measure. For a fixed $\omega \in \Omega$, the function $t \mapsto f(t|\omega)$ is a complex-valued function, called a *trajectory* (or a *realization*) of the stochastic process $\{f(t|\omega)| t \in \mathbb{R}, \omega \in \Omega\}$ corresponding to the event ω. A stochastic process is an *equivalence class* of trajectories with the same family of joint probability densities. The mean value m_f and the covariance function C_f of the stochastic process $\{f(t|\omega)| t \in \mathbb{R}, \omega \in \Omega\}$ are defined by

$$m_f(t) := \mathcal{E}\{f(t|\cdot)\} = \int_{\Omega} f(t|\omega)\mu(d\omega), \tag{19}$$

$$C_f(t,s) := \mathcal{E}\{f(t|\cdot)^* f(s|\cdot)\} - m_f(t)^* m_f(s)$$
$$= \int_{\Omega} f(t|\omega)^* f(s|\omega)\mu(d\omega) - m_f(t)^* m_f(s) \tag{20}$$

A stochastic process is said to be *stationary* if all joint probability densities are invariant under time translation. In this case, the mean value is time-independent, while the covariance function depends only on the difference of two times,

$$C_f(t,s) = C_f(t-s) \quad \forall t,s \in \mathbb{R}. \tag{21}$$

Khintchine (1934) proved that the covariance function of every stationary stochastic process can be represented in the form

$$C_f(t) = \int_{-\infty}^{\infty} e^{i\lambda t} dF_f(\lambda), \tag{22}$$

where $F_f : \mathbb{R} \mapsto \mathbb{R}$ is a real, never decreasing and bounded function, called the *spectral distribution function* of the stochastic process.

For ergodic stationary stochastic processes, Wiener's analytical representation theorem for a *single* function follows from Khintchine's ensemble representation theorem for stochastic processes: A realization $t \mapsto f(t|\omega)$ (ω fixed) of an ergodic stationary stochastic process $\{f(t|\omega)| t \in \mathbb{R}, \omega \in \Omega\}$ will be, with probability one, such a function that Wiener's individual autocorrelation function $t \mapsto A_f(t|\omega)$,

$$A_f(t|\omega) := \lim_{T\to\infty} \frac{1}{2T} \int_{-T}^{T} f(\tau|\omega)^* f(t+\tau|\omega)\, d\tau, \quad \omega \text{ fixed}, \tag{23}$$

exists and is equal to the covariance function $t \mapsto C_f(t)$. But it is important to realize that Wiener's theory is in no way probabilistic but applies to *single well defined functions* rather than to an ensemble of functions.[5]

Let $\Sigma(a,b) \subset \Sigma$ be the σ-field generated by the stochastic process $t \mapsto f_t := f(t|\cdot)$ in the time interval $a \leq t \leq b$, so that $\Sigma(-\infty,\infty) = \Sigma$. Since $\{\Sigma(-\infty,t)|t \in \mathbb{R}\}$ is a monotonically increasing, and $\{\Sigma(t,+\infty)|t \in \mathbb{R}\}$ a monotonically decreasing family of σ-fields, the remote past $\Sigma(-\infty)$ and the remote future $\Sigma(+\infty)$ are given by

$$\Sigma(-\infty) := \bigcap_{t \leq 0} \Sigma(-\infty,t), \quad \Sigma(+\infty) := \bigcap_{t \leq 0} \Sigma(t,+\infty). \tag{24}$$

These concepts are important for the prediction and the retrodiction of stochastic processes. First we consider the problem of *prediction*. Let $\mathcal{L}_t^2$ be the Hilbert space consisting of all $\Sigma(-\infty,t)$-measurable functions that are square-integrable with respect to the probability measure μ, $\mathcal{L}_t^2 := L^2\{\Omega, \Sigma(-\infty,t), \mu\}$. Having observed the past $\{f_s|s \leq t\}$ of a process, one wants to forecast $\tau \mapsto f_{t+\tau}$ using an element $g_{t,\tau}$ of $\mathcal{L}_t^2$. The predictor $g_{t,\tau}$ is in general a nonlinear function of the observed process $\{f_s|s \leq t\}$. If one adopts the least-square criterion, the prediction error is given by

$$\sigma^2(t,\tau) = \mathcal{E}\{|f_{t+\tau} - g_{t,\tau}|^2\}. \tag{25}$$

Since the process is assumed to be stationary, the error for the optimum predictor does not depend on $t, \sigma(t,\tau) = \sigma(\tau)$. The error for the optimal mean-square predictor is the conditional expectation of the process, given $\Sigma(-\infty,t)$ (see, e.g., Rosenblatt 1971, p. 164):

$$g_{t,\tau}^{\text{opt}} = \mathcal{E}\{f_{t+\tau}|\Sigma(-\infty,t)\}. \tag{26}$$

A process $\{f_t|t \in \mathbb{R}\}$ is called *forward deterministic* if the optimal predictor in terms of the past $\Sigma(-\infty,0)$ allows an error-free prediction. In this case the process is in fact already determined by the remote past $\Sigma(-\infty)$, and a perfect prediction can even be performed by a constructive algorithm (Scarpellini 1979a,b). If an error-free prediction is not possible, the process is called *forward nondeterministic*. Every process can be represented uniquely as the sum of a forward deterministic process and a so-called forward purely nondeterministic process (where, of course, one component may be absent).

---

[5] In his later work on *Extrapolation, Interpolation, and Smoothing of Stationary Time Series*, Wiener (1949) used individual functions $t \mapsto f(t)$, and not equivalence classes $t \mapsto \{f(t|\omega)|\omega \in \Omega\}$. This approach has been criticized as unnecessarily cumbersome (Kakutani (1950)). However, it has to be stressed that for the prediction of an individual time series only Wiener's approach is conceptually sound – for weather prediction or anti-aircraft fire control there is no ensemble of trajectories but just a single individual trajectory from whose past behavior one wants to predict something about its future behavior.

Hans Primas

A process is called *forward purely nondeterministic* if the unconditional expectation is the best forecast. In this case, the remote past $\Sigma(-\infty)$ is the trivial Borel field $\{\emptyset, \Omega\}$ consisting only of the impossible event $\emptyset$ and the certain event Ω. The present state of a forward deterministic process determines all its future states, while a forward purely nondeterministic process contains no components that can be predicted exactly from an arbitrarily long past record.

There is another extrapolation problem: *retrodiction*. Given the trajectory of a stochastic process on the positive real axis, can we retrodict the behavior of the process on the negative real axis? Of course, the answer is analogous to the problem of prediction; formally, t has just to be replaced by $-t$. Therefore we arrive at the following classification (cf. Krengel 1971, 1973):

A process $\{f_t \mid t \in \mathbb{R}\}$ is called

forward deterministic if $\Sigma(-\infty) = \Sigma$,
backward deterministic if $\Sigma(+\infty) = \Sigma$,
bidirectionally deterministic if $\Sigma(-\infty) = \Sigma(+\infty) = \Sigma$,
forward purely nondeterministic if $\Sigma(-\infty) = \{\emptyset, \Omega\}$,
backward purely nondeterministic if $\Sigma(+\infty) = \{\emptyset, \Omega\}$.

For a forward purely nondeterministic process the remote past does not contain any information that could be used for predictions. For backward purely nondeterministic processes the remote future does not contain any information that could be used for retrodictions. Bidirectionally deterministic processes are both forward and backward deterministic, they correspond to deterministic motions of time-reflection invariant Hamiltonian mechanics. It is important that in general forward determinism does not imply backward determinism. In fact, there are stationary processes which are forward purely nondeterministic and backward deterministic, and there are stationary processes which are forward deterministic and backward purely nondeterministic.

By breaking the time-reversal symmetry of the Hamiltonian dynamics one can derive one-sided processes. However, the spontaneous breaking of the symmetry of a group of order two gives *two* elementary realizations which have the same logical status. That is, if it is possible to derive backward deterministic and forward purely nondeterministic processes, then it is also possible to derive forward deterministic and backward purely nondeterministic processes. *The decision which of the two possibilities is appropriate can therefore not come from the first principles of physics.* So the conceptual problem is not the breaking of the time-reversal symmetry (though this may pose difficult mathematical questions), but the proper selection of one or the other one-sided realization.

To summarize: The theory of stochastic processes allows a precise description of all processes resulting from the breaking the time-reversal symmetry of a classical Hamiltonian system. Every process can be decomposed in a forward deterministic and a forward purely nondeterministic process. Since forward determinism does not imply backward determinism, a process can also be decomposed in a backward deterministic and a backward purely non-deterministic process. There exist mathematical models for all combinations of forward and backward determinism, and of forward and backward pure nondeterminism.

2.3   Forward Purely Nondeterministic Processes Generate Classical K-Flows

A flow $\{\tau_t | t \in \mathbb{R}\}$ on a probability space (Ω, Σ, μ) is said to be a K-flow (Cornfeld et al. 1982, p. 280) if there exists a σ-subalgebra of measurable sets $\Sigma_0 \subset \Sigma$ such that for $\Sigma_t := \tau_t \Sigma_0$ the following conditions hold (Cornfeld et al. 1982, p. 280):

- $\Sigma_0 \subset \Sigma_t$ for every $t > 0$,

- $\bigvee_{t=-\infty}^{\infty} \Sigma_t = \Sigma$,

- $\bigwedge_{t=-\infty}^{\infty} \Sigma_t$ is the trivial σ-algebra consisting of the sets of measure 0 and 1.

Here $\bigvee$ is the lattice sum and $\bigwedge$ is the intersection. Every K-flow is ergodic and has the mixing property of every degree. If we define $\Sigma_0 := \Sigma(-\infty, 0)$, every forward purely nondeterministic stationary process $\{f(\tau_t \omega) | t \in \mathbb{R}, \omega \in \Omega\}$ on a probability space (Ω, Σ, μ) generates a K-flow.

Every classical K-flow can be represented algebraically as a dynamical W*-system. The dynamical system $(\Omega, \Sigma, \mu, \tau_t)$ corresponds to the dynamical W*-system $(\mathcal{M}, \rho, \alpha_t)$ with the commutative W*-algebra $\mathcal{M} = L^\infty(\Omega, \mu)$. The automorphism group $\{\alpha_t | t \in \mathbb{R}\}$ on $L^\infty(\Omega, \mu)$ is defined by $\alpha_t \{A(\omega)\} = A(\tau_t \omega)$ for all $A \in L^\infty(\Omega, \mu)$. Let P_0 be the projection operator from $L^2\{\Omega, \Sigma(-\infty, +\infty), \mu\}$ onto $L^2\{\Omega, \Sigma(-\infty, 0), \mu\}$ and define a W*-algebra $\mathcal{M}_0 \subset \mathcal{M}$ by $\mathcal{M}_0 := P_0 \mathcal{M} P_0$. Then the K-flow generated by a forward purely nondeterministic stationary process is characterized by the W*-algebras $\mathcal{M}_t := \alpha_t \mathcal{M}_0$ with

- $\mathcal{M}_t \subset \mathcal{M}_s$ for every $t < s$,

- $\bigvee_{t=-\infty}^{\infty} \mathcal{M}_t = \mathcal{M}$,

- $\bigwedge_{t=-\infty}^{\infty} \mathcal{M}_t = \mathbb{C}\mathbb{1}$.

Here $\bigvee \mathcal{M}_t$ is the smallest W*-subalgebra of $\mathcal{M}$ which contains all $\mathcal{M}_t$, while $\bigwedge \mathcal{M}_t$ is the largest W*-subalgebra of $\mathcal{M}$ which is contained in all $\mathcal{M}_t$.

To summarize: Every forward purely nondeterministic stationary process generates a classical K-flow which can be represented as a commutative dynamical W*-system.

2.4 Linear Prediction

Since no systematic approach for nonlinear predictions for forward nondeterministic processes has been established so far, the discussion will be restricted
to the linear case. As long as linearity is retained, prediction theory is fairly
complete. (For an elementary introduction, compare for example Cramér and
Leadbetter (1967), Sect. 5.7 and 7.9.) This theory refers to weakly stationary processes which are characterized by their first and second moments. A
complex-valued process $\{f_t | t \in \mathbb{R}\}$ is said to be *weakly stationary* if the mo-
ments up to second order exist and are stationary, i.e., if for every $t, s, \tau \in \mathbb{R}$
we have

$$\mathcal{E}\{|f_t|^2\} < \infty, \quad \mathcal{E}\{f_{t+\tau}\} = \mathcal{E}\{(f_t)\}, \quad \mathcal{E}\{f_{t+\tau}^* f_{s+\tau}\} = \mathcal{E}\{f_t^* f_s\}. \tag{27}$$

Linear prediction theory is based on the Hilbert space $\mathcal{H}_t$ spanned by a
weakly stationary process $\{f_s | -\infty < s \le t\}$. The Hilbert space $\mathcal{L}_t^2 :=$
$L^2\{\Omega, \Sigma(-\infty, t), \mu\}$ is in general much larger than the Hilbert space $\mathcal{H}_t$.
When the nonlinear predictor $g_{t,\tau}$ is restricted to the subspace $\mathcal{H}_t \subseteq \mathcal{L}_t^2$, one
speaks of a *linear predictor*. For weakly stationary *Gaussian* processes we
have $\mathcal{H}_t = \mathcal{L}_t^2$ so that for Gaussian processes the best predictor is linear. A
weakly stationary second-order process is said to be *forward deterministic in
the linear sense* if the least-square prediction error vanishes for the optimum
linear predictor.

   In spite of the fact that a purely nondeterministic stochastic process exhibits irreversible and dissipative behaviour, it can be generated by an intrinsically conservative and reversible mechanical model. For example, it is well
known that every weakly stationary Gaussian process can be generated as
the output of a linear Hamiltonian system with an infinite-dimensional phase
space (Picci 1986, 1988).

   For every weakly stationary complex-valued process there exists a unique
orthogonal decomposition into a forward deterministic process (in the linear
sense) and a forward purely nondeterministic process (in the linear sense).[6]
Let $\mathcal{H}$ be the subspace of $L^2\{\Omega, \Sigma, \mu\}$ spanned by the variables f_t for all
$t \in \mathbb{R}$ and the constant function 1. The Hilbert space spanned by the process
$\{f_s - m | s \le t\}$ will be denoted by $\mathcal{H}_t$. Furthermore we define

$$\mathcal{H}_{-\infty} := \bigcap_{t \in \mathbb{R}} \mathcal{H}_t, \tag{28}$$

and the orthogonal projection $P_{-\infty}$ from $\mathcal{H}$ onto $\mathcal{H}_{-\infty}$, $\mathcal{H}_{-\infty} = P_{-\infty}\mathcal{H}$. Then
the Hilbert space $\mathcal{H}_t$ spanned by the stationary process $\{f_s | -\infty < s \le t\}$
decomposes into a direct sum

$$\mathcal{H}_t = \mathbb{C} \oplus \mathcal{H}_t^d \oplus \mathcal{H}_t^{nd}, \tag{29}$$

---

[6] This decomposition is due to Wold (1938) for the special case of discrete-time
stationary processes, and to Hanner (1950) for the case of continuous-time processes. The general decomposition theorem is due to Cramér (1961a, 1961b).

where $\mathcal{H}_t^d$ is the Hilbert space spanned by $\{f_s^d| - \infty < s \leq t\}$ with

$$f_s^d := P_{-\infty}\{f_s - m\}. \tag{30}$$

The Hilbert space $\mathcal{H}_t^{nd}$ is spanned by $\{f_s^{nd}| - \infty < s \leq t\}$, where

$$f_s^{nd} := (1 - P_{-\infty})\{f_s - m\}. \tag{31}$$

According to a criterion by Wiener and Krein (Wiener 1942, republished as Wiener 1949; Krein 1945a,b), a weakly stationary complex-valued process $\{f_t| t \in \mathbb{R}\}$ with the spectral distribution function $\lambda \mapsto F_f(\lambda)$ is purely nondeterministic in the linear sense if and only if the spectral distribution function is absolutely continuous and if

$$\int_{-\infty}^{\infty} \frac{|\ln\{\rho_f(\lambda)\}|}{1 + \lambda^2} d\lambda < \infty. \tag{32}$$

where $\lambda \mapsto \rho_f(\lambda)$ is the positive spectral density,

$$\rho_f(\lambda) := \frac{dF_f(\lambda)}{d\lambda}. \tag{33}$$

To summarize: A weakly stationary process is purely nondeterministic in the linear sense if and only if the spectral distribution function is absolutely continuous and satisfies the Paley-Wiener-Krein criterion. Such processes can be generated by classical Hamiltonian systems which are invariant under time-reversal if and only if the associated phase space is infinite-dimensional.

2.5   Purely Nondeterministic Processes and Time Operator

Let $\{f_s| t \in \mathbb{R}\}$ be a continuous weakly stationary purely nondeterministic process on the probability space (Ω, Σ, μ). Let $\mathcal{H}$ be the subspace of $L^2\{\Omega, \Sigma, \mu\}$ spanned by the variables f_t for all $t \in \mathbb{R}$. The time evolution for this process is given by a one-parameter group of the unitary shift operator U_t acting in the Hilbert space $\mathcal{H}$. They are defined by

$$f_{t+s} = U_t f_s, \quad f_{t+s}, f_s \in \mathcal{H}, \quad t, s \in \mathbb{R}. \tag{34}$$

The concept of a time operator for weakly stationary purely nondeterministic processes has been introduced by Tjøstheim (1975) (see also Hanner 1950; Tjøstheim 1976a,b; Gustafson and Misra 1976). Let $\mathcal{H}_t$ be the closed subspace of $\mathcal{H}$ spanned by $\{f_s^{nd}| - \infty < s \leq t\}$. Denote by P_t the projection operator from $\mathcal{H}$ onto $\mathcal{H}_t$. The set $\{P_t| t \in \mathbb{R}\}$ is a spectral family with $P_s \leq P_t$ for $s < t$, $P_{-\infty} = 0$ and $P_{+\infty} = 1$. Then the selfadjoint operator

$$T := \int_{-\infty}^{\infty} t \, dP_t \tag{35}$$

is called the time operator of the purely nondeterministic process $\{f_s | t \in \mathbb{R}\}$. It is shifted by t under the dynamics,

$$U_t^* T U_t = T + t\mathbb{1}, \quad t \in \mathbb{R}. \tag{36}$$

The unitary group $\{U_t | t \in \mathbb{R}\}$ can be represented by a selfadjoint generator Λ as $U_t = \mathrm{e}^{\mathrm{i}\Lambda t}$. If we introduce another unitary group $\{V_\lambda | \lambda \in \mathbb{R}\}$ by $V_\lambda = \mathrm{e}^{-\mathrm{i}\lambda T}$, then the unitary operators U_t and V_λ satisfy Weyl's canonical commutation relation

$$V_\lambda U_t = \mathrm{e}^{\mathrm{i}\lambda t} U_t V_\lambda, \quad \lambda, t \in \mathbb{R}. \tag{37}$$

On an appropriate domain, the time operator T and the generator Λ satisfy Heisenberg's canonical commutation relation

$$T\Lambda - \Lambda T = \mathrm{i}\mathbb{1} \tag{38}$$

The dynamical system associated with the weakly stationary purely nondeterministic process $\{f_s | t \in \mathbb{R}\}$ is characterized by the commutative algebra $\mathcal{M}$ of observables generated by the spectral family $\{P_t | t \in \mathbb{R}\}$ of projection operators P_t. Since $V_\lambda = \mathrm{e}^{-\mathrm{i}\lambda T} \in \mathcal{M}$, *the time operator T is an unbounded observable associated to the algebra of observables.* In contrast, the bounded functions of the generator Λ of the time evolution do not belong to $\mathcal{M}$ so that Λ cannot be considered as an observable. In particular, Λ does not represent the energy.

To summarize: To a weakly stationary process one can associate a time operator if and only if the process is purely nondeterministic in the linear sense. If the process is represented by commutative dynamical W*-system, the time operator is an unbounded observable associated to the commutative algebra of observables. The generator of the time evolution is canonically conjugate to the time operator but it is not an observable.

2.6   Forward Nondeterministic Processes and Nonanticipating Linear Filters

In communication theory, certain messages can be represented by a continuous stationary process. Let $\{f_t | t \in \mathbb{R}\}$ be a complex-valued weakly stationary process with zero mean and covariance

$$C_f(t) := \mathcal{E}\{f_t^* f_0\} = \int_{-\infty}^{\infty} \mathrm{e}^{\mathrm{i}\lambda t} dF_f(\lambda). \tag{39}$$

Suppose the signal $t \mapsto f_t$ passes through a time-invariant, linear filter with the response function $R : \mathbb{R} \mapsto \mathbb{R}$. Then the output of the filter is the zero mean process $t \mapsto g_t$,

$$g_t = \int_{-\infty}^{\infty} R(t - s) f_s ds, \tag{40}$$

with the covariance

$$C_g(t) := \mathcal{E}\{g_t^* g_0\} = \int_{-\infty}^{\infty} e^{i\lambda t} dF_g(\lambda). \tag{41}$$

The spectral distribution functions of the input and output processes are related via the frequency-response function H by

$$dF_g(\lambda) = |H(\lambda)|^2 dF_f(\lambda), \quad H(\lambda) := \int_{-\infty}^{\infty} e^{i\lambda t} R(t) dt. \tag{42}$$

According to the Wiener-Krein criterion, the output process $t \mapsto g_t$ is purely nondeterministic in the linear sense if $\lambda \mapsto F_g(\lambda)$ is absolutely continuous and if

$$\int_{-\infty}^{\infty} \frac{|\ln\{\rho_g(\lambda)\}|}{1+\lambda^2} d\lambda < \infty, \quad \rho_g(\lambda) := \frac{dF_g(\lambda)}{d\lambda}. \tag{43}$$

The relation

$$\int_{-\infty}^{\infty} \frac{\ln\{dF_g(\lambda)/d\lambda\}}{1+\lambda^2} d\lambda = \int_{-\infty}^{\infty} \frac{\ln|H(\lambda)|^2}{1+\lambda^2} d\lambda + \int_{-\infty}^{\infty} \frac{\ln\{dF_f(\lambda)/d\lambda\}}{1+\lambda^2} d\lambda \tag{44}$$

implies that the output process $t \mapsto g_t$ is purely nondeterministic in the linear sense if and only if the input process $t \mapsto f_t$ is purely nondeterministic in the linear sense and the linear filter satisfies the Paley-Wiener criterion. In particular, the output process is *forward* purely nondeterministic in the linear sense if and only if the input process is forward purely nondeterministic in the linear sense and the linear filter is *nonanticipating*.

The basic stochastic process is white noise. It is defined as the generalized derivative of the Wiener process $t \mapsto w_t$ with incremental covariance $\mathcal{E}\{dw_t dw_t\} = dt$. The Wiener process is a mathematically rigorous model for the Brownian motion $t \mapsto B_t$ in one dimension whose mean squared displacement $< B_t >^2$ has a linear dependence on time t, $\langle B_t \rangle^2 = 2Dt$, where D is the diffusion constant. White noise $t \mapsto n_t = dw_t/dt$ is a generalized weakly stationary real-valued process with zero mean whose power spectral density is constant at all frequencies,

$$\mathcal{E}\{n_t\} = 0, \quad C_n(t-s) = \mathcal{E}\{n_t n_s\} = \delta(t-s), \quad dF_n(\lambda)/d\lambda = 1/2\pi. \tag{45}$$

If the input process is white noise $t \mapsto n_t$, the output process $t \mapsto g_t$ has an absolutely continuous spectral distribution function with the density

$$\rho_g(\lambda) = \frac{1}{2\pi} |H(\lambda)|^2, \tag{46}$$

and satisfies the Wiener-Krein criterion.

For a given covariance function $t \mapsto C_g$ with the spectral density $\lambda \mapsto \rho_g(\lambda)$,

$$C_g(t) = \int_{-\infty}^{\infty} e^{i\lambda t} \rho_g(\lambda) \, d\lambda, \tag{47}$$

one can always find a linear input-output system with a response function $R : \mathbb{R} \mapsto \mathbb{R}$, such that the response to a white noise input is a real-valued weakly stationary process $t \mapsto g_t$ with the covariance $C_g(t - s) = \mathcal{E}\{g_t g_s\}$. But this representation is far from unique. The only condition one has to satisfy is

$$\rho_g(\lambda) = \frac{1}{2\pi} |H(\lambda)|^2 \quad \text{with} \quad H(\lambda) := \int_{-\infty}^{\infty} e^{i\lambda t} R(t) dt. \qquad (48)$$

If the Paley-Wiener criterion is met, the function $\lambda \mapsto |H(\lambda)|^2$ can be factorized. Then the spectral density ρ_g can be factorized into a boundary value of an analytic function which is holomorphic in the open upper half-plane and a boundary value of an analytic function which is holomorphic in the open lower half-plane. If we choose a filter of minimum phase type, these two functions are uniquely given by

$$H^{\pm}(\lambda) = \lim_{\epsilon \to 0} \exp \left(\frac{1}{\pi i} \int_{-\infty}^{\infty} \frac{1 + s(\lambda \pm i s \epsilon)}{s - (\lambda \pm i s \epsilon)} \frac{|\ln \rho_g(\lambda)|}{1 + (\lambda \pm i s \epsilon)^2} ds \right), \quad \lambda \in \mathbb{R}, \quad (49)$$

so that the canonical Wiener factorization of the spectral density is given by

$$2\pi \rho_g(\lambda) = H^+(\lambda) H^-(\lambda) = |H^+(\lambda)|^2 = |H^-(\lambda)|^2, \quad \lambda \in \mathbb{R}. \qquad (50)$$

Corresponding to these two canonical solutions, we get two moving average representations: the canonical backward moving average representation of a forward purely nondeterministic process $t \mapsto g_t^+$,

$$g_t^+ = \int_{-\infty}^{t} R^+(t - s) \, dw_s,$$
$$R^+(t) := \frac{1}{2\pi} \int_{-\infty}^{\infty} e^{-i\lambda t} H^+(\lambda) \, d\lambda,$$
$$\mathcal{E}\{g_t^+ g_s^+\} = C_g(t - s), \qquad (51)$$

and the canonical forward moving average representation of backward purely nondeterministic process $t \mapsto g_t^-$,

$$g_t^- = \int_{t}^{\infty} R^-(t - s) \, dw_s,$$
$$R^-(t) := \frac{1}{2\pi} \int_{-\infty}^{\infty} e^{-i\lambda t} H^-(\lambda) \, d\lambda,$$
$$\mathcal{E}\{g_t^- g_s^-\} = C_g(t - s). \qquad (52)$$

Since these stochastic processes are ergodic, the functions $t \mapsto g_t^+$ and $t \mapsto g_t^-$ can also be interpreted individually if the function $t \mapsto w_t$ is understood as the individual Wiener process.

To summarize: A nonanticipating, time-invariant, linear filter transforms forward purely nondeterministic input-processes into forward purely nondeterministic output-processes. Every weakly stationary process which is forward purely nondeterministic in the linear sense can be realized as the output of a nonanticipating linear filter with a white-noise input. The corresponding response function can be constructed by a Wiener factorization of the spectral density of the process.

2.7   Conclusion

The representation of facts requires the distinction of past and future. In classical physics it is possible to break the time-reversal symmetry of infinite, conservative, reversible Hamiltonian systems. The resulting irreversible and dissipative behavior can be described by nonanticipative measuring instruments or, equivalently, by forward purely nondeterministic stochastic processes. Both can be characterized by the Paley-Wiener criterion. The growth of the set of facts can be described by a time operator of the associated K-flow.

3   The Representation of Facts in Quantum Physics

3.1   Quantum K-Flows

In classical physics, K-flows are the paradigmatic model for the emergence of one-sided time evolutions from a bidirectionally deterministic dynamics. Within the algebraic formulation, the definition of a K-flow appropriate for classical physics can be generalized to quantum physics by replacing the commutative algebra by a noncommutative one. Let $(\mathcal{M}, \{\alpha_t | t \in \mathbb{R}\})$ be a dynamical W*-system consisting of a (in general noncommutative) W*-algebra $\mathcal{M}$ and a one-parameter group $\{\alpha_t | t \in \mathbb{R}\}$ of automorphisms α_t. The dynamical W*-system $(\mathcal{M}, \{\alpha_t | t \in \mathbb{R}\})$ is called a W*-K-flow if some subalgebra $\mathcal{M}_0 \subset \mathcal{M}$ exists such that for $\mathcal{M}_t := \alpha_t \mathcal{M}_0$ we have (Emch 1976):[7]

- $\mathcal{M}_t \subset \mathcal{M}_s$ for every $t < s$,
- $\bigvee_{t=-\infty}^{\infty} \mathcal{M}_t = \mathcal{M}$,
- $\bigwedge_{t=-\infty}^{\infty} \mathcal{M}_t = \mathbb{C}\mathbb{1}$.

Here $\bigvee \mathcal{M}_t$ is the smallest W*-subalgebra of $\mathcal{M}$ which contains all $\mathcal{M}_t$, while $\bigwedge \mathcal{M}_t$ is the largest W*-subalgebra of $\mathcal{M}$ which is contained in all $\mathcal{M}_t$.

Conclusion: The classical concept of K-flows representing forward purely nondeterministic processes can be generalized to noncommutative irreversible quantum processes. In contrast to the classical case, noncommutative K-systems are not sufficient to represent facts. The reason is that a representation of facts requires not only the existence of nonanticipative processes but also the impossibility of coherent superpositions of facts. In other words: *facts have to be represented by observables which commute with all observables of the system.*

---

[7] Emch's requirement of an invariant state is no longer called for. Compare, for example, Narnhofer and Thirring (1990); Benatti (1993), p. 129.

3.2   Classical Observables, Disjoint States, and Proper Mixtures

Nontrivial observables which commute with all observables are called *classical observables*. While in von Neumann's codification of traditional quantum theory there are no classical observables,[8] algebraic quantum mechanics allows to discuss classical observables from a fundamental point of view. All examples of rigorous derivations of classical observables refer to systems with infinitely many degrees of freedom. For example, it has been shown that in the thermodynamic limit the observables for temperature, the chemical potential, and the order parameters for ferromagnets, superfluids, and superconductors are classical observables.

The classical behavior of a quantum system is mirrored by the *center* of the algebra of observables. The center $\mathcal{Z}(\mathcal{M})$ of a W*-algebra $\mathcal{M}$ consists of all elements of $\mathcal{M}$ which commute with every element of $\mathcal{M}$,

$$\mathcal{Z}(\mathcal{M}) := \{Z | Z \in \mathcal{M}, ZM = MZ \ \forall \ M \in \mathcal{M}\} \tag{53}$$

The center $\mathcal{Z}(\mathcal{M})$ is a commutative W*-algebra. If the center consists only of the multiples of the identity, it is called trivial, $\mathcal{Z}(\mathcal{M}) = \mathbb{1}\mathbb{C}$. A W*-algebra is commutative if it is identical with its center, $\mathcal{M} = \mathcal{Z}(\mathcal{M})$. If the center of the algebra of observables is trivial, then the corresponding physical system is called a pure quantum system. If the algebra of observables is commutative, then the physical system is called classical. In general, physical systems are partially quantal and partially classical. That is, the algebra of observables is noncommutative and its center is nontrivial. The nontrivial selfadjoint operators of the center are called classical observables. If $\mathcal{M}$ describes a quantum system, then $\mathcal{Z}(\mathcal{M})$ describes its classical part. Although the classical part of a quantum system fulfills all requirements of a system of classical physics, all its quantities depend intrinsically on Planck's constant $\hbar$.

The central part of a quantum system allows an important classification of states. The support S_ρ of a normal state ρ of a W*-algebra $\mathcal{M}$ is defined as the smallest projection operator $S \in \mathcal{M}$ such that $\rho(S) = 1$. The central support C_ρ of a state ρ is defined as the smallest projection operator $C \in \mathcal{Z}(\mathcal{M})$ such that $\rho(C) = 1$. A state ρ on the W*-algebra $\mathcal{M}$ is pure if its support S_ρ is an atom. A state ρ is called a *factor state* if its central support is an atom. Two states ρ and ϕ are called *orthogonal* if their supports S_ρ and S_ϕ are orthogonal, $S_\rho S_\phi = 0$. Two states ρ and ϕ are called *disjoint* if their central supports C_ρ and C_ϕ are orthogonal, $C_\rho C_\phi = 0$. Disjointness implies orthogonality, but only in commutative algebras orthogonality implies disjointness. In general, disjointness is a much stronger condition than orthogonality.

---

[8] This is the irreducibility postulate introduced by von Neumann (1932) which implies that all selfadjoint operators acting on the Hilbert space of state vectors are observableless. However, we know empirically that von Neumann's irreducibility postulate is not valid in general.

A factor state ρ is distinguished by the fact that it is dispersion-free with respect to every classical observable,

$$\rho(Z^2) = \{\rho(Z)\}^2 \ \forall \ Z = Z^* \in \mathcal{Z}(\mathcal{M}). \tag{54}$$

If two states ρ and ϕ are disjoint, there exists a classical observable $Z \in \mathcal{Z}(\mathcal{M})$ such that $\rho(Z) \neq \phi(Z)$, so that *disjoint states can be distinguished and classified in a classical manner.* Two normal pure states ρ and ϕ are called *equivalent* if $\rho(Z) = \phi(Z)$ for *all* classical observables $Z \in \mathcal{Z}(\mathcal{M})$. Every state can be uniquely decomposed into a sum or an integral of disjoint factor states (for details compare, e.g., Takesaki 1979, Chap. IV.6). This so-called *central decomposition* represents the finest unique decomposition of a nonpure state into disjoint factor states.

In contrast to the classical case, for quantum systems the convex set of all states is not a simplex, hence a convex decomposition of a nonpure factor state is *never* unique.[9] A nonpure factor state ρ_α has infinitely many different decompositions into a convex sum of pure states. Such decompositions into pure states cannot be interpreted as a proper mixture.

If we speak of a classical mixture of two components (like a mixture of water and alcohol), then we tacitly presuppose that we can *distinguish* operationally between the two components. It makes no sense to speak of mixing indistinguishable entities. That is, *it must be possible to label every component of a proper mixture so that the components can be distinguished.* Since such a label must be determinable together with any other property of the component, it has to be characterized by a value of a *classical observable.*

The nonpurity of factor states cannot originate in some kind of mixing, it is always due to Einstein-Podolsky-Rosen correlations of the considered system with its environment. *Nonpure quantum states can be interpreted in terms of a proper mixture of pure states if and only if these pure states are mutually disjoint* (cf. Amann and Primas 1996). The finest possible decomposition which admits an ignorance interpretation is the central decomposition.

Conclusion: Disjoint states are of crucial importance as final states in *any* kind of processes – natural processes or measurement processes – which produce facts. This is decisive for a proper discussion of the notorious measurement problem of quantum theory. The measurement problem is not – as often asserted – the problem of how a pure state can be transformed into a nonpure state, or how the density operator can become diagonal in a preferred basis. This is a trivial task: appropriate dynamical linear semigroups and their Hamiltonian dilations can describe such a decoherence mechanism. A proper statistical description of the measurement process has to show that the dynamics of an isolated quantum system can create facts. That is, one has to show that a dynamics exists which transforms factor states into a classical mixture of *disjoint* factor states.

---

[9] The state space of a C*-algebra $\mathcal{A}$ is a simplex if and only if $\mathcal{A}$ is commutative. Compare Takesaki (1979), p. 251.

3.3   The Emergence of Facts in Quantum Systems

Classical observables and disjoint states exist only if the quantum system has infinitely many degrees of freedom. If the quantum system is coupled to the electromagnetic radiation field, there are always infinitely many inequivalent representations, so that in this case the existence of many disjoint states is the rule. Accordingly, there are many quantum systems with classical observables which can be used to represent facts. The creation of new facts by a dynamical system, however, is a hard nut to crack: a measurement process is a dynamical process which has to transform a factor state into disjoint final factor states. A very general result due to Klaus Hepp shows that such a process cannot be described by any automorphism of the algebra of observables (Hepp 1972, lemma 2, p. 246):

No-go theorem for any automorphic dynamics of measurement. *If ρ and ϕ are two disjoint states on a C*-algebra and if α is an automorphism of this C*-algebra, then the transformed states $\rho \circ \alpha$ and $\phi \circ \alpha$ are disjoint.*

This theorem also shows a crucial difference between commutative and noncommutative K-flows: A classical commutative K-flow maps equivalent states into disjoint states while a noncommutative K-flow cannot change the equivalence class. To the best of my knowledge, there are just two reasonable ways out of this situation:

1. The fundamental dynamics is invariant under time-reversal, but not given by a one-parameter group of automorphisms.
2. The fundamental dynamics is given by a one-parameter group of auto-morphisms, but the disjointness of the final states is reached only asymptotically.

I shall not consider the first possibility because no general theory is available for nonautomorphic time evolution invariant under time-reversal. But I would like to stress that the postulate of an automorphic dynamics has no sound physical basis. Many physically reasonable C*-algebraic systems without an automorphic dynamics are known.

The second point of view has been introduced by Hepp (1972). He proved the important result that for appropriate quantum systems there exists a one-parameter automorphism group $\{\alpha_t \,|\, t \in \mathbb{R}\}$ such that for equivalent initial states $\rho_1, \rho_2, \ldots$ the asymptotic limits $\rho_1 \circ \alpha_t, \rho_2 \circ \alpha_t, \ldots$ exist for $t \to \infty$ and are disjoint. Nevertheless, as shown by John Bell (1975), under Hepp's assumptions alone the measurement process is not well-posed since it is still possible to undo the measurement for any finite time by a quasilocal perturbation. As shown by Lockhart and Misra (1986), the key to the final resolution lies in recognizing the measurement apparatus as an irreversible dynamical system which breaks the time-reversal symmetry in such a way that it acts as a nonanticipative system.

In order to incorporate the irreversible and nonanticipative behavior of actual measurement apparatuses into the theory, one has to select a W*-algebra

$\mathcal{M}_0 \subset \mathcal{M}$ of observables which can be observed with a nonanticipative laboratory instrument. The requirement of nonanticipativity means that $\mathcal{M}_t \subseteq \mathcal{M}_s$ for $t < s$, where $\mathcal{M}_t := \alpha_t \mathcal{M}_0$. In terms of normal conditional expectations $\mathcal{E}_t := \alpha_t \mathcal{E}_0 \alpha_{-t}$ with $\mathcal{E}_0(\mathcal{M}) = \mathcal{M}_0$, one gets $\mathcal{E}_t \le \mathcal{E}_s$ for $t < s$. The family $\{\mathcal{E}_t | t \in \mathbb{R}\}$ is either constant (Hepp's case, where the time-reversal symmetry holds), or increasing as t increases (so that the time-reversal symmetry is broken). In the case of broken time-reversal symmetry, the automorphism group $\{\alpha_t\}$ and the W*-algebra $\mathcal{M}_0$ give rise to a generalized K-flow. The appropriate tools for the discussion of noncommutative K-flows are again the Wiener factorization and the theory of Hardy spaces.

Such symmetry-breaking K-flows cannot be classical and they do not generate classical observables in any finite time. However, they do produce asymptotically disjoint final states. That is, the emergence of disjoint states describing facts is gradual. It occurs progressively over finite amounts of time. The exact disjointness is reached only in the limit $t \to \infty$, but the objective irreversibility of the K-flow warrants that the process cannot be undone. In this sense we can speak of *approximately disjoint states* and *approximate facts* even for finite times.

In everyday life we usually idealize approximate facts. For example, the event of death is considered as a fact. It is characterized by the irreversible loss of the bodily attributes and functions that constitute life. This is a continuous physiological process which cannot be undone. Although this process can proceed very quickly, there is, however, no definite instant of death.

In this sense, the Hepp-Lockhart-Misra process is a very reasonable model for the emergence of facts. It also shows that the so-called "measurement problem of quantum mechanics" is neither a pseudoproblem nor a philosophical question. It is a well-posed problem of mathematical physics which can be discussed in the framework of algebraic quantum mechanics, provided we take the nonanticipative character of all laboratory instruments into account.

Conclusion: If we include the whole environment of an open quantum system and if we describe the resulting system by an automorphic dynamical C*-system, then this system cannot generate new facts in finite time. However such systems can, in a strictly irreversible manner, generate asymptotically disjoint final states which are described by noncommutative K-flows. Every K-flow has its own typical relaxation time. For finite times much larger than the relaxation time, asymptotically disjoint states describe approximate facts which correspond to the facts of everyday life.

Acknowledgments: I thank Harald Atmanspacher for many discussions about the topics presented in this article.

References

Amann A. and Primas H. (1996): What is the referent of a nonpure quantum state? In *Experimental Metaphysics – Quantum Mechanical Studies in Honor of Abner Shimony*, ed. by R.S. Cohen and J. Stachel. In press.

Bell J.S. (1975): On wave packet reduction in the Coleman-Hepp model. *Helv. Phys. Acta* **48**, 93–98.

Benatti F. (1993): *Deterministic Chaos in Infinite Quantum Systems* (Springer, Berlin).

Cornfeld I.P., Fomin S.V., and Sinai Ya.G. (1982): *Ergodic Theory* (Springer, Berlin).

Costa de Beauregard O. (1987): *Time, the Physical Magnitude* (Reidel, Dordrecht).

Cramér H. (1961a): On some classes of non-stationary stochastic processes. In *Proceedings of the Fourth Berkeley Symposium on Statistics and Applied Probability*, ed. by J. Neyman (University of California Press, Berkeley), 57–78.

Cramér H. (1961b): On the structure of purely non-deterministic stochastic processes. *Ark. Mat.* **4**, 249–266.

Cramér H. and Leadbetter M.R. (1967): *Stationary and Related Stochastic Processes* (Wiley, New York).

Einstein A. (1909): Zum gegenwärtigen Stand des Strahlungsproblems. *Physikalische Zeitschrift* **10**, 185–193.

Emch G.G. (1976): Generalized K-flows. *Commun. Math. Phys.* **49**, 191–215.

Feynman R.P. and Hibbs A.R. (1964): *Quantum Mechanics and Path Integrals* (McGraw-Hill, New York).

Fock V.A. (1957): On the interpretation of quantum mechanics. *Czechosl. J. Phys.* **7**, 643–656. Russian original: *Usp. Fiz. Nauk* **62**, 461–474 (1957).

Gustavson K. and Misra B. (1976): Canonical commutation relations of quantum mechanics and stochastic regularity. *Lett. Math. Phys.* **1**, 275–280.

Hanner O. (1950): Deterministic and non-deterministic stationary random processes. *Ark. Mat.* **1**, 161–177.

Havas P. (1965): Relativity and causality. In *Logic, Methodology, and Philosophy of Science*, ed. by Y. Bar-Hillel (North-Holland, Amsterdam), 347–362.

Hepp K. (1972): Quantum theory of measurement and macroscopic observables. *Helv. Phys. Acta* **45**, 237–248.

Kakutani S. (1950): Review of "Extrapolation, Interpolation and Smoothing of Stationary Time Series" by Norbert Wiener. *Bull. Amer. Math. Soc.* **56**, 378–381.

Kalman R.E., Falb P.L., and Arbib M.A. (1969): *Topics in Mathematical System Theory* (McGraw-Hill, New York).

Khintchine A. (1934): Korrelationstheorie der stationären stochastischen Prozesse. *Math. Ann.* **109**, 604–615.

Kolmogoroff A. (1933): *Grundbegriffe der Wahrscheinlichkeitsrechnung* (Springer, Berlin).

Krein M.G. (1945a): On a generalization of some investigations of G. Szeg, W.M. Smirnov, and A.N. Kolmogorov. *Dokl. Akad. Nauk SSSR* **46**, 91–94 (in Russian).

Krein M.G. (1945b): On a problem of extrapolation of A.N. Kolmogorov. *Dokl. Akad. Nauk SSSR* **46**, 306–309 (in Russian).

Krengel U. (1971): K-flows are forward deterministic, backward completely non-deterministic stationary point processes. *J. Math. Anal. Appl.* **35**, 611–620.

Krengel U. (1973): Recent results on generators in ergodic theory. In *Transactions of the Sixth Prague Conference on Information Theory, Statistical Decision Functions, Random Processes* (Academia, Prague), 465–482.

Lockhart C.M. and Misra B. (1986): Irreversibility and measurement in quantum mechanics. *Physica A* **136**, 47–76.

Narnhofer H. and Thirring W. (1990): Algebraic K-systems. *Lett. Math. Phys.* **20**, 231–250. Errata: *Lett. Math. Phys.* **22**, 81 (1991).

Neumann, J. von (1932): *Mathematische Grundlagen der Quantenmechanik* (Springer, Berlin).

Paley R.E.A.C. and Wiener N. (1934): *Fourier Transforms in the Complex Domain* (American Mathematical Society, Providence).

Pfaffelhuber E. (1971): Generalized impulse response and causality. *IEEE Transactions on Circuit Theory* **CT-18**, 218–223.

Picci G. (1986): Application of stochastic realization theory to a fundamental problem of statistical physics. In *Modelling, Identification, and Robust Control*, ed. by C.I. Byrnes and A. Lindquist (North Holland, Amsterdam), 211–258.

Picci G. (1988): Hamiltonian representation of stationary processes. In *Operator Theory: Advances and Applications*, ed. by I. Gohberg, J.W. Helton, and L. Rodman (Birkhäuser, Basel), 193–215.

Rosenblatt M. (1971): *Markov Processes: Structure and Asymptotic Behavior* (Springer, Berlin).

Scarpellini B. (1979a): Entropy and nonlinear prediction. *Z. Wahrscheinlichkeitstheorie verw. Gebiete* **50**, 165–178.

Scarpellini B. (1979b): Predicting the future of functions on flows. *Math. Systems Theory* **12**, 281–296.

Synge J.L. (1960): Classical dynamics. In *Principles of Classical Mechanics and Field Theory. Encyclopedia of Physics. Volume III/1*, ed. by S. Flügge (Springer, Berlin), 1–225.

Takesaki M. (1979): *Theory of Operator Algebras I* (Springer, Berlin).

Tjøstheim D. (1975): Multiplicity theory for multivariate wide sense stationary generalized processes. *J. of Multivariate Analysis* **5**, 314–321.

Tjøstheim D. (1976a): A commutation relation for wide sense stationary processes. *SIAM J. Appl. Math.* **30**, 115–122.

Tjøstheim D. (1976b): Spectral generating operators for non-stationary processes. *Adv. Appl. Prob.* **8**, 831–846.

Wiener N. (1923): Differential space. *Journal of Mathematics and Physics (MIT)* **2**, 131–174.

Wiener N. (1930): Generalized harmonic analysis. *Acta Math.* **55**, 117–258.

Wiener N. (1942): *Extrapolation, Interpolation, and Smoothing of Stationary Times Series, with Engineering Applications*. Report to Section D2, National Defence Research Committe (NDRC). Issued as Classified Report in February 1942.

Wiener N. (1949): *Extrapolation, Interpolation, and Smoothing of Stationary Times Series. With Engineering Applications* (MIT Technology Press and Wiley, New York).

Wold H. (1938): *A Study in the Analysis of Stationary Times Series* (Almquist and Wiksell, Stockholm).

Individual Complex Quantum Objects and Dynamics of Decompositions

Anton Amann

HIF E23.2, ETH-Hönggerberg, Postfach 164, CH-8093 Zürich, Switzerland

Abstract. At first sight, it seems that the usual system-theoretic approach cannot be applied to quantum theory. The reason is that an observer and the observed system cannot be separated due to the holistic structure of quantum mechanics. To escape this dilemma, one usually *enforces* a system-theoretic treatment ad hoc, declaring observer and observing apparatus to be classical. It is argued that this enforced separation of a quantum system and its environment should be *derived* from a stability requirement for the joint system {quantum object & environment}, and that the quantum fluctuations deviating from such a classical behavior should be quantitatively characterized by a *large-deviation entropy*. A mathematically rigorous derivation of system-theoretic behavior for quantum objects has not (yet) been achieved. Nevertheless, heuristic reasoning shows that one can expect to get a stochastic and usually nonlinear dynamical behavior of individual complex quantum systems. Experimentally, the stochastic behavior is the more pronounced the more the *individual* quantum behavior can be observed (instead of averaged results over many quantum systems, e.g., many molecules in a substance). Particularly interesting examples of individual stochastic behavior are investigated in single-molecule spectroscopy, where single molecules are embedded in a crystal or a polymorphic matrix. These examples are prototypes of individual complex quantum objects.

1 The Holistic Structure of Quantum Mechanics

One of the most interesting features of quantum theory is its holistic structure (Einstein et al. 1935, Aspect et al. 1982). It implies that a subsystem of a quantum object does not exist as an individual entity, i.e., that a subsystem need not be a quantum "object" a priori. Hence one must carefully distinguish between quantum *objects* and quantum *systems*, which do not have the same ontological status. Every quantum object is a quantum system; the converse is not true.

Explaining these matters in plain words leads almost inevitably to misunderstandings. In a quantum-mechanical formalism, instead, the situation is clearly characterized by the fact that a (pure) state with state vector Ψ of some joint quantum system is usually *not* of product form, i.e., *not* of the form

$$\Psi = \Psi_1 \otimes \Psi_2, \tag{1}$$

with Ψ_1 and Ψ_2 being state vectors of the subsystems. The subsystems can then only be described by non-pure states and not by state vectors anymore.[1] These non-pure states arise as restrictions of the overall state vector Ψ to the (observables of the) subsystems and are represented by density operators, say D_1 and D_2. In contrast to classical systems, the overall state vector Ψ of the joint system cannot be computed starting from the restricted information present in the density operators D_1 and D_2. In particular, the overall state vector Ψ is not specified as a product $D_1 \otimes D_2$ of the density operators.

In this sense, many states of a quantum joint system do not give rise to (pure) states of the subsystems. In this sense, the subsystems are not quantum objects. A quantum *object* is described by a (pure) state and the corresponding state vector, whereas a quantum (sub)*system* is only described by a density operator (incomplete information) or a state vector of a *larger* quantum object (complete information).

For a system-theoretic approach, this presents a serious problem if both measuring device and observed quantum system have quantum properties (which is usually the case, e.g., in spectroscopy). Therefore, it is not clear at all if the joint quantum object {measuring device & observed quantum system} allows a system-theoretic structure

$$input \rightarrow blackbox \rightarrow output,$$

where in- and output are well defined, without holistic correlations to the observed quantum system itself. Nevertheless, such a structure is usually taken for granted, because observers or observing tools are *declared* to have only classical (i.e., non-quantum) properties.

But even if such an input-output structure is legitimate, the observed quantum system can show holistic correlations to some quantum environment, as, e. g., to the quantum radiation field, which can never be completely screened. To be on the safe side, system-theoretic discussions of quantum systems are, therefore, based on density-operator dynamics and not on a dynamics of pure states. A typical example for such a density-operator dynamics is the Karplus-Schwinger dynamics specified by the differential equation

$$\dot{D}(t) = \frac{1}{i\hbar} [H_o, D(t)] - \Gamma \{D(t) - D_\beta\}, \quad t \geq 0. \tag{2}$$

Here the constant Γ determines the dissipation (and, finally, the line width in the context of spectroscopy). The solution of density operators $D(t)$ of the Karplus–Schwinger equation converges to the thermal state characterized by $D_\beta = \exp(-\beta H_o)/\mathrm{Tr}\,(\exp(-\beta H_o))$ for an inverse temperature β (i.e., the β-KMS state (Bratteli and Robinson 1981)) for long times t. Here H_o is the Hamiltonian of the unperturbed quantum system to be observed, $\hbar$ is

---

[1] Here state vectors always refer to an irreducible representation of the observables in question. In reducible representations, non-pure states can still be described by state vectors, but the intuitive picture (superposition principle, etc.) gets lost.

Planck's constant and Tr denotes the trace of an operator. An input $b(t)$ enters through perturbation of the Hamiltonian as in

$$H(t) = H_\circ - b(t)B \,. \tag{3}$$

Here B could, for example, be the dipole moment operator, whereas the scalar function b could be a sine input, $b(t) = \sin \omega t$, or an appropriate pulse (as in Fourier spectroscopy). As output, one usually takes the expectation value of some appropriate operator A

$$a(t) \stackrel{\text{def}}{=} \text{Tr}(D(t)A) \,. \tag{4}$$

Often only the linear part of the response/output is considered (Kubo et al. 1985, Fick and Sauermann 1986, Primas 1994).

2 Statistical Versus Individual Quantum Descriptions

A non-pure state with corresponding density operator D can have (at least) two different meanings. It can arise as a restriction of a pure state in a larger quantum system (see Sect. 1), or it can arise as a mixture of different pure states in the (sub)system considered. Hence, two different sorts of quantum mechanics can be envisaged:

(i) A description by non-pure states (i.e., density operators) which arise as restrictions of a larger quantum system. No relation between these density operators and pure states in the (sub)system itself is considered. This is the usual approach in "statistical" quantum mechanics.

(ii) A description by non-pure states, which arise as a particular (known or estimated) mixture of pure states in the (sub)system under discussion. This is called an *individual* approach to quantum mechanics, since pure states of the subsystem itself come into play. Via the particular mixture(s)/decomposition(s) for the density operators $D(t)$, it incorporates statistics of pure states, i.e., gives probabilities for pure states to arise.

A more detailed description of these two approaches to quantum mechanics can be found in Amann (1996) and Amann and Primas (1996). One must be quite careful, since there are quantum systems which cannot be turned into quantum objects by an appropriate dressing procedure (cf. Atmanspacher et al. 1995).

It is important to notice that the information contained in a density operator alone is rather restricted. I shall try to make plausible why chemists would never be satisfied with a density-operator description of molecules alone, though this would be sufficient to explain spectroscopic results of ensembles of molecules (but not for single-molecule spectroscopy). Similar situations arise in other fields of research.

The reason for the restricted information content of a given density operator is that it can be decomposed into pure states *in infinitely many different ways.* Consider, for example, a two-level system, describing the ground and excited state of ammonia, with respective state vectors Ψ_+, Ψ_- and energies E_+, E_-. The transition between these two states is the ammonia-maser transition with transition frequency $\nu = (E_- - E_+)/h = 23\,870\,110\,000\,\mathrm{s}^{-1}$. The thermal density operator is given as

$$D_\beta = \frac{\mathrm{e}^{-\beta E_+} |\Psi_+><\Psi_+| + \mathrm{e}^{-\beta E_-} |\Psi_-><\Psi_-|}{\mathrm{Tr}\,(\mathrm{e}^{-\beta E_+} |\Psi_+><\Psi_+| + \mathrm{e}^{-\beta E_-} |\Psi_-><\Psi_-|)}, \tag{5}$$

where Dirac bras and kets are used to describe the ground and excited states. (5) can be considered as a mixture of the ground and excited state with appropriate Boltzmann coefficients or as a decomposition of D_β into the pure states Ψ_+, Ψ_-. The important point already mentioned is that many other decompositions of D_β can be envisaged, for example, a decomposition into states which allow a nuclear structure (which is not the case for Ψ_+, Ψ_-, cf. Amann 1995, Pfeifer 1980, Quack 1989). "Nuclear structure" means that the nuclei in the molecule have unambiguous (though perhaps changing) positions. Hence, nuclear structure and other important structural features enter into quantum mechanics only through a particular decomposition of thermal density operators (into pure states), i.e., by an individual approach.

"Nuclear structure" is only one particular example, of enormous importance in chemistry. But in every quantum system, described by density-operators as in (2), one can use different decompositions of a thermal density operator D_β. Some of these decompositions may look weird, as, for a chemist, a decomposition into pure states without nuclear structure. Nevertheless, all possible decompositions can – in principle – give rise to a useful description in terms of pure states.

From the point of view of system theory, one may either be satisfied with a "statistical" description by density operators as in (2), or try to construct an *individual* description in terms of pure states of the (sub)system in question, i.e., give *some particular* decomposition of, say, the thermal state D_β into pure states. Note that in an individual description, different decompositions of D_β or the density operators $D(t)$ are *not* considered as being equivalent! A decomposition of D_β may be unstable under external perturbations and, therefore, only of limited physical relevance (Amann 1995).

Considering all possible decompositions of density operators as being equivalent would lead back to the usual "statistical" description ((i)). From a puristic statistical point of view, decompositions of density operators (into pure states) would not make physical sense at all, since here density operators are always thought to arise from a larger quantum system (as by restriction of a pure state).

3   Stochastic Input-Output Systems and Quantum Fluctuations

If a quantum (sub)system can reasonably be described in terms of pure states, it becomes a quantum *object* (see Sect. 1). Precise conditions for such an individual description to make sense have never been written down. Heuristically, it is necessary that there are only "small" holistic correlations between observables of the quantum subsystem and observables from the embedding system/environment.

Assume now that such an individual description makes sense. How could the "statistical" input-output description in (2) be replaced by an individual input-output description in terms of pure states?

The heuristic idea is as follows: *Strictly* isolated quantum objects exist only in theory. In practice, external perturbations by the object's environment have to be taken into account. Typical environments are the quantum radiation and gravitational fields, collisions with neighboring particles, or the coupling to some heat bath, etc. Hence,

- the dynamics of some given molecular initial state is not only governed by the Schrödinger equation, but also by external influences,
- depending on the (pure) state of the environment, different final molecular states can be reached,
- usually only probabilistic predictions (and no information about the precise trajectory of pure states) can be given.

Already in standard descriptions of spectroscopic experiments, a stochastic element comes in by "quantum jumps" between different eigenstates of the respective molecular Hamiltonian. Eigenstates are, of course, invariant under the usual Schrödinger dynamics, but change under the influence of the external radiation field. The transition probabilities are given by Fermi's Golden Rule (Fermi 1961).

Therefore, in an individual description, the usual Schrödinger dynamics (corresponding to the first term on the right-hand side of (2)) is supplemented by stochastic elements (roughly corresponding to the dissipative, second term on the right-hand side of (2)). Averaging over the stochastic part must reproduce the relevant density-operator dynamics. In this sense, Fermi's Golden Rule is compatible with the Karplus-Schwinger dynamics of (2).

It has already been stressed that density operators can be decomposed into pure states in infinitely many different ways. Correspondingly, *different* stochastic dynamics on the level of pure states are compatible with one and the same density-operator dynamics (cf. Wächter (1991)). Hence, a choice must be made among all the different possibilities. The approach using eigenstates and Fermi's Golden Rule to describe quantum jumps is interesting, but not the only relevant one. In particular, the higher molecular eigenstates are unstable under external perturbations (or small modifications of the molecular structure). Also, it has already been indicated above for ammonia, that molecules are not exclusively in eigenstates of the (unperturbed molecular) Hamiltonian. The superpositions $1/\sqrt{2}\,(\Psi_+ \pm \Psi_-)$, for example, describe pyramidal states of ammonia, which are non-stationary (tunnel back and forth) and more adapted to the chemical picture of ammonia. Actually, most molecular (pure) states admitting a nuclear structure are *not* strict eigenstates. Hence, more general stochastic dynamics must be taken into account (cf. Ghirardi et al. (1986), Gisin (1984), Primas (1990)). A simple stochastic dynamics has been discussed by Amann (1996).

Usual "statistical" quantum mechanics gives expectation values, interpreted as averages over many single measurements. In an individual approach, the *fluctuations* of the single-measurement values come into play. Hence a choice of some particular stochastic dynamics (compatible with a density-operator dynamics such as in (2)) gives rise to particular fluctuations which can be compared with experiments. A stochastic dynamics of the pure states of a quantum system is called *compatible* with a density-operator dynamics, if the latter arises from the first through averaging over the stochastic variable (i.e., over the diffusion part of the stochastic differential equation). The shift in perspective from *averages* to *fluctuations* is already interesting for deterministic mathematical sequences, as exemplified by Kalman (1995).

Experimental observations of fluctuations are interesting, but difficult. In single-molecule spectroscopy (Ambrose and Moerner 1991, Moerner and Basché 1993, Moerner 1994, Wild et al. 1994), for example, the fluctuations of migrating spectral lines have been observed, showing very different kinds of stochastic behavior, depending on the particular situation (of a single defect molecule in a crystal or matrix). Here "single molecule" is always to be understood as a single molecule embedded in a polymorphic matrix or a crystal. In the example of Fig. 1, the single molecule is a terrylene defect molecule embedded in a hexadecane matrix. Hence, the actual quantum system is the joint system {terrylene defect & hexadecane matrix}. Varying the laser frequency over a range of 800 MHz, the overall intensity of the emitted light (excluding the original laser frequency) is measured, so that it can be checked which laser frequencies have been absorbed. As asserted before, the absorption lines migrate stochastically. Some spectroscopic lines are stable over a time-interval of an hour, whereas others change rather quickly. Single-molecule spectroscopy should certainly fit into a future theory of individual quantum mechanics, incorporating fluctuations and, hence, going beyond the usual "statistical" formalism. In contrast to ensembles of molecules, where the stochastic behavior (migration of lines) averages out, single molecules are prototypes of individual stochastic behavior.

Replacing the "statistical" density-operator dynamics (2) by an individual (nonlinear) stochastic dynamics gives a more detailed description including dynamical fluctuations as well as (molecular, ...) individual structure. Formally, the scheme is as follows: Input $b(t)$ and output $a(t)$ are still treated classically, but the dynamics of the quantum object to be observed is stochas-

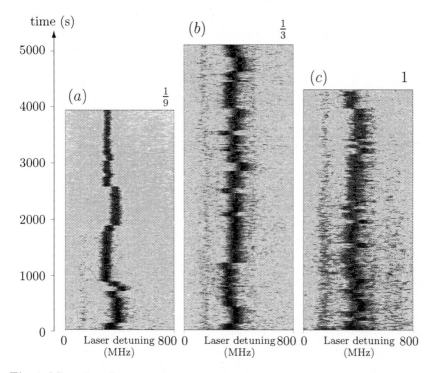

Fig. 1. Migrating absorption frequencies attributed to one single terrylene defect in a hexadecane matrix. The shown migrating absorption lines were measured at different relative laser intensities $\frac{1}{9} : \frac{1}{3} : 1$ (total time almost 4 hours). In other examples of the same material the stochastic behaviour is even more pronounced. The data were measured at February 18, 1994, by Mauro Croci, Viktor Palm, W. E. Moerner and Urs P. Wild. They can be downloaded from http://www.chem.ethz.ch/sms/. This figure is reproduced with permission by Prof. Urs Wild (ETH-Zürich).

tic due to coupling to its quantum environment (which can never be completely screened) *and* due to the measurement of the output observable A (see (4)), which does not leave the pure state of the quantum system unchanged. When an *ensemble* of quantum systems is observed (such as an ensemble of molecules), these stochastic terms average out and need, therefore, not be considered. It is only observation of *single individual* quantum objects, which reveals the interesting stochastic fluctuations, which are not treated in the usual "statistical" formalism of quantum mechanics.

Completely isolated *simple* quantum objects, described by a pure state and the usual Schrödinger dynamics, are, therefore, replaced by *complex* quantum objects, which are described by a pure state evolving under the Schrödinger time evolution *and* an additional stochastic dynamics due to external influences.

4   Approximate Classical Structures
and Approximate System-Theoretic Descriptions

Classical structures and classical descriptions abound: In system-theory, for example, input and output are treated classically, without any holistic correlations with one another and with the observed quantum system under consideration. Or different quantum systems (such as a molecule and the quantum radiation field), which could show holistic correlations, are treated as being entirely separated.

Such a classical way of thinking is, of course, useful; but it is not entirely correct. Strictly speaking, classical structures and descriptions arise only in an appropriate *limit* of infinitely many degrees of freedom, infinite nuclear masses, infinite volume, and so on (Bóna 1988, 1989, Bratteli and Robinson 1981, Primas 1983, Sewell 1986, Werner 1992). In practice, quantum systems have often only *finitely* many degrees of freedom, *finite* nuclear masses, *finite* volume, etc. Hence, quantum fluctuations "around" classical behavior can be expected and should be characterized quantitatively. Such quantum fluctuations "die out" with an increasing number of degrees of freedom, ..., because the corresponding states are *unstable* under external perturbations.

A relatively elementary example has been studied by Amann (1995), namely a Curie-Weiss model consisting of N spins below the Curie temperature, with Hamiltonian

$$H_N = -\frac{J}{2N} \sum_{i,j=1}^{N} \sigma_{z,i}\,\sigma_{z,j} \,. \tag{6}$$

For finite N, the superposition principle of quantum mechanics is fully valid. Therefore, superpositions of two pure states with positive and negative permanent magnetization can – in theory – be written down and also – in principle – be prepared experimentally. Similarly, superpositions of molecular states with different nuclear structure can be experimentally prepared (Cina and Harris 1995, Quack 1986, 1989). Nevertheless, one expects that such superpositions become increasingly unstable with an increasing number of spins (in the magnet) or with increasing nuclear masses in a molecule, etc. For the Curie-Weiss magnet, such stability considerations can be quantitatively characterized by a large-deviation entropy s_β. For precise definitions of the stability requirement, its connection to the dynamics, etc., see Amann (1995). Here only a heuristic version will be given: In an *individual* description, the magnet is in a particular pure state Ψ_t at time t, and this pure state develops in time under the Schrödinger dynamics and additional external perturbations (in a thermal environment with inverse temperature β, chosen in Fig. 2 as three times the *inverse* Curie temperature, $\beta = 3\beta_{\text{Curie}}$). Then the probability $\text{Prob}_\beta[m_1, m_2]$ to find the expectation values

$$\langle \Psi_t | \hat{m}_N \Psi_t \rangle \tag{7}$$

of the specific magnetization operator

$$\hat{m}_N \stackrel{\text{def}}{=} \frac{1}{N} \sum_{j=1}^{N} \sigma_{z,j} \tag{8}$$

for large times t in some interval $[m_1, m_2]$ is given by

$$\text{Prob}_\beta[m_1, m_2] \sim \exp\left\{ -N \inf_{m \in [m_1, m_2]} (s_\beta(m)) \right\}. \tag{9}$$

The infimum in (9) is taken over the values $s_\beta(m)$, where the expectation value m of the specific magnetization operator $\hat{m}_N$ lies in the interval $[m_1, m_2]$. Hence, depending on the chosen interval $[m_1, m_2]$ of specific magnetizations, the probability $\text{Prob}_\beta[m_1, m_2]$ will either go to 0 *or* to 1 exponentially. The latter situation arises if and only if the interval $[m_1, m_2]$ contains one (or both) of the specific magnetization values $+m_\beta$ or $-m_\beta$ of positive or negative permanent magnetization at inverse temperature β. These two values are expectation values of a *classical* magnetization operator $\hat{m}_\infty := \lim_{N \to \infty} \hat{m}_N$ in the *limit of infinitely many spins* (cf. Bóna 1988, 1989).

Therefore, the entropy s_β describes the *quantum fluctuations around classical behavior*. In a *strict* sense, classical behavior arises only in the limit of *infinitely* many spins. Hence, the specific magnetization operator $\hat{m}_N$ is only *approximately* classical for finite N, but converges to a strictly classical observable $\hat{m}_\infty$ in the limit $N \to \infty$.

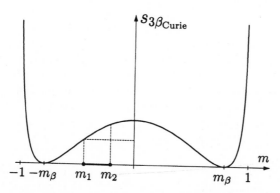

Fig. 2. An entropy function in the sense of large-deviation theory, describing how fast the mean magnetization of a spin system becomes classical with an increasing number of spins. The figure is based on an approximate calculation for the Curie-Weiss model. The temperature is fixed and has been taken here as one third of the critical (Curie) temperature. Above the Curie temperature, the respective entropy s_β would only have one minimum, namely at $m = 0$.

Similarly, all sorts of strictly classical descriptions (such as, e.g., classical input and output in system theory) entail quantum fluctuations which deviate from it. And similarly, any strict exclusion of holistic correlations between two systems (e.g., between the system to be observed and the observing tools, or between a molecule and the radiation field, or between different molecules in a vessel) entail holistic quantum fluctuations, which should be taken into account and characterized quantitatively (by a large-deviation entropy).

Hence, a derivation and description of a *quantum* input-output system needs two different components. First, an appropriate limit of infinitely many degrees of freedom is required to derive classical behavior without any holistic entanglement between input/output-observer and quantum system or without any holistic entanglement between quantum system and its quantum environment. In a second step, the infinite limit should be replaced by a *large, but finite* number of degrees of freedom, thus leading to a description of quantum fluctuations. These quantum fluctuations should be characterized by large-deviation techniques.

Acknowledgment: The author gratefully acknowledges an APART grant from the Austrian Academy of Sciences.

References

Amann A. (1995): Structure, dynamics, and spectroscopy of single molecules: a challenge to quantum mechanics. *J. Math. Chem.* **18**, 247–308.

Amann A. (1996): Can quantum mechanics account for chemical structures? In *Fundamental Principles of Molecular Modeling*, ed. by W. Gans, J. Boeyens, and A. Amann) (Plenum, New York), 55–97.

Amann A. and Primas H. (1996): What is the referent of a non-pure quantum state? In *Experimental Metaphysics – Quantum Mechanical Studies in Honor of Abner Shimony*, ed. by R.S. Cohen and J. Stachel (Kluwer, Dordrecht), in press.

Ambrose W.P. and Moerner W.E. (1991): Fluorescence spectroscopy and spectral diffusion of single impurity molecules in a crystal. *Nature* **349**, 225–227.

Aspect A., Grangier P., and Roger G. (1982): Experimental realization of Einstein-Podolsky-Rosen-Bohm- Gedankenexperiment: a new violation of Bell's inequalities. *Phys. Rev. Lett.* **49**, 91–94.

Atmanspacher H., Wiedenmann G., and Amann A. (1995): Descartes revisited – the endo-exo distinction and its relevance for the study of complex systems. *Complexity* **1**(3), 15–21.

Bóna P. (1988): The dynamics of a class of quantum mean-field theories. *J. Math. Phys.* **29**, 2223–2235.

Bóna P. (1989): Equilibrium states of a class of quantum mean-field theories. *J. Math. Phys.* **30**, 2994–3007.

Bratteli O. and Robinson D.W. (1981): *Operator Algebras and Quantum Statistical Mechanics, Vol. 2* (Springer, Berlin).

Cina J.A. and Harris R.A. (1995): Superpositions of handed wave functions. *Science* **267**, 832–833.

Einstein A., Podolsky B., and Rosen N. (1935): Can quantum-mechanical description of physical reality be considered complete? *Phys. Rev.* **47**, 777–780.

Fermi E. (1961): *Notes on Quantum Mechanics* (The University of Chicago Press, Chicago).

Fick E. and Sauermann G. (1986): *Quantenstatistik dynamischer Prozesse. Band IIa: Antwort- und Relaxationstheorie* (Verlag Harri Deutsch, Thun).

Ghirardi G.C., Rimini A., and Weber T. (1986): Unified dynamics for microscopic and macroscopic systems. *Phys. Rev. D* **34**, 470–491.

Gisin N. (1984): Quantum measurements and stochastic processes. *Phys. Rev. Lett.* **52**, 1657–1660.

Kalman R.E. (1995): Randomness and probability. *Math. Jap.* **41**, 41–58.

Kubo R., Toda M., and Hashitsume N. (1985): *Statistical Physics II* (Springer, Berlin).

Moerner W.E. (1994): Examining nanoenvironments in solids on the scale of a single, isolated impurity molecule. *Science* **265**, 46–53.

Moerner W.E. and Basché T. (1993): Optical spectroscopy of single impurity molecules in solids. *Angew. Chem. Int. Ed.* **32**, 457–476.

Pfeifer P. (1980): *Chiral Molecules - a Superselection Rule Induced by the Radiation Field* (Thesis ETH-Zürich No. 6551, ok Gotthard S+D AG, Zürich).

Primas H. (1983): *Chemistry, Quantum Mechanics, and Reductionism.* (Springer, Berlin).

Primas H. (1990): Induced nonlinear time evolution of open quantum objects. In *Sixty-Two Years of Uncertainty*, ed. by A.I. Miller (Plenum, New York), 259–280.

Primas H. (1994): *Physikalische Chemie A* (Course at the Swiss Federal Institute of Technology, Zürich).

Quack M. (1986): On the measurement of the parity violating difference between enantiomers. *Chem. Phys. Lett.* **132**, 147–153.

Quack M. (1989): Structure and dynamics of chiral molecules. *Angew. Chem. Int. Ed. Engl.* **28**, 571–586.

Sewell G.L. (1986): *Quantum Theory of Collective Phenomena* (Clarendon, Oxford).

Wächter M. (1991): *Zur Lokalisierung makroskopischer Systeme* (Diplomarbeit ETH Zürich).

Werner R.F. (1992): Large deviations and mean-field quantum systems. In *Quantum Probability and Related Topics, Vol. VII*, ed. by L. Accardi (World Scientific, Singapore), 349–381.

Wild U.P., Croci M., Güttler F., Pirotta M., and Renn A. (1994): Single molecule spectroscopy: Stark-, pressure-, polarization-effects and fluorescence lifetime measurements. *J. Luminescence* **60 & 61**, 1003–1007.

Decoherence and Quantum-Classical Correspondence in Chaotic Systems

F. Tito Arecchi[1,2]

[1] Department of Physics, University of Florence, I–50125 Florence, Italy,
[2] Istituto Nazionale di Ottica, Largo E. Fermi 6, I–50125 Florence, Italy

1 Introduction

Given a system with a limited number of degrees of freedom, together with its interaction with the environment; can one evaluate a critical time (decoherence time) within which the system behaves fully quantum mechanically (including Einstein-Podolsky-Rosen (EPR)–entanglements), and after which it displays quasi-classical features? And if the system is isolated, but its classical limit displays chaotic behavior, would this be a sufficient replacement for the environment? A general positive answer to these questions would settle many open issues.

First of all, quantum vs. classical (Q vs. C) would not be a matter of logical interpretation, but of real physical behavior on different time scales. Second, it is not the size of the system which decides about Q vs. C, but rather the time scale, hence, even macroscopic objects can display Q behavior on suitable time ranges, as put forward by Leggett and Garg (1985) and Leggett (1987). Third, even in isolation from a thermal environment, an intrinsically chaotic dynamics could introduce some classical aspects. In other words, decoherence may play a role even without environment.

This paper aims at answering this last question, which, in the spirit of the title of this volume, could be reformulated as "The time edge between quantum and classical behavior in the presence of chaos". It is organized as follows. In Sects. 2 and 3, I review the meaning and the physical implications of "decoherence", following the approach of Zurek (Zurek 1982, 1991, Paz et al. 1993, Zurek and Paz 1994). Sections 4 and 5 carry the central message of this paper. As shown recently by my research group (Farini et al. 1996), whenever a quantum system has classically chaotic dynamics, this introduces a new characteristic time scale represented by the coherence time of the local direction of maximum expansion. For a quantum system whose classical limit yields chaotic dynamics, the ratio between this new time scale and the decorrelation time (of the order of the reciprocal of the maximum Ljapunov exponent) rules the ratio between non-classical (Moyal) and classical (Liouville) terms in the evolution of the density matrix. However, such a ratio does not provide a complete criterion for quantum-classical correspondence, since a very small amount of Moyal contributions, interpreted in time, provides an asymptotic deviation from purely classical behavior. This fact represents a problem which has yet to be settled.

2   Quantum Superposition and Decoherence

Quantum theory allows many more states to be realized than we encounter. Moreover, quantum dynamics takes simple, localized initial states of individual systems into complicated nonlocal superpositions. We do not perceive most of such superpositions. Macroscopic objects always appear to us in a small classical subset of a much larger quantum menu which is in principle available in Hilbert space. This point was made by Einstein, who, in a letter to Born, wrote in 1954 (Einstein 1969): "Let ψ_1 and ψ_2 be solutions of the same Schrödinger equation. ... When the system is macroscopic and ψ_1 and ψ_2 are narrow with respect to the macrocoordinates, then ... (typically) this is no longer the case for $\psi = \psi_1 + \psi_2$. Narrowness with respect to macrocoordinates is not only independent of the principles of quantum mechanics, but, moreover, incompatible with them...." Hence, predictions of quantum theory seem to be in conflict with our perceptions.

The key point of the decoherence approach is simple: there is a basic difference between the predictions of quantum theory for quantum systems which are closed (isolated) and open (interacting with their environments). In case of a closed system, the Schrödinger equation and the superposition principle apply literally. In contrast, the superposition principle is not valid for open quantum systems. Here the relevant physics is quite different, as has been shown by many examples in the context of condensed matter physics, quantum chemistry, etc. The evolution of open quantum systems has to be described in a way violating the assumption that each state in the Hilbert space of a closed system is equally significant. Decoherence is a negative selection process which dynamically eliminates non-classical states.

The distinguishing feature of classical systems, the essence of "classical reality", is the persistence of their properties – the ability of systems to exist in predictably evolving states, to follow a trajectory which may be chaotic, but is deterministic. This suggests the relative stability – or, more generally, predictability – of the evolution of quantum states as a criterion which decides whether they will be repeatedly encountered by an observer and can be used as ingredients of a "classical reality". The characteristic feature of the decoherence process is that a generic initial state will be dramatically altered on a characteristic decoherence time scale: only certain stable states will be left on the scene.

Quantum measurement is a classic example of a situation in which a coupling of a macroscopic quantum apparatus A and a microscopic measured system S forces the composite system into a correlated, but usually exceedingly unstable state. In a notation where $|A_0\rangle$ is the initial state of the apparatus and $|\psi\rangle$ the initial state of the system, the evolution establishing an A - S correlation is described by:

$$|\psi\rangle|A_0\rangle = \sum_k \alpha_k |\sigma_k\rangle|A_0\rangle \rightarrow \sum_k \alpha_k |\sigma_k\rangle|A_k\rangle = |\Phi\rangle. \tag{1}$$

An example is the Stern-Gerlach apparatus. There the states $|\sigma_k\rangle$ describe orientations of the spin and the states $|A_k\rangle$ are the spatial wavefunctions centered on the trajectories corresponding to different eigenstates of the spin. When the separation of the beams is large, the overlap between them tends to zero ($\langle A_k|A_{k'}\rangle \sim \delta_{kk'}$). This is a precondition for a good measurement. Moreover, when the apparatus is not consulted, A - S correlations would lead to a mixed density matrix for the system S:

$$\rho_s = \sum_k |\alpha_k|^2 |\sigma_k\rangle\langle\sigma_k| = \mathrm{Tr}\,|\Phi\rangle\langle\Phi|. \tag{2}$$

However, this pre-measurement quantum correlation does not provide a sufficient foundation to build a correspondence between the quantum formalism and the familiar classical reality. It only allows for Einstein-Podolsky-Rosen quantum correlations between A and S, which imply the entanglement of an arbitrary state – including non-local, non-classical superpositions of the localized status of the apparatus (observer) – with the corresponding relative state of the other system. This is a prescription for a Schrödinger cat, not a resolution of the measurement problem. What is needed is a fixed set of states in which classical systems can safely exist. What is needed, therefore, is an effective superselection rule which "outlaws" superpositions of these preferred "pointer states". This rule cannot be absolute: there must be a time scale sufficiently short, or an interaction strong enough, to render it invalid, for otherwise measurements could not be performed at all. Superselection should become more effective when the size of the system increases. It should apply, in general, to all objects (not just idealized models of a quantum apparatus) and allow us to deduce elements of our familiar reality – including the spatial localization of macroscopic systems – from Hamiltonians.

Environment-induced decoherence has been proposed to fit these requirements. The transition from a pure state $|\Phi\rangle\langle\Phi|$ to the effectively mixed ρ_{AS} can be accomplished by coupling the apparatus A to the environment ϵ. The requirement to get rid of unwanted, excessive, EPR-like correlations (1) is equivalent to the demand that the correlations between the pointer states of the apparatus and the measured system ought to be preserved in spite of an incessant measurement-like interaction between the apparatus pointer and the environment. In simple models of the apparatus this can be assured by postulating the existence of a pointer observable with eigenstates (or, more precisely, eigenspaces) which remain unperturbed during the evolution of the open system. This "nondemolition" requirement will be exactly satisfied when the pointer observable O commutes with the total Hamiltonian generating the evolution of the system:

$$[(H + H_{\mathrm{int}}), O] = 0. \tag{3}$$

For an idealized quantum apparatus this condition can be assumed to be satisfied, and – provided that the apparatus is in one of the eigenstates of O – leads to an uneventful evolution:

$$|A_k\rangle|\epsilon_0\rangle \rightarrow |A_k\rangle|\epsilon_k(t)\rangle. \tag{4}$$

However, when the initial state is a superposition corresponding to different eigenstates of O, the environment will evolve into an $|A_k\rangle$-dependent state:

$$\left(\sum_k \alpha_k|A_k\rangle\right)|\epsilon_0\rangle \rightarrow \sum_k \alpha_k|A_k\rangle|\epsilon_k(t)\rangle. \tag{5}$$

The decay of the interference terms is inevitable. The environment causes decoherence only when the apparatus is forced into a superposition of states, which are distinguished by their effect on the environment. The resulting continuous destruction of the interference between the eigenstates of O leads to an effective environment-induced superselection. Only states which are stable in spite of decoherence can exist long enough to be accessed by an observer so that they can count as elements of our familiar, reliably existing reality.

Effective reduction of the state vector follows immediately. When the environment becomes correlated with the apparatus,

$$|\Phi\rangle|\epsilon_0\rangle \rightarrow \sum_k \alpha_k|A_k\rangle|\sigma_k\rangle|\epsilon_k(t)\rangle = |\Psi\rangle, \tag{6}$$

but the apparatus is not consulted (so that it must be traced out), we have:

$$\rho_{AS} = \mathrm{Tr}|\Psi\rangle\langle\Psi| = \sum_k |\alpha_k|^2|A_k\rangle\langle A_k||\sigma_k\rangle\langle\sigma_k|. \tag{7}$$

Only correlations between the pointer states and the corresponding relative states of the system retain their predictive validity. This form of ρ_{AS} follows, provided that the environment becomes correlated with the set of states $\{|A_k\rangle\}$ (it could be any other set) and that it has acted as a good measuring apparatus, so that $\langle\epsilon_k(t)|\epsilon_{k'}(t)\rangle = \delta_{kk'}$ (the states of the environment and the different outcomes are orthogonal).

3   Reduction of the Wavepacket Due to the Environment

Can a similar process be responsible for the classical behavior of systems which cannot be idealized as simply as an abstract apparatus? The crucial difference arises from the fact that, in general, there will be no observable which commutes with both parts of the total Hamiltonian $H + H_{\mathrm{int}}$. Thus, all states – and all correlations – will evolve on some time scale. The distinction between various states will now have to be quantitative rather than qualitative: the majority of states will deteriorate on the decoherence time scale. This is the time required for the "reduction of the wavepacket". For non-classical states of macroscopic objects, it is many orders of magnitude shorter than the dynamical time scale – so short that, from the point of view of an observer, responding on his dynamical time scale, it can be regarded as instantaneous.

The interaction with the environment will continue to play a crucial role: monitoring of the classical observable by the environment is still the process responsible for decoherence, and H_{int} determines the set of states which leave distinguishable imprints in the environment. For example, the commutation condition (3) for the interaction Hamiltonian alone explains the approximate localization of classical states of macroscopic objects: the environment is coupled almost always through the coordinate x (interactions depend on distance) and, therefore, states which are localized will be favored: this feature of preferred states follows from the form of the interaction alone; it does not need to be put in ad hoc.

However, the kinetic term $(p^2/2m)$ in the Hamiltonian does not commute with the position observable. Therefore, exact position eigenstates – which have to be completely "nonlocal in momentum" – are also unstable. We need a more systematic procedure to filter out non-classical states. A natural generalization of absolutely stable pointer states of the apparatus are the most predictable states of less idealized open quantum systems. An algorithm for "trying out" all the states in the Hilbert space can be readily outlined. For each candidate of a pure initial state we can

1. calculate the density matrix which is provided by its evolution in contact with the environment,
2. compute its entropy as a function of time, and
3. at some instant much in excess of a typical decoherence time scale, construct a list of the pure initial states ordered according to how much entropy was generated, or how much predictability was lost in the process.

The most predictable states near the top of the list would be, in effect, most classical. A similar procedure can be used to compare the predictability of mixed initial states.

Such a "predictability sieve" was recently proposed (Zurek 1993) and implemented for a harmonic oscillator with the resulting evolution of the reduced density matrix generated by the appropriate master equation (Caldeira and Leggett 1983). For a weakly damped harmonic oscillator, pure states selected by the predictability sieve turn out to be the familiar coherent states. This is true in spite of the fact that the stability of the states is crucially influenced by their dispersion in position. For instance, for pure states the rate of purity loss (quantified by the linear entropy given by $\operatorname{Tr} \rho^2$) is proportional to their dispersion in x,

$$\frac{d}{dt}\mathrm{Tr}\,\rho^2 \sim \left(\langle\psi|x^2|\psi\rangle - |\langle\psi|x|\psi\rangle|^2\right) . \tag{8}$$

Most predictable mixtures turn out to be Gaussian and – for weakly damped oscillators – symmetric in x and p.

An issue raised by critics is the applicability of the decoherence approach to the Universe as a whole and its relation to the consistent histories approach (Griffiths 1984, Omnès 1992, Gell-Mann and Hartle 1993). The Universe is a closed system, so it does not have an environment. However, macroscopic subsystems within it (including recording apparatuses and observers) do have environments, and the decoherence program can be implemented in such a setting. The projection operators which define sequences of events in the consistent histories approach would then have to satisfy more than just the probability sum rules. Rather, the process of decoherence singles out events and observables which become "recorded" as a result of environmental monitoring. For example, when a well-defined pointer basis exists, the histories consisting of sequences of pointer states are consistent. Thus, the additivity of probabilities of histories expressed in terms of the "usual" observables appears to be guaranteed by the efficiency with which unstable states and the corresponding off-diagonal terms of the density matrix in the preferred pointer basis representation are removed by a coupling with the environment.

Let us consider a system S ruled by a Hamiltonian H_0 and coupled to the environment through the term

$$H' = \nu x E, \tag{9}$$

where ν is the coupling strength, x is a coordinate of the system, and E is an environment operator. As we trace the overall density operator over an ensemble of environments with temperature T, the system's density matrix in the coordinate representation, $\rho(x, x')$, evolves according to the following master equation (Caldeira and Leggett 1983):

$$\frac{d\rho}{dt} = \frac{1}{i\hbar}[H_0, \rho] - \gamma(x - x')(\partial_x - \partial_{x'})\rho - \eta\frac{kT}{\hbar^2}(x - x')^2\rho, \tag{10}$$

where $\eta := \nu^2/2$, and $\gamma := \eta/2m$ is the drift coefficient which rules the evolution of the first moments.

"Negative selection" consists in the rapid decay of the off-diagonal elements of $\rho(x, x')$. Indeed, for $\hbar \to 0$, the last term on the right hand side of (10) prevails, providing the solution

$$\rho(x, x', t) = \rho(x, x', 0)\exp\left(-\eta\frac{kT}{\hbar^2}(x - x')^2 t\right). \tag{11}$$

With $\Delta x = x - x'$, we see that an initial offset $\rho(x, x', 0)$ decays after a decoherence time

$$\tau_D = \frac{1}{\gamma} \left(\frac{\lambda_{DB}}{\Delta x} \right)^2, \tag{12}$$

where

$$\lambda_{DB} = \frac{\hbar}{p} = \frac{\hbar}{\sqrt{2mkT}} \tag{13}$$

is the thermal deBroglie length. At length scales $\Delta x \gg \lambda_{DB}$, we have $\tau_D \gg 1/\gamma$, such that the system decoheres rapidly and then continues with the standard Brownian decay on the time scale $1/\gamma$.

4   Decoherence and Chaos

A quantum classical correspondence (QCC) implies an answer to the question of the limitations imposed by the quantum nature of the system to the measuring process (von Neumann 1983). In particular, it has recently been shown (Zurek and Paz 1994) that the ratio between non-classical and classical terms in the evolution equation of the phase space density diverges for an unstable motion, whereas it decays to zero if one accounts for a coupling with the environment. This decay was taken as a definition of QCC. In the following section we present an example in which

1. the mentioned ratio remains confined to very small values even for an isolated system (i.e., in the absence of an environment) because of the intrinsic spread due to chaotic motion, and yet
2. a QCC, defined more rigorously by the absence of appreciable differences between classical and quantum phase space densities, is not obtained since for long times the quantum phase space density shows appreciable deviations from the classical one.

Claim (1) is based on the introduction of a novel indicator for chaos, not considered previously in classical chaos. Claim (2) is supported by numerical evidence. More precisely, we refer to a classically chaotic, nonautonomous system (i.e., a system with a time dependent forcing term), which models a hydrogen atom in a Rydberg state excited by a microwave field (Bayfield and Koch 1974). The quantum evolution is described by the quasi-probability Wigner function (Wigner 1932, Hillery et al. 1984)

$$W(q, p) = \frac{1}{2\pi\hbar} \int_{-\infty}^{\infty} dy \, e^{\frac{i}{\hbar}py} \left\langle q - \frac{y}{2} \middle| \hat{\rho} \middle| q + \frac{y}{2} \right\rangle, \tag{14}$$

where $\hat{\rho}$ is the density operator of the system, p and q are the conjugate variables and $\langle \ldots \rangle$ stands for the expectation value.

The Wigner function can take negative values, thus it is not a probability function (Wigner 1932, Hillery et al. 1984). However, it represents a good tool for inspecting the classical or quantum nature of the system. Its time evolution is given by

$$\frac{dW}{dt} = \{H, W\}_{\text{PB}} + \sum_{n \geq 1} \frac{\hbar^{2n}(-1)^n}{2^{2n}(2n+1)!} \partial_x^{2n+1} V \partial_p^{2n+1} W, \qquad (15)$$

where H is the Hamiltonian of the system written in the Wigner form (Degroot 1974, Takahashi 1989), V is the potential, $\{\ldots\}_{\text{PB}}$ stands for the Poisson bracket, and ∂_x, ∂_p denote $\partial/\partial x, \partial/\partial p$. Notice that the right hand side of (15) contains two different terms which contribute to the evolution of W. The Poisson bracket generates the ordinary Liouville flow corresponding to the classical evolution of W, while the sum represents the Moyal contribution (Moyal 1949), reflecting the quantum evolution of W.

In Zurek and Paz (1994), the Moyal terms are comparable to the Poisson bracket after a time τ_1 proportional to the reciprocal of the maximum Ljapunov exponent Λ. In their example, the growth of Moyal contributions is quenched by the coupling to a thermal bath, which involves further diffusive and dissipative terms to be added to (2). Hence, dissipation and diffusion can guarantee that the Moyal terms vanish on long time scales, thus establishing a correspondence between quantum and classical descriptions. In the following, we shall discuss sufficient conditions for the confinement of Moyal contributions within a limited range for isolated systems.

5   The Twisting Rate as a New Indicator of Chaos, and Its Role in QCC

Let us introduce one of the simplest chaotic dynamics which gives rise to non-zero Moyal contributions. The Hamiltonian of the system is

$$H = \frac{p^2}{2m} + V(x, t), \qquad (16)$$

where m is the mass of the system and $V(x,t) = V_1(x,t) + V_2(x,t)$; $V_1 = -\alpha x^2/2 + \beta x^4/4$ is the unperturbed potential, $V_2 = \varepsilon x \cos \omega t$ is the driving potential, and $\alpha, \beta, \varepsilon, \omega$ are real parameters to be specified later. For the Hamiltonian (16), (15) becomes:

$$\frac{dW}{dt} = (-\alpha x + \beta x^3 + \varepsilon \cos \omega t)\frac{\partial W}{\partial p} - \frac{p}{m}\frac{\partial W}{\partial x} - \frac{\hbar^2 \beta x}{4}\frac{\partial^3 W}{\partial p^3}. \qquad (17)$$

The classical behavior is ruled by the Hamilton-Jacobi equations

$$\frac{dx}{dt} = \frac{\partial H}{\partial p} = \frac{p}{m},$$

$$\frac{dp}{dt} = -\frac{\partial H}{\partial x} = \alpha x - \beta x^3 - \varepsilon \cos \omega t. \qquad (18)$$

These equations can give rise to chaotic dynamics. Furthermore, choosing $\omega \gg \sqrt{\alpha/m}$ (the forcing frequency is much higher than the proper frequency of the autonomous part) preserves (on average) the conservative nature of the system.

The guideline of our approach is that, for a bounded chaotic system, another characteristic time scale in addition to τ_1 arises due to the continuous stretching and folding process which causes a twisting of the direction of the eigenvector corresponding to the local maximum Ljapunov exponent Λ. As a consequence, the system does not always expand in the same direction of phase space, but each variable experiences both contraction and expansion depending upon the local position along the trajectory. Thus, besides the decorrelation time τ_1, there is an additional time scale τ_2, proportional to the reciprocal of the averaged frequency $\langle \Omega \rangle$ of twisting of the local direction corresponding to the maximum expansion rate.

Depending on the values of τ_1 and τ_2, we can classify chaotic dynamics as follows. In systems for which $\tau_1 \leq \tau_2$ ($\Lambda \geq \langle \Omega \rangle$), the direction of the maximal Ljapunov exponent remains essentially invariant on the time scale which is sufficient for Moyal contributions to become comparable with the Liouville term. Hence, these systems are intrinsically quantum. A transition from quantum to classical description can be achieved only through coupling to an environment (Zurek and Paz 1994).

If $\tau_1 \gg \tau_2$ ($\Lambda \ll \langle \Omega \rangle$), the growth of Moyal terms within the decorrelation time is limited by the change of the direction of maximum expansion. Indeed, the growth of Moyal terms is due to the fact that – since they are proportional to powers of the reciprocal of $\Delta x (\Delta p)$ – if one of the two conjugate variables is locally expanding, the other one is contracting to preserve the volume in phase space. As a consequence, those terms containing powers of the contracting variable in the denominator tend to explode. To apply these general considerations to specific chaotic dynamics, we first identify the range of control parameters for which the system belongs to one of the two cases, respectively.

For the classical evolution of the driven double-well system (18), Fig. 1 shows Λ and $\langle \Omega \rangle$ as a function of the control parameter ε. To determine $\langle \Omega \rangle$, we have introduced the following strategy. At each time t_n we consider a circle of radius η centered at the intersection of the actual trajectory with a plane transversal to the flow. In our case, this plane coincides with the two dimensional (x, p) projection of the phase space for any time. We let all the points on the circle evolve to time t_{n+1} ($t_{n+1} - t_n = t_{RK}$, where t_{RK} is the Runge-Kutta integration interval). At t_{n+1}, the points will form an ellipse around the evolved actual trajectory. The direction of the local eigenvector corresponding to the maximal Ljapunov exponent then coincides with the direction of the major axis of this ellipse. For fixed $\varepsilon = 0.7$, the inset of Fig. 1 shows the temporal evolution of the angle ϑ formed by the major axis of the local ellipse with the x axis. The twisting process determines a characteristic

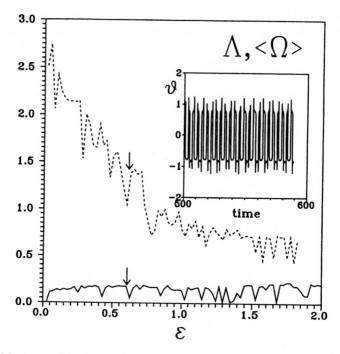

Fig. 1. Maximum Ljapunov exponent Λ (solid line) and averaged twisting frequency $\langle \Omega \rangle$ (dashed line) as functions of the forcing amplitude ε for the driven double-well system. Λ and $\langle \Omega \rangle$ have been evaluated for the classical system (18) with $\alpha = 1, \beta = 2, m = 1, \omega = 2.5$. Inset: temporal evolution of the angle ϑ formed by the eigenvector corresponding to the local direction of maximum expansion with respect to the x axis. The angle ϑ has been calculated for $\varepsilon = 0.7$ (arrows) and the above values of the other parameters.

oscillation of ϑ, the averaged frequency of which is $\langle \Omega \rangle$. Figure 1 clearly shows that $\tau_1 \gg \tau_2$ ($\Lambda \ll \langle \Omega \rangle$) in the range of small control parameters ε.

To discuss the quantum behavior of the system in more detail, let us focus on the case $\tau_1 \gg \tau_2$ to show that the Moyal terms are limited so that their growth cannot be made responsible for the lack of QCC. For this purpose, (17) has been numerically integrated over a two-dimensional $N \times N$ (x, p)-grid, where the maximum and minimum values of x and p were selected according to the choice of α and β. At each time t_n, the Liouville terms have been evaluated using the Lax-Wendroff scheme (Mitchell and Griffiths 1980), while Moyal contributions have been evaluated using a finite difference method for the third order partial derivatives (Davis and Polonsky 1970). Boundary conditions have been chosen to imply vanishing derivatives at the borders of the grid. The numerical integration has been performed over a global time much larger than $1/\omega$ to assure the conservative nature of the quantum solution.

A further useful parameter for our purpose is the logarithmic time (Berry et al. 1979)

$$t_{\log} = \frac{1}{\Lambda} \ln \left(\frac{S}{\hbar} \right). \tag{19}$$

In our case, taking S as the action per bounce in the double well potential at the threshold orbit and $\hbar = 1/200\pi$ (Helmkamp and Browne 1994), the argument of the logarithm is of the order of $200 - 300$. So, for $1/\Lambda$ as in Fig. 1 we obtain $t_{\log} = 50 - 60$.

Due to an argument by Berry et al. (1979), integration over a time much longer than $t_{\log}$ should yield appreciable differences between quantum and classical features of the system. Denoting by M (L) the Moyal (Liouville) term of the evolution equation for $\tau_1 \gg \tau_2$, i.e., for $\Lambda \ll \langle \Omega \rangle$, one should expect that M (L) be confined within an amplitude range $M_0 \exp\{3\Lambda/\langle\Omega\rangle\}$ ($L_0 \exp\{\Lambda/\langle\Omega\rangle\}$), where M_0 (L_0) is the initial amplitude of the Moyal (Liouville) term. The reason is that the expansion of M (L) due to the contraction of Δx or Δp lasts only for a time $\tau_2 = 1/\langle\Omega\rangle$. Thus, the ratio

$$\frac{|M|}{|L|} \sim \frac{|M_0|}{|L_0|} \exp \left(\frac{2|\Lambda|}{\langle\Omega\rangle} \right) \tag{20}$$

can be limited by a suitable choice of the ratio $\Lambda/\langle\Omega\rangle$. Notice that the ratio $|M_0|/|L_0|$ can be arbitrarily chosen by selecting a suitable classical initial state. Equation (20) shows that, for $\Lambda \ll \langle\Omega\rangle$, the ratio of Moyal to Liouville terms can be adjusted to a pre-assigned value. This limitation is guaranteed for all values of $\hbar$ which give rise to negligible ratios $|M_0|/|L_0|$.

Figure 2 shows the maxima of Moyal and Liouville terms as a function of time. Although there are no diffusive and dissipative terms in (17), Moyal contributions are always much smaller than the Liouville terms. This result is at variance with the expectation of Zurek and Paz (1994). Note that this limitation is obtained in a range of ε (extracted from Fig. 1), for which the classical chaotic loss of directional coherence of the maximum expansion is much faster than the decorrelation time.

Yet the small size of the Moyal term does not necessarily imply a small dynamical effect, i.e., it is not a sufficient condition for a QCC in the sense of an "absence of appreciable differences between classical and quantum phase space densities". To compare quantum and classical systems, we have calculated *at each integration time* the differences δW between the quantum Wigner function W_Q, obtained by integration of (17), and the classical Wigner function W_C, obtained from (17) without the Moyal term,

$$\delta W(x, p, t) = W_Q(x, p, t) - W_C(x, p, t). \tag{21}$$

We integrate δW over the grid, obtaining $\int \delta W(x, p, t) \, dx \, dp$, and normalize with respect to the grid integral of the Wigner distribution. Denoting such a

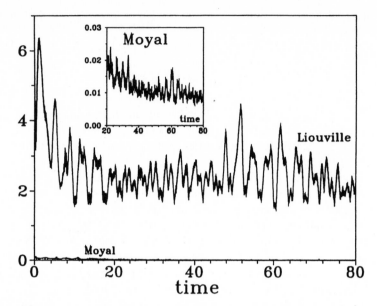

Fig. 2. Maxima of Liouville and Moyal terms as a function of time for $\varepsilon = 0.7$, $\hbar = 1/200\pi$, other parameters as in Fig. 1. A Gaussian wave packet centered at $x = 0.3$ and $p = 0.1$ with $W(q,p) = (1/\pi\hbar)\exp\{(x - x_0)^2/\delta^2 - (p - p_0)^2\delta^2/\hbar^2\}$ ($\delta = 0.1$) has been chosen as initial condition. Boundary conditions are as discussed in the text. Inset: vertically magnified plot of Moyal contributions. Notice that the window confining the Moyal terms remains limited even for $t > \tau_1 \approx (1/\Lambda)\ln(S/\hbar) \approx 50$.

ratio by ΔW, we plot its temporal evolution in Fig. 3. If $\langle M \rangle$ is the average of M over the grid, then ΔW is proportional to $\int \langle M \rangle dt$.

Since $\langle \Omega \rangle / \Lambda > 1$, the maxima of the Moyal term ($M_{\max}$) remain limited over a time longer than any relevant time scale of the system. Thus, ΔW has an upper limit smaller than $M_{\max} t$. Indeed, the temporal evolution of ΔW is approximately fitted by a linear increase (Fig. 3).

For more complex systems, even though $\langle \Omega \rangle$ is still a relevant indicator, the direction of maximum expansion generally does not lie in the (x, p)-plane. Our argument – that dynamical aspects have to be considered together with the quantum conjugacy of x and p – is then limited to a single degree of freedom x, p with a forcing term. Nevertheless, it is descriptive for a large test class of candidates for quantum chaos (Bayfield and Koch 1974).

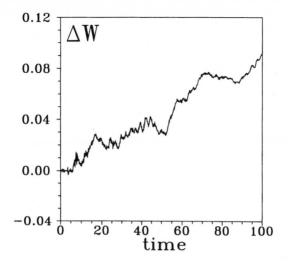

Fig. 3. Differences ΔW (see text) between quantum and classical integrated Wigner functions extracted from the solution of (17) with $\varepsilon = 0.7$ and $\hbar = 1/200\pi$. Boundary and initial conditions as in Fig. 2.

References

Bayfield J.E. and Koch P.M. (1974): Multiphoton ionization of highly excited hydrogen atoms. *Phys. Rev. Lett.* **33**, 258–261.

Berry M.V., Balazs N.L., Tabor M., and Voros A. (1979): Quantum maps. *Ann. Phys.* **122**, 26–63.

Caldeira A.O. and Leggett A.J. (1983): Path integral approach to quantum Brownian motion. *Physica A* **121**, 587–616.

Davis P.J. and Polonsky I. (1970): Numerical interpolation, diiferentiation, and integration. In *Handbook of Mathematical Functions*, ed. by M. Abramowitz and I.A. Stegun (Dover, New York), p. 884.

Degroot S.R. (1974): *La transformation de Weyl et la fonction de Wigner: une forme alternative de la méchanique quantique* (Les Presses Universitaires de Montréal, Montréal).

Einstein A. (1969): In *The Born-Einstein letters*, ed. by M. Born, translated by I. Born (Walker, New York), p. 213.

Farini A., Boccaletti S., and Arecchi F.T. (1996): Quantum-classical comparison in chaotic systems. *Phys. Rev. E* **53**, 4447–4450.

Gell-Mann M. and Hartle J.B. (1993): Classical equations for quantum systems. *Phys. Rev. D* **47**, 3345–3382.

Griffiths R.B. (1984): Consistent histories and the interpretation of quantum mechanics. *J. Stat. Phys.* **36**, 219–272.

Helmkamp B.S. and Browne D.A. (1994): Structures in classical phase space and quantum chaotic dynamics. *Phys. Rev. E* **49**, 1831–1839.

Hillery M., O'Connell R.F., Scully M.O., and Wigner E.P. (1984): Distribution functiosn in physics: fundamentals. *Phys. Rep.* **106**, 121–167.

Leggett A.J. (1987): Experimental approaches to the quantum measurement paradox. *Found. Phys.* **18**, 939–952.

Leggett A.J. and Garg A. (1985): Quantum mechanics versus macroscopic realism: is the flux there when nobody looks? *Phys. Rev. Lett.* **54**, 857–860.

Moyal J.E. (1949): Quantum mechanics as a statistical theory. *Proc. Cambridge Phil. Soc.* **45**, 99–124.

Mitchell A.R. and Griffiths D.F. (1980): *The Finite Difference Method in Partial Differential Equations* (Wiley, New York).

Neumann J. von (1983): "Measurement and reversibility" and "The measuring process". In *Quantum Theory and Measurement*, ed. by J.A. Wheeler and W.H. Zurek (Princeton University Press, Princeton), 549–647 (originally chapters V and VI in J. von Neumann, *Mathematische Grundlagen der Quantenmechanik*, Springer, Berlin 1932).

Omnès R. (1992): Consistent interpretations of quantum mechanics. *Rev. Mod. Phys.* **64**, 339–382.

J.P. Paz J.P., Habib S., and Zurek W.H. (1993): Reduction of the wave packet: preferred observables and decoherence time scale. *Phys. Rev. D* **47**, 488–501.

Takahashi K. (1989): Distribution functions in classical and quantum mechanics. *Prog. Theor. Phys. Supp.* **98**, 109–156.

Wigner E.P. (1932): On the quantum correction for thermodynamic equilibrium. *Phys. Rev.* **40**, 749–759.

Zurek W.H. (1982): Environment-induced superselection rules. *Phys. Rev. D* **26**, 1862–1880.

Zurek W.H. (1991): Decoherence and the transition from quantum to classical. *Physics Today*, October 1991, 36–44. See also comments on this article by J. Anderson, G.C. Ghirardi *et al.*, N. Gisin, D. Albert and G. Feinberg, P. Holland, V. Ambegaokar, and K.J. Epstein together with Zurek's reply in *Physics Today*, April 1993, 13–15, 81–90.

Zurek W.H. (1993): Preferred states, predictability, classicality, and the environment-induced decoherence. *Progr. Theor. Phys.* **89**, 281–312.

Zurek W.H. and Paz J.P. (1994): Decoherence, chaos, and the second law. *Phys. Rev. Lett.* **72**, 2508–2511.

Spectral Decomposition and Extended Formulation of Unstable Dynamical Systems

Ioannis Antoniou[1,2] and Zdzislaw Suchanecki[1,3]

[1] International Solvay Institute for Physics and Chemistry, CP 231, ULB,
Campus Plaine, Boulevard du Triomphe, 1050 Brussels, Belgium
[2] Theoretische Natuurkunde, Free University of Brussels, 1050 Brussels, Belgium
[3] Hugo Steinhaus Center and Institute of Mathematics,
Wroclaw Technical University, Wroclaw, Poland

Abstract. The work of the Brussels-Austin groups over the last six years has demonstrated that for unstable systems, classical or quantum, there exist spectral decompositions of the evolution in terms of resonances and resonance states which appear as eigenvalues and eigenprojections of the evolution operator. These new spectral decompositions are non-trivial only for unstable systems and define an extension of the evolution to suitable dual pairs. Duality here is the states/observables duality and the spectral decompositions extend the algebraic approach to dynamical systems to an intrinsically probabilistic and irreversible formulation. The extended formulation allows for probabilistic prediction and control beyond the traditional local techniques.

1 Introduction

The well-known problem of irreversibility versus dynamics (Prigogine 1980) means that the conventional formulation of classical and quantum dynamics is incomplete. The question is therefore, how to extend the conventional formulation of dynamics in order to include irreversible processes like diffusion, heat conduction, chemical reactions, decay of unstable states, etc. The conventional formulation of unstable dynamical systems is also incomplete with respect to the question of integration. Not only integrability is undecidable but the actual integration of integrable systems is also not an algorithmically computable problem (Pour-el and Richards 1989, da Costa and Doria 1991). Recent work of the Brussels-Austin groups (Petrosky and Prigogine 1991, Petrosky et al. 1991, Prigogine 1992, Antoniou and Prigogine 1993, Antoniou and Tasaki 1992, 1993a, 1993b) has shown that both problems share a common solution, namely the existence and construction of complex spectral decompositions of the dynamical evolution operators.

These new spectral decompositions are actually decompositions in terms of resonances and include correlation decay rates, lifetimes, and Lyapunov times. In this sense they provide a natural representation for the evolution of unstable dynamical systems. The decompositions acquire meaning in suitable dual pairs of functional spaces beyond the conventional Hilbert space framework. For invertible systems, the reversible evolution group, once extended to the functional space, splits into two distinct semigroups. Irreversibility emerges therefore naturally as the selection of the semigroup corresponding to future observations. The resonances are the singularities of the extended resolvent of the evolution operator while the resonance states are the corresponding Riesz projections computed as the residues of the extended resolvent at the singularities (Antoniou et al. 1996a, Bandtlow et al. 1996a).

In this article we present earlier as well as some new results concerning this novel approach to the problem of irreversibility through the spectral theory of extension of unstable dynamics. In Sect. 2 we present the basic concepts of the operator theoretic algebraic approach to dynamical instability and the spectrum of dynamical systems. We define the extension to dual pairs and the spectral decomposition in terms of resonances. and apply them to the example of the Renyi map. In Sect. 3, we discuss the precise meaning of spectral decompositions and the problem of how to choose suitable rigged Hilbert spaces allowing such a decomposition. Section 4 addresses the issue of quantum systems with diagonal singularities, their algebras of observables, their state spaces, and their temporal evolution. Finally, the logic of unstable dynamical systems is described in Sect. 5.

2   Dynamical Instability and the Spectrum of Dynamical Systems

Liouville established in 1855 (Arnold 1978) that a Hamiltonian system with N degrees of freedom can be integrated through a canonical transformation to a solvable system, if N constants of motion in involution exist. Despite the fact that all one-dimensional systems can be integrated, the integration program faces difficulties for systems with $N \geq 2$ degrees of freedom. Attempts to construct canonical transformations using perturbation theory formulated as in the one-dimensional case lead to divergences, due to the famous problem of small denominators. Most interesting problems of mechanics starting from the three-body problem (sun, earth and moon or jupiter) do not possess analytic constants of motion. Poincaré's theorem (Poincaré 1892) made clear that the reason for the non-existence of analytic canonical transformations to a solvable system, is the presence of resonances.

Poincaré's theorem was a turning point in the development of mechanics. The role of resonances became clear, however, only in the late 1950's from the Kolmogorov, Arnold and Moser theorem (Arnold 1978). It should be emphasized however, that the KAM theorem did not solve the problem of integration of Poincaré's non-integrable systems, but only established conditions which guarantee that sufficiently far from resonances, reasonable answers may be found through suitable perturbation schemes. Let us also remark that the resonance divergence is a manifestation of the general fact (da Costa and Doria 1991) that not only the integrability question is undecidable but also the actual integration of integrable systems is a non-computable problem.

Furthermore, KAM's result cannot be generalized to large systems because the small denominators are not necessarily different from zero as the number N of degrees of freedom tends to infinity (Pöschell 1989), in contrast to the case of finite degrees of freedom. In large systems, resonances are present everywhere, as the frequencies become arbitrarily close to resonance and Fourier sums are replaced by Fourier integrals. Large Poincaré systems are a special class of dynamical systems with continuous spectrum for which the resonant denominators in the secular perturbation terms are arbitrarily close to zero for a continuous set of spectral values. We emphasize that the term "large" refers to the continuous spectrum while the term "Poincaré" refers to the difficulties in any perturbational treatment of the problem, expressed by Poincaré in his famous theorem (Poincaré 1892).

A dynamical system S_t, where t labels the time which is continuous for flows and discrete for cascades, on the phase space Y has continuous spectrum if the Frobenius-Perron operator U_t, which is the adjoint to the Koopman operator V_t,

$$U_t = V^\times, \tag{1}$$
$$V_t \rho(y) = \rho(S_t y), \quad y \in Y, \tag{2}$$

has continuous spectrum on the Hilbert space of square integrable phase functions ρ with respect to the invariant equilibrium measure.

The idea of using operator theory for the study of dynamical systems is due to Koopman (1931) and was extensively used thereafter in statistical mechanics (Prigogine 1962, Goodrich et al. 1980) and ergodic theory (Koopman and von Neumann 1932, Cornfeld et al. 1982, Lasota and Mackey 1985) because the dynamical properties are reflected in the spectrum of the density evolution operators. Koopman showed that for any measure preserving dynamics S_t, the operator (2) is isometric on the space of square integrable phase functions. In the case of invertible dynamics the Koopman operator is unitary. The Koopman operator carries all dynamical information. This important result was proved by Goodrich, Gustafson and Misra (1980) who showed the converse, namely that any positive isometry is implementable by a point transformation.

Dynamical systems with continuous spectrum evolve in a mixing manner, as first pointed out by Koopman and von Neumann (1932). They observed that for dynamical systems with continuous spectra, "the states of motion corresponding to any set, become more and more spread out into an amorphous everywhere dense *chaos*. Periodic orbits, and such like, appear only as very special possibilities of negligible probability". This result was refined later into the ergodic hierarchy of mixing, Lebesgue, Kolmogorov, exact, and Bernoulli systems (Cornfeld et al. 1982, Lasota and Mackey 1985), which is the basis of the modern theory of "chaos". To our knowledge, the paper by Koopman and von Neumann (1932) was the first to use the term "chaos" in the context of dynamical systems. The Renyi maps and the baker's tranformation are simple representative examples of exact and Kolmogorov systems, respectively.

Non-integrability due to resonances appears also in large quantum systems with continuous spectrum. Here, divergences due to resonances appear in the constructive perturbative solutions of the eigenvalue problem (Petrosky et al. 1991). These divergences are manifestations of the general non-computability of the eigenvalue problem for self-adjoint operators in Hilbert space (Pour-el and Richards 1989). Large Poincaré systems, classical or quantum, are quite common in our dynamical models of natural phenomena, including kinetic theory and Brownian motion, the interaction between matter and light, as well as interacting fields and many body quantumn systems involving collisions. The algebraic formulation of dynamical systems allows us to formulate the essentials for both classical and quantum systems as well as stochastic processes in a unified way.

Any dynamical evolution is described by the semigroup of endomorphisms V_t of the algebra of observables $\mathcal{A}$. The time t is discrete for cascades and continuous for flows. For invertible evolutions time takes values over the whole integers and reals, respectively. The states ρ are normalized positive linear functionals over the algebra of observables so that the expectation value of any observable A in the state ρ as a function of time t is given by the bilinear form

$$(\rho|V_t A) \equiv \langle A; t \rangle_\rho \ . \tag{3}$$

This is a generalization of the Heisenberg picture of evolution. The states ρ evolve in terms of the dual evolution which is just the Schrödinger picture,

$$U_t = V^\times , \tag{4}$$

$$(U_t \rho | A) = (\rho | V_t A). \tag{5}$$

The precise mathematical structure of the dual pair of states and observables depends upon the specific types of systems considered. To illustrate the concepts we compare them with the traditional Hilbert space formulation.

For classical systems the states/observables duality is usually the Hilbert space duality $\mathcal{A} = L_\Gamma^2$ with $\rho \in (L_\Gamma^2)^\times \cong L_\Gamma^2$ (i.e., the observables are the square integrable functions), or the $L_\Gamma^1, L_\Gamma^\infty$ duality (i.e., the observables are the essentially bounded functions and the states ρ are the normalized positive integrable densities). Observables evolve according to the Koopman operator V_t and states evolve according to the Frobenius-Perron operator U_t. For quantum systems, the observables form a C^*-algebra $\mathcal{A}$ of bounded operators on a Hilbert space $\mathcal{H}$ and states are elements of the predual $\mathcal{A}^\times$ (tracial operators). In the Heisenberg picture the observables A evolve according to the group of automorphisms of the algebra $\mathcal{A}$ induced by the Schrödinger equation

$$V_t A \equiv e^{iHt} A e^{-iHt} = e^{iLt} A \ . \tag{6}$$

The Heisenberg evolution equation is

$$\partial_t A = i[H, A] = i(HA - AH) = iLA , \tag{7}$$

where L is the Liouville operator. In the Schrödinger picture, the states ρ evolve according to the dual evolution

$$U_t = V_t^\times,\tag{8}$$

or

$$(U_t\rho|A) = (\rho|V_tA),\tag{9}$$

which follows from the Liouville equation

$$\partial_t\rho = i[H,\rho] = iL\rho\tag{10}$$

for any density operator ρ.

We emphasize that, following the tradition of physicists we used the same symbol for the Liouville operators L in (7) and (10). However, this is mathematically and conceptually wrong and led to a lot of confusion in the past. The two Liouville operators are identical only in case of Hilbert-Schmidt observables. In all other cases the Liouville operators should be understood as the generators of (6) and (8). Therefore the correct way to write equations (7) and (10) is

$$\partial_t A = iLA,\tag{11}$$

$$\partial_t\rho = -iL^\times\rho,\tag{12}$$

where $L^\times$ is the dual of L.

The formalism that we develop in the following differs radically from the standard approach to dynamical systems in terms of Hilbert spaces. Depending on the problem considered, we reduce or enlarge the spaces of states and, conversely (due to the duality of states and observables), enlarge or reduce the space of observables.

The evolution of probability densities associated with a dynamical system is described by the Frobenius-Perron operator for a given invariant measure (Cornfeld et al. 1982, Lasota and Mackey 1985). The Frobenius-Perron operator of unstable systems on the space L^2 of square integrable densities has a continuous spectrum, which does not contain characteristic time scales of irreversible changes such as decay rates of the correlation functions. However, isolated point spectra corresponding to the decay rates of the correlation functions emerge if the Frobenius-Perron operator is restricted to suitable locally convex subspaces of L^2. For example, for expanding maps, as the differentiability of the domain functions increases, the essential spectrum (i.e., the spectrum excluding isolated point spectra) of the Frobenius-Perron operator decreases from the unit disk to a smaller one and isolated point spectra appear in the annulus between the two disks. This fact has been shown by Tangermann (1986), Pollicott (1990), and Ruelle (1990). The logarithms of these isolated point eigenvalues of the Frobenius-Perron operator are known as the Pollicott-Ruelle resonances (Pollicott 1985, Ruelle 1986).

It turns out that in some particular cases it is possible to restrict the Frobenius-Perron operator to a dense subspace in such a way that its spectrum becomes discrete. In such a case the Hilbert space L^2 can be replaced by the dual pair $(\Phi, \Phi^\times)$, where Φ is an invariant with respect to U in a locally convex subspace of L^2, and $\Phi^\times$ is its topological dual. Here the Koopman operator V is extended to the space $\Phi^\times$. Moreover, in many cases it is possible to obtain a generalized spectral decomposition of V (and also of U) of the form

$$V = \sum_i z_i |\varphi_i)(F_i| \,, \tag{13}$$

where z_i are the eigenvalues of U and $|\varphi_i)$, and $(F_i|$ is a biorthogonal family of the corresponding eigenvectors which are elements of Φ and $\Phi^\times$, respectively. Following Dirac's notation (Dirac 1958) we denote the linear and antilinear functionals by bras (| and kets |), respectively. Equation (13) has to be understood as

$$(\phi|Vf) = \sum_i z_i(\phi|\varphi_i)(F_i|f) = (U\phi|F)\,, \tag{14}$$

for any state ϕ and observable f from a suitable dual pair. This procedure is referred to as *rigging*, and the triple

$$\Phi \subset L^2 \subset \Phi^\times \tag{15}$$

is called a *rigged Hilbert space*. (For details see Bohm and Gadella (1989), Maurin (1968), Gelfand and Shilov (1967), and Gelfand and Vilenkin (1964).)

Summarizing for the reader's convenience: a dual pair $(\Phi, \Phi^\times)$ of linear topological spaces constitutes a rigged Hilbert space for the linear endomorphism V of the Hilbert space $\mathcal{H}$ if the following conditions are satisfied:

1. Φ is a dense subspace of $\mathcal{H}$,
2. Φ is complete and its topology is stronger than the one induced by $\mathcal{H}$,
3. Φ is stable with respect to the adjoint $V^\dagger$ of V, i.e., $V^\dagger \Phi \subset \Phi$,
4. the adjoint $V^\dagger$ is continuous on Φ.

The extension V_{ext} of V to the dual $\Phi^\times$ of Φ is then in the standard way defined as

$$(\phi|V_{\text{ext}}f) = (V^\dagger \phi|f)\,, \tag{16}$$

for every $\phi \in \Phi$. The choice of the test function space Φ depends on the specific operator V and on the physically relevant questions to be asked about the system.

Let us now present the spectrum of the Renyi map as an example for the generalized spectral decomposition of the Frobenius-Perron operator. Corresponding results have been obtained for the tent map (Antoniou and Qiao 1996a), the logistic map (Antoniou and Qiao 1996b), the Chebyshev map (Antoniou and Qiao 1996c), the baker's transformation (Antoniou and

Tasaki 1992), the Arnold cat map (Antoniou et al. 1996b), and for simple one-dimensional, piecewise-linear maps admitting fractal repellers (Tasaki et al. 1993, 1994).

The β-adic Renyi map S on the interval $[0,1)$ is the multiplication, modulo 1, by the integer $\beta \geq 2$:

$$S : [0,1) \to [0,1) \; : \qquad x \mapsto Sx = \beta x \quad (\text{mod } 1). \tag{17}$$

The probability densities $\rho(x)$ evolve according to the Frobenius-Perron operator U (Lasota and Mackey 1985)

$$U\rho(x) \equiv \sum_{y, S(y)=x} \frac{1}{|S'(y)|}\rho(y) = \frac{1}{\beta}\sum_{r=0}^{\beta-1}\rho\Big(\frac{x+r}{\beta}\Big) \; . \tag{18}$$

According to (1) and (2) the Frobenius-Perron operator is a partial isometry on the Hilbert space L^2 of all square integrable functions over the unit interval as the dual of the isometric Koopman operator V.

The Koopman operator admits the following generalized spectral decomposition (Antoniou and Tasaki 1993a, 1993b)

$$V = \sum_{n=0}^{\infty} \frac{1}{\beta^n}\,|\tilde{B}_n)(B_n| \; , \tag{19}$$

where $B_n(x)$ is the n-degree Bernoulli polynomial defined by the generating function (Boas and Buck 1964, §9):

$$\frac{ze^{zx}}{e^z - 1} = \sum_{n=0}^{\infty} \frac{B_n(x)}{n!}z^n \tag{20}$$

and

$$|\tilde{B}_n) = \begin{cases} |1) \,, & n = 0 \\ |\frac{(-1)^{(n-1)}}{n!}\{\delta^{(n-1)}(x-1) - \delta^{(n-1)}(x)\}) & n = 1, 2, \dots \end{cases} \tag{21}$$

Equation (19) defines a spectral decomposition for the Koopman and Frobenius-Perron operators according to

$$(\rho|Vf) = (U\rho|f) = \sum_{n=0}^{\infty} \frac{1}{\beta^n}(\rho|\tilde{B}_n)\,(B_n|f), \tag{22}$$

for any density function ρ and observable f in the appropriate pair $(\Phi, \Phi^\times)$. Consequently, the Frobenius-Perron operator acts on density functions as

$$U\rho(x) = \int_0^1 dx'\rho(x') + \sum_{n=1}^{\infty} \frac{\rho^{(n-1)}(1) - \rho^{(n-1)}(0)}{n!\beta^n}B_n(x) \; . \tag{23}$$

The orthonormality of the system $|\tilde{B}_n)$ and $(B_n|$ follows immediately, while the completeness relation is just the Euler–MacLaurin summation for the Bernoulli polynomials (Boas and Buck 1964, §9)

$$\rho(x) = \int_0^1 dx' \rho(x') + \sum_{n=1}^{\infty} \frac{\rho^{(n-1)}(1) - \rho^{(n-1)}(0)}{n!} B_n(x) . \qquad (24)$$

The precise meaning of this spectral decomposition and the choice of the test function space will be discussed in the next section.

3   "Tight" Rigging for the Spectral Decompositions

Now we discuss the problem of "tight" rigging for the generalized spectral decompositions of the Koopman operator for the Renyi map. We call a rigging "tight" if the test function space is the (set-theoretically) largest possible space within a chosen family of test function spaces, such that the physically relevant spectral decomposition is meaningful.

In the case of the Renyi map various riggings exist (Bandtlow et al. 1996b). For example, we can consider the restrictions of the Frobenius-Perron operator to a series of test function spaces: the inductive limit $\mathcal{P}$ of polynomials in one variable, the Banach space $\mathcal{E}_c$ of entire functions of exponential type c, the inductive limit $\tilde{\mathcal{E}}_c$ of entire functions of exponential type less than c, the Fréchet space $\mathcal{H}(D_r)$ of functions analytic in the open disk with radius r, the Fréchet space C^∞ of infinitely differentiable functions on the closed unit interval, the Banach space C^m of m-times continuously differentiable functions on the closed unit interval, and finally the Hilbert space L^2. Each of these spaces is densely and continuously embedded according to

$$\mathcal{P} \hookrightarrow \mathcal{E}_c \hookrightarrow \tilde{\mathcal{E}}_c \hookrightarrow \mathcal{E}_{c'} \hookrightarrow \tilde{\mathcal{E}}_{c'} \hookrightarrow \mathcal{H}(D_r) \hookrightarrow C^\infty \hookrightarrow C^m \hookrightarrow L^2 , \qquad (25)$$

(where $c < c'$) and should therefore be a suitable test function space. We do now explain how suitable they are.

Let us observe that the Bernoulli polynomials are the only polynomial eigenfunctions as any polynomial can be uniquely expressed as a linear combination of the Bernoulli polynomials.

The spectral decomposition (19) has no meaning in the Hilbert space L^2, as the derivatives $\delta^{(n)}(x)$ of Dirac's delta function appear as right eigenvectors of V. A natural way to give meaning to formal eigenvectors of operators which do not admit eigenvectors in Hilbert space is to extend the operator to a suitable rigged Hilbert space. A suitable test function space is the space $\mathcal{P}$ of polynomials. The space $\mathcal{P}$ fulfills the following conditions:

1. $\mathcal{P}$ is dense in L^2 (Treves 1967, Chap. 15]),
2. $\mathcal{P}$ is a nuclear LF-space (Treves 1967, Chap. 51) and, thus, complete and barreled,

3. $\mathcal{P}$ is stable with respect to the Frobenius-Perron operator U,
4. U is continuous with respect to the topology of $\mathcal{P}$, because U preserves the degree of polynomials.

$\mathcal{P}$ therefore provides an appropriate rigged Hilbert space, which gives meaning to the spectral decomposition of V. In order to obtain *tight* rigging, the test functions should at least provide a domain for the Euler-MacLaurin summation (24). The requirement of absolute convergence of the series (24) means that

$$\sum_{n=1}^{\infty} \left| \frac{\phi^{(n-1)}(y)}{n!} B_n(x) \right| < \infty \qquad (y = 0, 1).\qquad (26)$$

This implies (Antoniou and Tasaki 1993a) that the appropriate test functions are restrictions on the interval $[0,1)$ of entire functions of exponential type c with $0 < c < 2\pi$. For simplicity we identify the test functions space with the space $\mathcal{E}_c$ of entire functions $\phi(z)$ of exponential type $c > 0$ such that

$$|\phi(z)| \leq K e^{c|z|}, \quad \forall z \in \mathbf{C}, \text{ for some } K > 0.\qquad (27)$$

Each member of the whole family $\mathcal{E}_c$, $0 < c < 2\pi$ is a suitable test function space, since properties (1)-(4) are fulfilled. Indeed, each space $\mathcal{E}_c$ is a Banach space with norm (Treves 1967, Chap. 22)

$$\|\phi\|_c \equiv \sup_{z \in \mathbf{C}} |\phi(z)| e^{-c|z|},\qquad (28)$$

which is dense in the Hilbert space L^2, as $\mathcal{E}_c$ includes the polynomial space $\mathcal{P}$. Each $\mathcal{E}_c$ is stable under the Frobenius-Perron operator U, and it is easily verified that U is continuous on $\mathcal{E}_c$. Now, observe that the spaces are ordered,

$$\mathcal{E}_c \subset \mathcal{E}_{c'}, \qquad c < c',\qquad (29)$$

and consider the space

$$\tilde{\mathcal{E}}_{2\pi} \equiv \bigcup_{c < 2\pi} \mathcal{E}_c.\qquad (30)$$

The space $\tilde{\mathcal{E}}_{2\pi}$, also preserved by U, is the (set-theoretically) largest test function space in our case. Since $\tilde{\mathcal{E}}_{2\pi}$ is a natural generalization of the space $\mathcal{P}$ of polynomials, we want to equip it with a topology which is a generalization of the topology of $\mathcal{P}$.

Recall that $\mathcal{P}$ was given the strict inductive limit topology of the spaces $\mathcal{P}^n$ of all polynomials of degree $\leq n$. A very important property of this topology is that the strict inductive limit of complete spaces is complete. Moreover, it is exceptionally simple to describe convergence in this topology. For example, a sequence $\{w_n\}$ of polynomials converges in $\mathcal{P}$ if and only if the degrees of all w_n are uniformly bounded by some n_0 and $\{w_n\}$ converges in $\mathcal{P}^{n_0}$.

We cannot, however, define the strict inductive topology on $\tilde{\mathcal{E}}_{2\pi}$, because for $c < c'$ the topology on $\mathcal{E}_c$ induced by $\mathcal{E}_{c'}$ is essentially stronger than the initial one. Nevertheless, it is possible to define a topology on $\tilde{\mathcal{E}}_{2\pi}$, which is a natural extension of the topology on $\mathcal{P}$ in the following sense (Suchanecki et al. 1996):

Theorem. *There is a locally convex topology $\mathcal{T}$ on $\tilde{\mathcal{E}}_{2\pi}$ for which it is a nuclear, complete Montel space. Moreover, a sequence $\{f_n\} \subset \tilde{\mathcal{E}}_{2\pi}$ is convergent in the $\mathcal{T}$ topology if and only if there is $c_0 \in (0, 2\pi)$ such that*

1. *f_n, $n = 1, 2, \ldots$, are of exponential type c_0*
2. *$\{f_n\}$ converges in $\| \cdot \|_{c_0}$ - norm.*

It follows from the above considerations that the generalized spectral decomposition of the Koopman operator of the Renyi map is valid only in the spaces (25) up to and including $\tilde{\mathcal{E}}_{2\pi}$. Moreover, it has been proven (Bandtlow et al. 1996b) that the eigenvectors form a complete system in these spaces.

Together with Bandtlow (Bandtlow et al. 1996a), we have recently shown how it is possible to produce the same results as those of the spectral decomposition technique in terms of an analytic continuation of the resolvent of the Frobenius-Perron operator. Following Antoniou et al. (1996a), we understand resonances of dynamical systems as certain singularities of the extended resolvent of the evolution operator on a dual pair or rigged Hilbert space. A sufficient condition for resonances to arise is that the Frobenius-Perron operator is a Fredholm-Riesz operator on a rigged Hilbert space. For the Renyi map, the possible space are given by (25).

This approach is based on the work of Ruelle (1986) who showed that the Fourier transform of the correlation function, i.e., the power spectrum, of certain chaotic dynamical systems can be described by suitably chosen evolution operators of these systems. The resonances of the power spectrum, i.e. the complex poles of the meromorphic extension of the power spectrum coincide with the logarithms of the eigenvalues of the Frobenius-Perron operator (Eckmann 1989). The resonances and the resonance projections depend on the choice of observables, i.e., on the experimentally accessible information about the system. In the case of a piecewise expanding interval transformation, for example, Keller (1984) showed that the L^2-spectrum of the associated Frobenius-Perron operator fills the unit disk, whereas the BV-spectrum, i.e. the spectrum of the restriction of the Frobenius-Perron operator to the space of functions of bounded variation, consists of two parts: eigenvalues filling the interior of a closed disk, centered at the origin and strictly contained in the unit disk, and isolated eigenvalues of finite multiplicity outside this disk, but contained in the unit disk. This means that there are square-integrable functions whose power spectrum cannot be meromorphically continued to a strip containing the real line, whereas this is always the case for functions of bounded variation. In particular, a generic "L^2-observation" will not have poles in its power spectrum, and if there is a pole it might not coincide with that of another generic L^2-observation. The choice of the test function space on which the Frobenius-Perron operator is considered is therefore not a purely academic problem.

Another interesting result of our studies is that the eigenvectors and the spectrum of chaotic maps in the same rigged Hilbert space may be different although they have the same Kolmogorov-Sinai entropy (Prigogine 1992). For instance, the spectrum and the eigenvectors of the Renyi map (17) are different from those of the tent map (Antoniou and Qiao 1996c). This means that isomorphic shifts may have significantly different extensions beyond the Hilbert space. Since these extensions give the physically interesting spectral properties of unstable dynamical systems, the Hilbert space formulation of ergodic theoretic properties is not sufficient to answer all interesting questions like the rate of approach to equilibrium. We propose therefore that the conventional isomorphism of dynamical systems in terms of Kolmogorov-Sinai entropy should be refined by a natural isomorphism in terms of the extension of the evolution operator to a suitable Hilbert spaces where the resonance spectra appear. Two dynamical systems are physically isomorphic if and only if they have the same Kolmogorov-Sinai entropy *and* the same resonance spectrum. Isomorphic systems in this sense have the same rates of approach to equilibrium.

4 Quantum Systems with Diagonal Singularity

4.1 Algebra of Observables

The diagonal singularity of operators in large quantum systems was discovered by van Hove (1955, 1956) in his perturbation analysis of such systems. The very large size of the system manifests itself in the continuous nature of the unperturbed energies. Van Hove supplemented the diagonal singularity property with the initial random phase condition in order to derive the Pauli master equation (van Hove 1957, 1959). The importance of states with diagonal singularity in non-equilibrium statistical physics of both quantum and classical systems was emphasized at the same time by Prigogine and coworkers (Prigogine 1962, Philippot 1961, Prigogine and Balescu 1959, Brout and Prigogine 1956) in their establishment of the master equation in the weak coupling approximation. The irreversible equations like Boltzmann, Fokker-Planck and Pauli equations arise as first order approximations in the subdynamics decomposition of the Liouville operator (Prigogine et al. 1973, George 1973, Grecos et al. 1975, Balescu 1975, Petrosky and Hasegawa 1989).

Moreover, the subdynamics decomposition leads to a spectral representation of the evolution of large Poincaré (non-integrable) systems which includes the resonances and diffusion parameters in the spectrum (Petrosky et al. 1991, Petrosky and Prigogine 1991, Antoniou and Prigogine 1993, Antoniou and Tasaki 1993b). These spectral decompositions are meaningless in the conventional Hilbert space topology but they acquire meaning in terms

of weaker topologies like those of rigged Hilbert spaces (Antoniou and Prigogine 1993, Antoniou and Tasaki 1993b). The key point in the subdynamics
construction is the projection onto the diagonal part of states or observables.
This projection is usually performed through the thermodynamic limit because the direct calculation of the continuous spectrum in the conventional
formulation suffers from divergences. The reason for the divergences is that
the continuum labeled basis $|\alpha\rangle$ of the space of wave functions

$$\langle \alpha | \, \alpha' \rangle = \delta(\alpha - \alpha') \,, \tag{31}$$

cannot be lifted to a product basis for the operators. Indeed, the dyadic basis
$|\alpha\rangle\langle\alpha'|$ gives rise to meaningless expressions in the continuous case because

$$\mathrm{tr}\{(\,|\alpha\rangle\langle\alpha|\,)^{\dagger}(\,|\alpha'\rangle\langle\alpha'|\,)\} \equiv (\alpha\alpha|\,\alpha'\alpha')_{\mathrm{HS}} = \delta(\alpha - \alpha')\delta(\alpha - \alpha') \,, \tag{32}$$

where $(\,|\,)_{\mathrm{HS}}$ is the scalar product in the Hilbert-Schmidt space $\mathcal{H} \otimes \mathcal{H}^{\times}$.

Furthermore, the conventional Hilbert-Schmidt formulation does not allow us to define the microcanonical equilibrium and therefore the approach
to equilibrium. The reason is that the identity operator, although bounded,
is not in the Hilbert-Schmidt space $\mathcal{H} \otimes \mathcal{H}^{\times}$, since

$$\mathrm{Tr}I = \sum_{\nu}\langle u_{\nu} | \, u_{\nu} \rangle = +\infty \,, \tag{33}$$

for any orthonormal basis (u_{ν}) of the Hilbert space $\mathcal{H}$.

However, the extension of the Hilbert-Schmidt algebra to an algebra with
identity (Naimark 1972, Dixmier 1977)

$$\mathbf{C} \cdot I \oplus (\mathcal{H} \otimes \mathcal{H}^{\times}) \tag{34}$$

together with the diagonal singularity property holds the clue for a proper
formulation of quantum theory which allows for

1. a clear definition of states and observables with the diagonal singularity,
2. a definition of the projections onto the diagonal and off-diagonal parts
   without resorting to the thermodynamic limit,
3. the construction of continuous biorthonormal bases for states and observables.

We have shown (Antoniou and Suchanecki 1995, 1996) that following
the algebraic approach to dynamical systems we can construct a natural
extension of quantum theory which fulfills the requirements (1)-(3) and gives
meaning to the resulting spectral decompositions of the Liouville operator.
The main idea of the algebraic approach is that the states ρ correspond to the
normalized positive linear functionals $(\rho|$ over a suitably chosen algebra of
operators representing the observables of the dynamical system. The value of
the functional $(\rho|$ for an operator A is the expectation value of the observable
A in the state ρ

$$(\rho|A) \equiv \langle A \rangle_{\rho}. \tag{35}$$

The algebraic generalization of quantum theory was pioneered by Segal (1947) who assumed that the observables form a C*-algebra of bounded operators. In the 1960s, C*-algebraic methods were applied to local quantum field theory and statistical physics. For a general introduction to the methodology and results, we refer to Emch (1972), Bratteli and Robinson (1979, 1981), and Haag (1992). In fact, Emch (1972) concludes his paper with the remark that the algebraic approach can be of definite use in the problem of Poincaré invariants.

However the algebra of operators with the diagonal singularity turns out to be an involutive Banach algebra which is not a C*-algebra. This more general structure allows us to incorporate states with diagonal singularity, which cannot be described in terms of the conventional (von Neumann 1955, Prugovecki 1981) or generalized (Dirac 1958, Bohm 1986, Antoniou and Prigogine 1993) formulation of quantum theory.

We consider quantum systems associated with a separable Hilbert space $\mathcal{H}$ with a left conjugate linear and right linear scalar product $\langle \ | \ \rangle$. The Hamilton operator H can be decomposed into a diagonal part H_0 and an off-diagonal part or perturbation V

$$H = H_0 + \lambda V \tag{36}$$

The operator H_0 is diagonal with respect to the non-interacting entities $|\alpha\rangle$. The eigenvectors $|\alpha\rangle$ are the common eigenvectors of a complete system of commuting observables and provide a representation of the system. The actual choice of the complete set of commuting observables is based upon physical considerations. The labeling quantum numbers α are discrete as the case of atoms and oscillators or continuous in the case of scattering, fields and large systems.

The perturbation λV (λ is a dimensionless parameter) couples the non-interacting degrees of freedom $|\alpha\rangle$. In the case of integrable systems the perturbation just gives rise to new independent conserved entities associated with the total Hamiltonian H

$$H = \sum_a z_a \, |a\rangle\langle a| \ . \tag{37}$$

However, if resonances are present, the interaction cannot be re-incorporated into any conventional non-interacting conserved entities. The main idea of the Brussels school is that in such cases the natural non-interacting entities are not conserved. In fact, the subdynamics decomposition amounts to a generalized spectral decomposition of the evolution operator with complex eigenvalues associated with resonances (Petrosky et al. 1991, Petrosky and Prigogine 1991, Antoniou and Prigogine 1993, Antoniou and Tasaki 1993b). The construction is particularly appropriate for large Poincaré (non-integrable) systems, classical or quantum mechanical.

In the representation provided by continuous bases like $|\alpha\rangle$, van Hove (1955, 1956) encountered operators represented by the kernels

$$A_{\alpha\alpha'} \equiv \langle \alpha | \, A \, | \alpha' \rangle = A_\alpha^d \delta(\alpha - \alpha') + A_{\alpha\alpha'}^c \, . \tag{38}$$

The operator A is said to have a diagonal singularity with respect to the basis $|\alpha\rangle$ if the diagonal part A_α^d does not vanish. The diagonal function A_α^d has no singularities while the off-diagonal part $A_{\alpha\alpha'}^c$ may have certain weaker singularities as in the case of continuous degeneracies appearing in scattering or in coupled fields. In such cases one may decompose further the off-diagonal part $A_{\alpha\alpha'}^c$. Here we shall assume the simplest case where $A_{\alpha\alpha'}^c$ are kernels of compact operators. The reason is that the states on the algebra of compact operators correspond to trace class operators, because the dual of the algebra $\mathcal{B}_\mathcal{H}^\infty$ of compact operators is isomorphic to the trace class operators $\mathcal{B}_\mathcal{H}^1$. Tracial states are the traditional states of quantum statistical physics (von Neumann 1955, Prugovecki 1981).

A reasonable choice for the algebra of operators with diagonal singularity is the sum

$$\mathcal{A} \equiv \mathcal{A}^d + \mathcal{B}_\mathcal{H}^\infty \, . \tag{39}$$

The diagonal part is the maximal Abelian von Neumann algebra generated by the chosen complete system of commuting observables (Jauch 1960, de Dormale and Gautrin 1975). The algebra $\mathcal{A}^d$ is a Banach algebra with respect to the norm

$$\|A^d\| \equiv \sup_{\psi \neq 0} \frac{\|A\psi\|_H}{\|\psi\|_H} \quad \forall \, A^d \in \mathcal{A}^d \, . \tag{40}$$

If some observables correspond to unbounded self-adjoint operators, as in the case of potential scattering, the algebra $\mathcal{A}^d$ is generated by their spectral projections which are bounded operators.

The operator H_0 as well as any other unbounded operator are affiliated with the algebra $\mathcal{A}^d$, if their spectral projections are included in $\mathcal{A}^d$. In the case of unbounded operators, Antoine et al. (1983) considered more general structures like V^*-algebras instead of maximal Abelian von Neumann algebras.

The non-diagonal part $\mathcal{B}_\mathcal{H}^\infty$ is the C*-algebra (Dixmier 1977) of compact operators on the Hilbert space $\mathcal{H}$. It was proven by Antoniou and Suchanecki (1995) that the algebras $\mathcal{A}^d$ and $\mathcal{B}_\mathcal{H}^\infty$ have no common elements. Consequently, any operator A in the algebra is uniquely represented as the sum of the diagonal and off-diagonal parts

$$A = A^d + A^c \, . \tag{41}$$

A natural norm on $\mathcal{A}$ which captures the physical properties of these operators can be defined by

$$\|A\|_\mathcal{A} = \|A^d\| + \|A^c\| \, , \tag{42}$$

where $\| \cdot \|$ is the operator norm defined in (40).

The properties of the algebra $\mathcal{A}$ are summarized (Antoniou and Suchanecki 1995) by the following

Theorem.

1. $\mathcal{A}$ *is an involutive Banach algebra of bounded operators which includes the identity operator.*
2. $\mathcal{A}$ *is not a C^*-algebra.*

The generalized basis $|\alpha\rangle$ for $\mathcal{H}$ leads to a natural basis for the algebra $\mathcal{A}$ constructed as follows. The diagonal part $\mathcal{A}^d$ can be expanded in terms of operators $|\alpha) \equiv |\alpha)^d \equiv |\alpha\rangle\langle\alpha|$

$$A^d = \int_\sigma d\alpha A^d_\alpha \, |\alpha) \, , \tag{43}$$

where A^d_α is an essentially bounded measurable function (Jauch and Misra 1965, de Dormale and Gautrin 1975, Prugovecki 1981) on the spectrum σ.

Physicists usually expect that the off-diagonal compact part can be expanded in terms of the operators $|\alpha\alpha') \equiv |\alpha\rangle\langle\alpha'|$. If this is the case, for example if A^c is Hilbert-Schmidt, we have

$$A = \int_\sigma d\alpha A_\alpha \, |\alpha) + \int_{\sigma\times\sigma} d\alpha d\alpha' A^c_{\alpha\alpha'} \, |\alpha\alpha') \, , \tag{44}$$

where $A^c_{\alpha\alpha'}$ is the kernel function function on the product space $\sigma \times \sigma$.

The basis $\{|\alpha), |\alpha\alpha')\}$ may be used for the expansion of unbounded operators which are not in the algebra $\mathcal{A}$. In such cases, we have more general structures like the partial involutive algebra (Antoine and Karwowski 1985, Epifanio and Trapani 1988). For example, in the case of kernel operators, which of course include the the Hilbert-Schmidt operators, one can show that the decomposition (44) is unique (Antoniou and Suchanecki 1996). Moreover, a different choice for the complete system of commuting observables gives rise to a different generalized basis $|\alpha\rangle$ for the Hilbert space $\mathcal{H}$ and therefore to a different algebra $\mathcal{A}$. However, both bases are related by a unitary operator which induces a Banach algebra isomorphism.

4.2   States with Diagonal Singularity

The states ρ of the system correspond to the normalized positive linear functionals $(\rho|$ over the algebra $\mathcal{A}$, i.e., they satisfy the properties

1. Linearity:
$$(\rho|z_1 A_1 + z_2 A_2) = z_1(\rho|A_1) + z_2(\rho|A_2) \, . \tag{45}$$

2. Positivity:
$$(\rho|A^\dagger A) \geq 0 \, . \tag{46}$$

3. Normalization:

$$(\rho|I) = 1 . \tag{47}$$

As $\mathcal{A}$ is an involutive Banach algebra, the states ρ are bounded with norm (Naimark 1972, Dixmier 1977)

$$\|\rho\| = (\rho|I) = 1 . \tag{48}$$

We shall adopt here the following bra/ket notation in connection with the states/observables duality.

$$\rho(A) \equiv (\rho|A) \equiv \langle A \rangle_\rho . \tag{49}$$

The bracket $(\rho|A)$ represents the expectation value of the observable A in the state ρ. The bras $(\rho|$ are to be understood as linear functionals over the algebra $\mathcal{A}$ of observables. The reader should not confuse the bras and kets $(\cdot|$ and $|\cdot)$ expressing the state/observables duality with the Dirac bracket notation $\langle\cdot|$ and $|\cdot\rangle$ for the linear and antilinear functionals representing the wave functions (Bohm 1986). Our bras $(\cdot|$ and kets $|\cdot)$ notation is just Dirac's notation in the Liouville space.

The states ρ form a convex subset of the Banach dual $\mathcal{A}^\times$ of the algebra $\mathcal{A}$,

$$\mathcal{A}^\times = \mathcal{A}^{d^\times} \oplus \left(\mathcal{B}_H^\infty\right)^\times = \mathcal{A}^{d^\times} \oplus \mathcal{B}_H^1 . \tag{50}$$

Any state ρ has therefore a diagonal and an off-diagonal part, ρ^d and ρ^c, such that

$$\rho = \rho^d + \rho^c , \tag{51}$$

with

$$\begin{aligned}(\rho^d|A) &= (\rho^d|A^d) , \\ (\rho^c|A) &= (\rho^c|A^c) ,\end{aligned} \tag{52}$$

for any operator $A \in \mathcal{A}$.

As the dual space of $\mathcal{B}_H^\infty$ is the space of trace class $\mathcal{B}_H^1$, the off diagonal states ρ correspond to the tracial functionals

$$(\rho^c|A) = \text{Tr}\hat{\rho}\hat{A}, \tag{53}$$

where $\hat{\rho}$ is the density operator representing the state ρ.

The dual space $\left(\mathcal{L}_\sigma^\infty\right)^\times$ is isomorphic to the space of complex additive measures of bounded total variation and absolutely continuous with respect to a suitable (Dunford and Schwartz 1988) extension $(\widetilde{\mathcal{B}}, \widetilde{\mu})$ of the Lebesgue measure structure $(\mathcal{B}, \mu)$ associated with the spectrum σ. The diagonal states ρ^d are therefore represented in terms of these normalized positive measures or their Radon-Nikodym derivatives ρ_α^d.

The diagonal elements ρ_α^d represent the probability densities associated with the states $|\alpha\rangle$ while the off-diagonal elements $\rho_{\alpha\alpha'}^c$ describe correlations or coherent superpositions of the states $|\alpha\rangle$.

For observables A admitting a kernel representation, the expectation value in the state ρ can be written

$$(\rho|A) = \langle A \rangle_\rho = \int_\sigma d\alpha\, \rho_\alpha^d A_\alpha^d + \int_{\sigma \times \sigma} d\alpha d\alpha'\, \rho_{\alpha\alpha'}^{c*} A_{\alpha\alpha'}^c . \tag{54}$$

In view of the duality relation (54) we define the action of the states ρ on the basis $|\alpha)$, $|\alpha\alpha')$ as

$$\begin{aligned}(\rho|\alpha) &\equiv \rho_\alpha^{0*} = \rho_\alpha^0 , \\ (\rho|\alpha\alpha') &\equiv \rho_{\alpha\alpha'}^{c*} .\end{aligned} \tag{55}$$

These equations suggest the definition of a dual basis $(\alpha|$, $(\alpha\alpha'|$ for the space of states

$$\begin{aligned}(\alpha|A) &\equiv (\alpha|A^d) \equiv A_\alpha^d \\ (\alpha\alpha'|A) &\equiv (\alpha\alpha'|A^c) \equiv A_{\alpha\alpha'}^c\end{aligned} \tag{56}$$

The duality relation (54) suggests the expansion of any state ρ

$$(\rho| = \int_\sigma d\alpha \rho_\alpha^d (\alpha| + \int_{\sigma \times \sigma} d\alpha d\alpha' \rho_{\alpha\alpha'}^{c*} (\alpha\alpha'| \tag{57}$$

The normalized positive linear functionals over the algebra $\mathcal{A}$ include not only the conventional quantum states but also the states with diagonal singularity which cannot be discussed in terms of the Hilbert space formulation. In fact, the reason for the algebraic formulation of quantum theory (Segal 1947, Emch 1972, Bratteli and Robinson 1979, 1981, and Haag 1992) is the possibility to discuss states of physical interest which cannot be formulated in terms of the conventional Hilbert space language. For example, the dual $\mathcal{B}_\mathcal{H}^\times$ of the von Neumann algebra $\mathcal{B}_\mathcal{H}$ of bounded operators on the Hilbert space $\mathcal{H}$ includes not only the density operators which exist in the trace class space $\mathcal{B}_\mathcal{H}^1$ but also states which vanish with respect to the compact operators $\mathcal{B}_\mathcal{H}^\infty$. In fact we have the isomorphism (Dixmier 1981, p. 51)

$$\mathcal{B}_\mathcal{H}^\times \cong \mathcal{B}_\mathcal{H}^1 \oplus (\mathcal{B}_\mathcal{H}^\infty)^\perp \tag{58}$$

The norm of the state ρ extends in a natural way the trace concept because

$$\|\rho\| = (\rho|I) = \int_\sigma d\alpha\, \rho_\alpha^d . \tag{59}$$

The microcanonical and canonical equilibrium states for the evolution generated by H_0 have a natural description within our formalism

$$(\rho_{\text{micro}}| = \frac{1}{\mathcal{N}_{\text{micro}}} \int d\alpha \delta(\omega_\alpha - E)(\alpha| , \tag{60}$$

$$(\rho_{\text{can}}| = \frac{1}{\mathcal{N}_{\text{can}}} \int d\alpha e^{-\beta\omega_\alpha}(\alpha| , \tag{61}$$

where $\mathcal{N}_{\text{micro}}$, $\mathcal{N}_{\text{can}}$ are the normalization constants determined by the condition

$$(\rho_{\text{micro}}|I) = (\rho_{\text{can}}|I) = 1 . \tag{62}$$

We see that ρ_{micro} and ρ_{can} cannot be described in terms of the conventional formulation of quantum physics because they are purely diagonal states.

Respecting the tradition of statistical physics we work with states although, whenever convenient, some calculations may be performed with observables using the duality of states and observables. The action of a linear operator U on the state ρ satisfies the duality relation

$$(U\rho|A) = (\rho|VA) \tag{63}$$

for all observables A. The operator V acting on the observables A is the predual of U, i.e. $V = U^+$ or equivalently U is the dual of V, i.e. $U = V^\times$. This is just the duality between the Schrödinger and the Heisenberg pictures.

4.3   Time Evolution

The Liouville-von Neumann evolution equation for states is given by

$$(\partial_t \rho|A) = (\rho|i[H, A]) = (-iL^\times \rho|A) . \tag{64}$$

If the state ρ is tracial then the evolution is implemented by the evolution on the Hilbert space and $L^\times$ is the Liouville-von Neumann operator of the conventional formulation of quantum statistical physics (von Neumann 1955, Bratteli and Robinson 1979, 1981, Haag 1992). However, for simplicity we shall use the symbol L instead of $L^\times$, keeping in mind that it acts on states.

The Liouville operator may lead out of the algebra $\mathcal{A}$. Indeed for Hamiltonians of the form

$$H = H_0 + V , \tag{65}$$

with

$$H_0 = \int_\sigma d\alpha \omega_\alpha \, |\alpha\rangle\langle\alpha| , \tag{66}$$

$$V = \int_{\sigma \times \sigma} d\alpha d\alpha' V_{\alpha\alpha'} \, |\alpha\rangle\langle\alpha'| , \tag{67}$$

the Liouville operator acts on the algebra as

$$\begin{aligned}
LA = [H, A] &= [H_0, A^c] + [V, A^d] + [V, A^c] \\
&= \int_{\sigma^2} d\alpha d\alpha' (\omega_\alpha - \omega_{\alpha'}) |\alpha\alpha')(\alpha\alpha'| \\
&\quad + \int_{\sigma^3} d\alpha d\alpha' d\alpha'' \Big[V_{\alpha''\alpha} |\alpha''\alpha')(\alpha\alpha'| - V_{\alpha'\alpha''} |\alpha\alpha'')(\alpha\alpha'| \Big] \\
&\quad + \int_{\sigma^2} d\alpha d\alpha' \Big[V_{\alpha'\alpha} |\alpha'\alpha)(\alpha| - V_{\alpha\alpha'} |\alpha\alpha')(\alpha| \Big] .
\end{aligned} \tag{68}$$

It is now clear that if $V_{\alpha'\alpha}$ contains singularities as in case of the van Hove condition discussed above, then L leads out of the algebra and we have to extend the algebra accordingly.

The solution of the eigenvalue problem may lead to a spectral decomposition of the Liouville operator

$$L = \sum_{\nu} z_{\nu} |F_{\nu})(f_{\nu}| , \tag{69}$$

with f_{ν}, F_{ν} a biorthonormal dual pair.

The evolution of states can be decomposed in terms of the eigencomponents

$$(e^{-iL^{\times}t}\rho| = \sum_{\nu} e^{iz_{\nu}t}(\rho|F_{\nu})(f_{\nu}| . \tag{70}$$

Our extended formulation allows z_{ν} to be complex. We expect that the imaginary part represents decay rates or diffusion coefficients. In the conventional Hilbert-Schmidt formulation, of course, z_{ν} is real and $f_{\nu} \equiv F_{\nu}$ because L is a self-adjoint operator on the Hilbert-Schmidt space.

5   The Logic of Unstable Systems

The logic of a physical system consists of the elementary statements (propositions, questions) about the physical system, equipped with the relations of implication and negation. In classical systems each elementary statement is identified with a subset of the phase space, corresponding to the question whether an observed point or trajectory lies in the set. Implication and negation correspond to set inclusion and complement, respectively. This leads to the commonly accepted assumption that a Boolean algebra of subsets of the phase space is the logic of classical physical system.

The logic of quantum systems turns out to be more complicated. A quantum system cannot be associated with any phase space. The observables are operators on a Hilbert space and, consequently, the most elementary observables must also be operators (in fact, they are projectors).

Although a similar logic as in classical systems can be easily constructed by identifying projectors on a Hilbert space with appropriate subspaces, there is an essential difference – quantum logic is non-distributive and therefore non-Boolean. Non-distributivity expresses the fact that in quantum systems not all possible (Boolean) statements are experimentally verifiable (for example, simultaneous measurements of two incompatible observables).

Similar, although more complex, is the problem of a logic of quantum systems with diagonal singularities. Here the set of operators at our disposal is restricted. Consequently, one may expect that the logic is also more restrictive, and this raises the question: What are the most elementary observables (questions) in systems with diagonal singularities and what are the differences between their logic and traditional quantum logic?

In a recent paper (Antoniou and Suchanecki 1994) we answered this question, giving the full analytical characterization of the elements of this logic and an analysis of its axioms. In particular, one essential difference concerning the completeness of the logic is crucial. We review these results in the next subsection.

Yet another type of logic holds for classical chaotic dynamical systems as discussed in Sect. 2. The logic of chaotic systems turns out to be a fuzzy logic and is strictly connected with the extended formulation in rigged Hilbert spaces discussed in Sects. 3 and 4. It is related to the fact that in prediction and computation we can never obtain sharp results due to the sensitivity of such systems on the initial conditions and computational imprecision.

5.1   The Logic of Systems with Diagonal Singularity

Following Birkhoff and von Neumann's (1936) pioneering work, the Hilbert space formulation of quantum mechanics (von Neumann 1955, Prugovecki 1981) is related to the logic of the relevant propositions/questions (Jauch 1973, Beltrametti and Cassinelli 1981). As a result of these investigations, the logic of quantum systems was shown to be non-distributive. Therefore quantum logic is non-Boolean. The minimal properties of a non-Boolean logic, which guarantee that it is essentially a Hilbert space logic, have been finally clarified by Piron (1964). The result is that the logic of conventional quantum mechanics is a complete orthomodular, orthocomplemented lattice. Completeness means here that any countable combination of propositions leads to a meaningful proposition within the logic.

The algebra of operators with diagonal singularity specifies a logic different from conventional Hilbert space logic, since only finite combinations of propositions/questions are allowed. This fact is a manifestation of the underlying instability of the dynamics and shows another aspect of our extended quantum theory.

The logic of a system results from the partial order relations between the elementary propositions or questions relevant for the system. The partial order relation between two propositions is denoted by

$$A \leq B, \tag{71}$$

which means: if A is true then B is also true. This relation satisfies the following conditions:

1. $A \leq A$,
2. $A \leq B$ and $B \leq A$ implies $A = B$,
3. $A \leq B$ and $B \leq C$ implies $A \leq C$.

In the Hilbert space formulation of quantum mechanics, the observables are bounded linear operators on a Hilbert space $\mathcal{H}$ and, consequently, the propositions are the projectors on $\mathcal{H}$. Moreover, in such an approach it is *ad hoc* assumed (see Mackey 1963, Axiom VII) that all possible projectors on $\mathcal{H}$ are propositions. In the set of projectors on $\mathcal{H}$, (71) means that the range $\mathcal{H}_A$ of A is a subspace of the range $\mathcal{H}_B$ of B.

In a partially ordered set (poset) we can combine propositions by the operations $\wedge$ and $\vee$. Recall that $A \wedge B$ (if it exists) is the proposition with the properties

1. $A \wedge B \leq A$ and $A \wedge B \leq B$,
2. if C is another proposition such that $A \leq C$ and $B \leq C$ then $A \wedge B \leq C$.

This definition extends naturally on an arbitrary set of propositions. The operation $A \vee B$ is defined in a similar way (see, e.g., Beltrametti and Cassinelli 1981).

The existence of the operations $\wedge$ and $\vee$ in a poset is not automatically guaranteed. If they, however, exist for any two elements (thus for any finite number of propositions) we call the poset a *lattice*. If the operations $\wedge$ and $\vee$ are admissible for a countable number of propositions the lattice is said to be *complete* (strictly speaking σ-complete).

In the set of all projections on a Hilbert space $\mathcal{H}$, the operation $A \wedge B$ is defined as the projection onto the (closed) space

$$\mathcal{H}_A \cap \mathcal{H}_B , \tag{72}$$

where $\mathcal{H}_A$ and $\mathcal{H}_B$ denote the ranges of A and B, respectively. The operation $\vee$ is the sum of the projections A, B, i.e., the projection onto the space

$$\overline{\mathcal{H}_A + \mathcal{H}_B} , \tag{73}$$

where the bar denotes the closure in the norm topology of $\mathcal{H}$. One can show that the set of *all* projections on $\mathcal{H}$ is a complete lattice (Beltrametti and Cassinelli 1981, §10).

The negation of a proposition gives additional structure to the poset of propositions. Namely, we say that a poset of propositions is orthocomplemented if there is a map, denoted here by $^\perp$, of the set of propositions onto itself such that:

1. $(A^\perp)^\perp = A$,
2. $A^\perp \wedge A = 0$ (0 is the absurd proposition),
3. $A \leq B$ implies $B^\perp \leq A^\perp$.

The orthocomplement of the projector A on a Hilbert space is defined as $I - A$, i.e., as the projection on the space $\mathcal{H} \ominus \mathcal{H}_A$.

An orthocomplemented poset is distributive if

$$A \vee (B \wedge C) = (A \vee B) \wedge (A \vee C) . \tag{74}$$

Distributivity is, according to Birkhoff and von Neumann (1936), "a law in classical, but not in quantum mechanics". Distributivity fails if we replace sets by the closed subspaces of a Hilbert space.

For finite dimensional Hilbert spaces a weaker "version" of distributivity holds. It is called *modularity*:

$$A \leq C \quad \text{implies} \quad A \vee (B \wedge C) = (A \vee B) \wedge C \tag{75}$$

Modularity does not hold for infinite dimensional Hilbert spaces (see Birkhoff and von Neumann (1936) for a counterexample). Instead, we have the property of *weak modularity* (or *orthomodularity*; Jauch 1973) which in the case of projections can be (equivalently) written as

$$A \leq C \quad \text{implies} \quad AB = BA. \tag{76}$$

These are the basic properties of the logic of quantum mechanics which we are now going to verify for systems with diagonal singularities.

Denote by L the set of projectors of the algebra $\mathcal{A}$ with diagonal singularity and recall that the algebra $\mathcal{A}$ in the simplest case is the direct sum

$$\mathcal{A} = \mathcal{A}^d \oplus \mathcal{B}_{\mathcal{H}}^{\infty} , \tag{77}$$

where $\mathcal{A}^d$ is a commutative subalgebra of $\mathcal{B}_{\mathcal{H}}$. The algebra of all bounded operators on $\mathcal{H}$ and $\mathcal{B}_{\mathcal{H}}^{\infty}$ is the algebra of all compact operators on $\mathcal{H}$.

In the following we assume for simplicity that the Hilbert space $\mathcal{H}$ is isomorphic to L_{σ}^2, where σ is a subset of $\mathbb{R}$ and μ is the Lebesgue measure. This implies that the diagonal part $\mathcal{A}^d$ is isomorphic to L_{σ}^{∞}.

Since $\mathcal{L}$ is a subset of $\mathcal{A}$, each projector $P \in \mathcal{L}$ is uniquely represented as the sum of the diagonal and off-diagonal parts

$$P = P^d + P^c . \tag{78}$$

Since $\mathcal{L}$ is a subset of $\mathcal{B}_{\mathcal{H}}$, operations such as multiplication, adjointness, etc. are the ordinary multiplication of operators, the operator adjoint, etc., respectively. Therefore, $\mathcal{L}$ is a poset (partially ordered set) with respect to the natural ordering (ordering on projections), and all the projections from $\mathcal{L}$ can be characterized by the conditions:

(a) $P^+ = P$,
(b) $P^2 = P$.

This definition remains consistent when we restrict $\mathcal{L}$ to elements of $\mathcal{A}^d$ or $\mathcal{B}_{\mathcal{H}}^{\infty}$, respectively.

Condition (a) implies that for an element $P \in \mathcal{L}$, $P = P^d + P^c$, where $P^d \in \mathcal{A}^d$ and $P^c \in \mathcal{B}_{\mathcal{H}}^{\infty}$, we have

$$P^d + P^c = (P^d + P^c)^{\dagger} = P^{d\dagger} + P^{c\dagger}. \tag{79}$$

Because of the uniqueness of the decomposition of elements of the algebra $\mathcal{A} = \mathcal{A}^d \oplus \mathcal{B}_{\mathcal{H}}^{\infty}$, we have therefore

$$P^{d\dagger} = P^d \quad \text{and} \quad P^{c\dagger} = P^c . \tag{80}$$

Condition (b) implies that

$$P^d + P^c = (P^d + P^c)^2 = P^{d^2} + P^d P^c + P^c P^d + P^{c^2}. \tag{81}$$

The uniqueness of the decomposition implies that

(c) $P^{d^2} = P^d$,
(d) $P^c = P^d P^c + P^c P^d + P^{c^2}$.

Consequently, P^d is a projector and P^c is a selfadjoint operator.

The assumption that $\mathcal{A}^d$ is isomorphic to L_σ^∞ implies that P^d is also a multiplication operator, say

$$P^d f(x) = q(x) f(x), \quad \text{for almost all } x \in \sigma, \tag{82}$$

where $q(\cdot)$ is a function from L_σ^∞. Consequently $q(\cdot)$ must be a real function such that $q^2(x) = q(x)$, for almost all x. This implies that $q(\cdot)$ can only be the indicator of a measurable subset Δ of σ

$$q(x) = \mathbf{1}_\Delta(x). \tag{83}$$

Since $\mathcal{L}$ is not closed with respect to countable sums of projections, $\mathcal{L}$ *cannot be a complete lattice* (see Antoniou and Suchanecki (1994) for a counterexample).

Orthocomplementation on $\mathcal{L}$ can be defined as in case of ordinary projections, i.e.,

$$P^\perp \equiv I - P. \tag{84}$$

If $P \in \mathcal{L}$ then $P^\perp$ also belongs to $\mathcal{L}$. Indeed, let $P = P^d + P^c$, where $P^d \in \mathcal{A}^d$ and $P^c \in \mathcal{B}_\mathcal{H}^\infty$. Then $I - P = (I - P^d) + (-P^c)$, but we have also $I - P^d \in \mathcal{A}^d$ and $-P^c \in \mathcal{B}_\mathcal{H}^\infty$. Hence, $\mathcal{L}$ is orthocomplemented.

The axiom of orthomodularity or weak modularity in the sense of Jauch (1973) can be equivalently written as

$$P_1 \leq P_2 \Rightarrow P_2 = P_1 + (P_2 - P_1). \tag{85}$$

To check that $\mathcal{L}$ is orthomodular let us take $P_1, P_2 \in \mathcal{L}$, $P_1 \leq P_2$ (what is equivalent that $P_1 P_2 = P_1$). Then taking the operator adjoints in $\mathcal{B}_\mathcal{H}$ we obtain

$$P_2 P_1 = P_2^\dagger P_1^\dagger = (P_1 P_2)^\dagger = P_1^\dagger = P_1 P_2, \tag{86}$$

meaning that P_1 and P_2 are "compatible" (see Jauch 1973, p. 81 and problem 1 on p. 86).

5.2   The Fuzzy Logic of Chaos

We stressed already that the logic of a physical system is the class of its most elementary observables. Therefore, before we begin the discussion about the logic in classical chaotic system we must be clear about the class of observables for such systems. This is an easy problem in the traditional Hilbert space approach where both observables and states are square integrable functions. We may also consider states as integrable and observables as bounded functions. Analogous to quantum systems, the construction of dual pairs restricts the state space (the domain of the Frobenius-Perron operator) and extends the observable space (the domain of the Koopman operator). Though it is very convenient to use this duality from the point of view of application of functional analysis, one may ask the the question: is the extended domain of the Koopman operator indeed the space of physical observables?

In the study of chaotic dynamical systems the only experimentally accessible quantities are time correlation functions. They show the decay rate of the system and are formally equivalent with the power spectrum of resonances of the system. For a dynamical system corresponding to a flow $t \mapsto S_t$ on a phase space Y with invariant measure μ, the correlation function of two observables f and g is defined by

$$C_{f,g}(t) = \int_Y \mu(dy) f(S_t y) g(y) - \left(\int_Y \mu(dy) f(y) \right) \left(\int_Y \mu(dy) g(y) \right). \quad (87)$$

For $f = g$ we obtain the autocorrelation function of the observable f

$$C_f(t) \equiv C_{f,f}(t) = \int_Y \mu(dy) f(S_t y) f(y) - \left(\int_Y \mu(dy) f(y) \right)^2. \quad (88)$$

It is therefore natural to denote as "physical observables" those functions f for which $C_f(t)$ is at least correctly defined.

Equation (87) cannot be easily extended on arbitrary dual pairs $(\Phi, \Phi^\times)$. Of course we can write $\int_Y \mu(dy) f(S_t y) g(y) = (V_t f | g)$, where V_t is the Koopman operator, $f \in \Phi^\times$ and $g \in \Phi$. However, $\int_Y \mu(dy) f(y)$ cannot be in general defined. Trying to extend (88) is even more difficult. The only reasonable way to extend the definition of the correlation function $C_f(t)$ on rigged Hilbert spaces is to restrict f to the common domain of the Koopman and the Frobenius-Perron operator, i.e., to $\Phi \cap \Phi^\times$. We shall call this space the space of *physical observables*. For chaotic systems as the Renyi map, we have $\Phi \subset \Phi^\times$. Therefore, in such a case the space of physical observables coincides with the state space.

Let us consider the case when the set $\mathcal{A}$ of all possible observables of a physical system is represented by a specific topological vector space of real or complex valued functions defined on a phase space X. We assume that $\mathcal{A}$ is, at least partially, of algebraic structure, i.e., $\mathcal{A}$ is a vector space and the multiplication is correctly defined for those observables which we regard as elementary.

One possibility to verify the truth value of a statement about an observable is: an (elementary) statement about $a \in \mathcal{A}$ is true (or false), if the repetition of the same observation of $a(\cdot)$ gives (or does not give) the same result as a single observation $a(\cdot)$:

$$a^2 = a \tag{89}$$

If $a(\cdot)$ is a function then $a(\cdot)$ can assume only two possible values – zero or one. If $\mathcal{A}$ is a space of measurable functions, such as $L^p(X)$, $0 \le p \le \infty$, $a(\cdot)$ must be the indicator of a measurable subset of X. In general, the topological structure imposed on the phase space X implies that $\mathcal{A}$ is a subspace of $\mathcal{C}(X)$ – the space of all continuous function on X, such as the space of infinitely differentiable function with compact supports and many others. In all these cases, (89) has only trivial – 0 or 1 – solutions as elements of $\mathcal{A}$.

However, trivial solutions are insufficient for the logical structure of our algebra since one expects that the propositions should generate, in some sense, the whole space of observables. It is therefore impossible to establish a classical logic for such systems and we have to look for a possibility that enables us to take into account those observables which are "close" to indicators of sets, thus replacing sets by fuzzy sets. By a fuzzy set A we mean a function

$$m_A : X \longrightarrow [0, 1] , \tag{90}$$

called the *membership function*. For a given $x \in X$ $m_A(x)$ can be interpreted as the degree of the membership that x is an element of the set A. The basic relations between fuzzy sets can be defined in the same way as, expressed by indicators, relations between ordinary sets. For example, a fuzzy set A is a subset of a fuzzy set B if

$$m_A(x) \le m_B(y) \text{ for each } x \in X . \tag{91}$$

Similarly, the complement of a fuzzy set A is

$$1 - m_A(x). \tag{92}$$

For reasons to be discussed below, we cannot, however, assume a universal definition of the union and intersection of fuzzy sets.

The most natural way to extend a given definition to a larger class of objects is to specify the class of axioms which we want to be satisfied. This does also apply to the generalization of the ordinary set operations to fuzzy sets, which are to satisfy a set of axioms concerning fuzzy connectives. We are not going to present all these "natural axioms" which are discussed extensively in Klir and Folger (1988). All these axioms are satisfied by the following choice of the union and the intersection of two fuzzy sets:

$$m_{A \cup B}(x) = \max\{m_A(x), \ m_B(x)\}, \tag{93}$$

$$m_{A \cap B}(x) = \min\{m_A(x), \ m_B(x)\}. \tag{94}$$

We can assume these definitions in the case where the space of observables is $\mathcal{C}(X)$. However, if we demand that the observables should be at least differentiable functions then (93) and (94) lead to difficulties since these elementary operations lead outside the space of observables. Therefore we propose to use other fuzzy connectives which belong to the Yager class:

$$m_A(x) \cup_F m_B(x) \equiv m_{A\cup B}(x) = m_A(x) + m_B(x) - m_A(x)m_B(x), \quad (95)$$

$$m_A(x) \cap_F m_B(x) \equiv m_{A\cap B}(x) = m_A(x)m_B(x). \quad (96)$$

These connectives do not satisfy the axiom of idempotency since

$$m_A \cup_F m_A = 2m_A - m_A^2 \quad \text{and} \quad m_A \cap m_A = m_A^2. \quad (97)$$

However, the violation of idempotency concerns only the "purely fuzzy" part of m_A, i.e., those x for which $0 < m_A(x) < 1$. Moreover, performing the same operation on a fuzzy set m_A we decrease its fuzziness in the case of union and enhance it in the case of intersection. This seems to be in perfect agreement with the intuitive point of view on repeated observations of a physical system. Another interesting consequence of equalities (97) is the impossibility to perform infinitely many operations. This means that we do not have an analog of σ-completeness, i.e.:

$$\text{if } \ m_{A_1} \leq m_{A_2} \leq \dots \text{ then the fuzzy join } \bigcup_n m_{A_n} \text{ exists in } \mathcal{A} \quad (98)$$

or

$$\text{if } \ m_{A_1} \geq m_{A_2} \geq \dots \text{ then the fuzzy meet } \bigcap_n m_{A_n} \text{ exists in } \mathcal{A} . \quad (99)$$

The given arguments concerning the impossibility of performing infinitely many operations on the same fuzzy set and the related incompleteness cannot be applied for a space $\mathcal{A}$ with the operations (93) and (94) since in this case the axiom of idempotency is fulfilled. Nevertheless it can be proven that the space $\mathcal{C}(X)$ with these operations is still incomplete. Note also that, if the space $\mathcal{A}$ contains the identity, then both types of connectives (93)-(94) and (95)-(96) satisfy DeMorgan's laws with respect to the fuzzy complement.

Let us now return to the question how to describe the set of the most elementary observables for a given dynamical system. Having introduced a fuzzy logic we must answer the question whether this new logic can be identified with the set of the most elementary observables. As we already know the indicators of sets cannot be considered as observables. Therefore we would like to call those observables "most elementary" which are sufficiently close to indicators. Of course, not all fuzzy subsets of X can be considered to be most elementary since the correspondence

$$\text{set } A \longmapsto \text{ fuzzy set } m_A \quad (100)$$

does not mean any particular relation between A and m_A. The latter can be an arbitrary function; for example, it can be the same for several different sets. Thus the question is: how to measure the fuzziness?.

Several measures of fuzziness have been proposed in the literature but none of them is, as we shall see below, appropriate to determine the class of the fuzzy sets which we can intuitively accept as most elementary observables. Indeed, these measures of fuzziness depend more or less explicitly on the maximal or, most commonly, on an average distance between the functions 1_A and m_A. The first type does not distinguish between fuzzy sets which differ insignificantly from 1_A and are, say, identically 0 or 1. Measures of fuzziness which are based on an average distance are inappropriate for another reason. Let us recall that in our approach to unstable dynamical systems the space $\mathcal{A}$ of observables is the test function space for all physically relevant states. Now consider, for example, the functional δ_{x_0}, where x_0 is a given element of X, $\langle f, \delta_{x_0} \rangle = f(x_0)$, defined on the space $\mathcal{C}_b(X)$ – the space of all continuous bounded functions. It extends, as a Radon measure, on indicators of measurable subsets of X. In particular, taking as A an open set which contains x_0 we have

$$\langle 1_A, \delta_{x_0} \rangle = 1 . \tag{101}$$

Equality (101) is no longer true if we replace 1_A by a fuzzy set m_A – it can be an arbitrary number within the interval [0,1]. However, we expect that (101) will be (at least approximately) true for most elementary observables. In other words, most elementary observables should play the role of indicators, giving reliable results. This condition cannot be satisfied when the most elementary observables are classified according to the average distance between 1_A and m_A.

In order to avoid these problems, we postulate that the most elementary observables have to satisfy the following property:

1. For each functional ν on $\mathcal{A}$ which has a correctly defined support such that sup $\nu \subset A$ and each $\varphi \in \mathcal{E}$ such that $\varphi(x) = 1$ on A we have

$$\langle m_A, \nu \rangle = \langle \varphi, \nu \rangle . \tag{102}$$

This imposes a significant reduction of the class $\mathcal{E}$ of those fuzzy sets which can be regarded as most elementary observables. This class must not, however, be too small. For this reason we postulate further

2. that the class $\mathcal{E}$ separates the elements of the dual to $\mathcal{A}$, i.e., if μ and ν are two elements of the dual space then there is an $m_A \in \mathcal{E}$ such that

$$\langle m_A, \mu \rangle \neq \langle m_A, \nu \rangle . \tag{103}$$

3. and that each function from $\mathcal{A}$ can be approximated by linear combinations of elements from $\mathcal{E}$.

We have briefly referred to the basic problems related to the fuzzy logic in unstable system. The full discussion and the description of the spaces of most elementary observables for concrete spaces and their properties will be the subject of a forthcoming paper.

6   Concluding Remarks

1. The Koopman operators of chaotic systems are shift operators of uniform countably infinite multiplicity on the Hilbert space $L^2 \ominus [1]$ of square integrable deviations from equilibrium. This observation follows if one compares the discussion of Cornfeld et al. (1982) and Eckmann and Ruelle (1985) with the terminology of Sz-Nagy and Foias (1970), Halmos (1982), and Hoffmann (1988). The Koopman operator is a unilateral shift for exact endomorphisms (Renyi map, tent map, logistic map, Chebyshev map) and a bilateral shift for Kolmogorov automorphisms (baker's map, Arnold cat map). The spectral decomposition presented in Sect. 2 shows that the extended unilateral shifts have spectral decompositions although they are not decomposable and have no spectral theorem in Hilbert space (Hoffmann 1988).

2. In case of chaotic maps, the admissible initial densities exclude Dirac's delta functions. This means that the trajectories $\delta(y' - S^n y)$, $n = 0, 1, 2, \ldots$, are excluded from the domain of the Frobenius-Perron operator. The spectral decomposition (13) can be used for probabilistic predictions with initial densities representing the experimentally accessible knowledge of initial conditions which are never sharply localized to a point. This reflects the intrinsically probabilistic character of unstable dynamical systems.

3. Apart from predictability questions, the extension of the evolution operator to suitable rigged Hilbert spaces leads to a natural classification of dynamical systems. For stable systems the extension has the same spectrum within the Hilbert space formulation, but for unstable systems the extension leads to resonances which are not visible in the Hilbert space spectrum. Two unstable systems are physically isomorphic if they have the same decay rates, i.e., the same resonance spectra.

4. The extended evolution of invertible unstable systems like Kolmogorov systems or quantum unstable systems like the Friedrichs model (Antoniou and Prigogine 1993) split into two distinct semigroups. In this way the problem of irreversibility is resolved in a natural way without any additional assumptions like ad hoc coarse graining or asymptotic approximations. As Prigogine (1980, 1992) puts it, irreversibility should be considered as an intrinsic property of certain dynamical systems and manifests itself without additional assumptions in an extended formulation of the dynamics. The suggestion to use the rigged Hilbert spaces to give meaning to the extended dynamics is due to Bohm (1981) who conjectured that "the rigged Hilbert space with a suitably defined topology can accomplish it". Bohm and Gadella (1989) constructed the rigged Hilbert spaces of Hardy class for scattering resonances and Gamov

vectors. In fact, the Gamov vectors of Bohm and Gadella define the Riesz projection associated with resonances (Petrosky and Prigogine 1991, Antoniou et al. 1996). The arrow of time associated with the measurement process coincides with the arrow of time associated with resonance scattering (Bohm et al. 1995).

5. The logic of unstable quantum systems with diagonal singularity reflects the nonlocality of classical chaotic systems, as it is not possible to combine propositions/questions ad infinitum. The restriction to Hilbert space logic is not a physical necessity but a mathematical postulate (Mackey 1963) in order to force the lattice of propositions to be essentially a lattice of Hilbert space projections. As Varadarajan (1993) has pointed out, "the result of von Neumann, Mackey, Gleason and their successors have pretty much closed the door for the discovery of any non trivial situation violating the canonical interpretations". We see now for the first time that systems which are responsible for irreversibility provide physical examples of a logic beyond conventional Hilbert space logic.

6. We wish to conclude by emphasizing that the results presented here are the first exploration towards a new systematic approach to complex systems on the basis of intrinsically irreversible and intrinsically probabilistic extensions of dynamics to suitable functional spaces. The origin of this approach lies in the work of Prigogine and his collaborators almost 30 years ago (Prigogine 1980).

Acknowledgments: This work originated from the persistent and stimulating discussions with Prof. I. Prigogine. Parts of the algebraic formulation were worked out together with Drs. S. Tasaki and R. Laura. We are also grateful to Profs. J.-P. Antoine, A. Bohm, E. Brändas, M. Gadella, C. George, K. Gustafson, L. Horwitz, B. Misra, T. Petrosky and A. Weron for several critical remarks. We thank Dr. H. Atmanspacher who suggested to us to write this paper and pointed out a confusion of the standard notation in the Liouville equations for states and observables. This work received financial support from the European Commission ESPRIT P9282 ACTCS and the Belgian Government through the Interuniversity Attraction Poles.

References

Antoine J.-P., Epifanio G., and Trapani C. (1983): Complete sets of unbounded observables. *Helv. Phys. Acta* **56**, 1175–1186.

Antoine J.-P. and Karwowski J. (1985): Partial *-algebras of closed linear operators in Hilbert space. *Publ. R.I.M.S. Kyoto Univ.* **21**, 205–236.

Antoniou I., Dimitrieva L., Kuperin Yu., and Melnikov Yu. (1996a): Resonances and the extension of dynamics to rigged Hilbert spaces. *Computers and Mathematics with Applications*. In press.

Antoniou I. and Prigogine I. (1993): Intrinsic irreversibility and integrability of dynamics. *Physica A* **192**, 443–464.

Antoniou I., Qiao B., and Suchanecki Z. (1996b): Generalized spectral decomposition and intrinsic irreversibility of the Arnold cat map. *Chaos, Solitons and Fractals*. In press.

Antoniou I. and Qiao B. (1996a): Spectral decomposition of the tent maps and the isomorphism of dynamical systems. *Phys. Lett. A* **215**, 280–290.

Antoniou I. and Qiao B. (1996b): Spectral decomposition of the chaotic logistic map. *Nonlinear World*. In press.

Antoniou I. and Qiao B. (1996c): Spectral decomposition of Chebyshev maps. *Physica A*. In press.

Antoniou I. and Tasaki S. (1992): Generalized spectral decomposition of the β-adic baker's transformation and intrinsic irreversibility. *Physica A* **190**, 303–329.

Antoniou I. and Tasaki S. (1993a): Spectral decomposition of the Renyi map. *J. Phys. A* **26**, 73–94.

Antoniou I. and Tasaki S. (1993b): Generalized spectral decompositions of mixing dynamical systems. *Int. J. Quant. Chem.* **46**, 425–474.

Antoniou I. and Suchanecki Z. (1994): The logic of quantum systems with diagonal singularities. *Found. Phys.* **24**, 1439–1457.

Antoniou I. and Suchanecki Z. (1995): Quantum systems with diagonal singularity: observables, states, logic, and intrinsic irreversibility. In *Nonlinear, Deformed and Irreversible Systems*, ed. by H.-D. Doebner, V.K. Dobrev, and P. Nattermann (World Scientific, Singapore), 22–52.

Antoniou I. and Suchanecki Z. (1996): Quantum systems with diagonal singularity. *Adv. Chem. Phys.* In press.

Arnold V. (1978): *Mathematical Methods of Classical Mechanics* (Springer, Berlin).

Balescu R. (1975): *Equilibrium and Non-Equilibrium Statistical Mechanics* (Wiley, New York).

Bandtlow O.F., Antoniou I., and Suchanecki Z. (1996a): Resonances of dynamical systems and Fredholm-Riesz operators on rigged Hilbert spaces. *Computers and Mathematics with Applications*. In press.

Bandtlow O.F., Antoniou I., and Suchanecki Z. (1996b): On some properties of the Frobenius-Perron operator of the Renyi map. *J. Stat. Phys.* In press.

Beltrametti E.G. and Cassinelli G. (1981): *The Logic of Quantum Mechanics*. In *Encyclopedia of Mathematics and its Applications* (Addison-Wesley, Reading).

Birkhoff G. and von Neumann J. (1936): The logic of quantum mechanics. *Ann. Math.* **37**, 823–843.

Boas R.P. and Buck R.C. (1964): *Polynomial Expansions of Analytic Functions* (Berlin, Springer).

Bohm A. (1981): Resonance poles and Gamow vectors in the rigged Hilbert space formulation of quantum mechanics. *J. Math. Phys.* **22**, 2813–2823.

Bohm A. (1986): *Quantum Mechanics: Foundations and Applications* (Springer, Berlin).

Bohm A., Antoniou I., and Kielanowski P. (1995): A quantum mechanical arrow of time and the semigroup time evolution of Gamow vectors. *J. Math. Phys.* **36**, 2593–2504.

Bohm A. and Gadella M. (1989): *Dirac Kets, Gamow Vectors, and Gelfand Triplets*. Springer Lecture Notes on Physics **348** (Springer, Berlin).

Bratteli O. and Robinson D. (1979, 1981): *Operator Algebras and Quantum Statistical Mechanics, Vols. I, II* (Springer, Berlin).

Brout R. and Prigogine I. (1956): Statistical mechanics of irreversible processes, part VII: a general theory of weakly coupled systems. *Physica* **22**, 621–636.

Cornfeld I.P., Fomin S.V., and Sinai Ya.G. (1982): *Ergodic Theory* (Springer, Berlin).

da Costa N. and Doria F. (1991): Undecidability and incompleteness in classical mechanics. *Int. J. Theor. Phys.* **30**, 1041–1073.

de Dormale B. and Gautrin H.-F. (1975): Spectral representation and decomposition of self-adjoint operators. *J. Math. Phys.* **16**, 2328–2332.

Dirac P.A.M. (1958): *The Principles of Quantum Mechanics* (Clarendon Press, Oxford).

Dixmier J. (1977): *C*-Algebras* (North-Holland, Amsterdam).

Dixmier J. (1981): *Von Neumann Algebras* (North-Holland, Amsterdam).

Dunford N. and Schwartz J. (1988): *Linear Operators I: General Theory* (Wiley, New York).

Eckmann J.-P. (1989): Resonances in dynamical systems. In *Proc. IX Intl. Congr. on Mathematical Physics*, ed. by B. Simon et al. (Adam Hilger, Bristol).

Eckmann J.-P. and Ruelle D. (1985): Ergodic theory of chaos and strange attractors. *Rev. Mod. Phys.* **57** 617–656.

Emch G. (1972): An introduction to C*-algebraic methods in physics. *Adv. Chem. Phys.* **22**, 315–364.

Epifanio G. and Trapani C. (1988): Partial *-algebras of matrices and operators. *J. Math. Phys.* **29**, 536–540; see also: Partial *-algebras and Volterra convolution of distribution kernels. *J. Math. Phys.* **32**, 1096–1101 (1991).

Gelfand I. and Shilov G. (1967): *Generalized Functions, Vol. 3: Theory of Differential Equations* (Academic Press, New York).

Gelfand I. and Vilenkin N. (1964): *Generalized Functions, Vol. 4: Applications of Harmonic Analysis* (Academic Press, New York).

George C. (1973): Subdynamics and correlations. *Physica* **65**, 277–302.

Goodrich K., Gustafson K., and Misra B. (1980): On converse to Koopman's lemma. *Physica A* **102**, 379–388.

Grecos A., Guo T., and Guo W. (1975): Some formal aspects of subdynamics. *Physica A* **80**, 421–446.

Haag R. (1992): *Local Quantum Physics* (Springer, Berlin).

Halmos P. (1982): *A Hilbert Space Problem Book* (Springer, Berlin).

Hoffmann K. (1988): *Banach Spaces of Analytic Functions* (Dover, New York).

Jauch J.M. (1960): Systems of observables in quantum mechanics. *Helv. Phys. Acta* **33**, 711–726.

Jauch J.M. (1973): *Foundations of Quantum Mechanics* (Addison-Wesley, Reading).

Jauch J.M. and Misra B. (1965): The spectral representation. *Helv. Phys. Acta* **38**, 30–52.

Keller G. (1984): On the rate of convergence to equilibrium in one-dimensional systems. *Commun. Math. Phys.* **96**, 181–193.

Klir G.J. and Folger T.A. (1988): *Fuzzy Sets, Uncertainty, and Information* (Prentice-Hall, Englewood Cliffs).

Koopman B. (1931): Hamiltonian systems and transformations in Hilbert space. *Proc. Natl. Acad. Sci. USA* **17**, 315–318.

Koopman B. and Neumann J. von (1932): Dynamical systems of continuous spectra. *Proc. Natl. Acad. Sci. USA* **18**, 255–263.

Lasota A. and Mackey M.C. (1985): *Probabilistic Properties Deterministic Systems* (Cambridge University Press, Cambridge).

Mackey G.W. (1963): *The Mathematical Foundations of Quantum Mechanics* (Benjamin, New York)

Maurin K. (1968): *General Eigenfunction Expansions and Unitary Representations of Topological Groups* (Polish Scientific Publishers, Warsaw).

Naimark M. (1972): *Normed Algebras* (Wolfer-Noordhoff, Groningen).

Petrosky T. and Hasegawa H. (1989): Subdynamics and nonintegrable systems. *Physica A* **160**, 351–385.

Petrosky T. and Prigogine I. (1991): Alternative formulation of classical and quantum dynamics for non-integrable systems. *Physica A* **175**, 146–209.

Petrosky T., Prigogine I., and Tasaki S. (1991): Quantum theory of non-integrable systems. *Physica A* **173**, 175–242.

Philippot J. (1961): Initial conditions in the theory of irreversible processes. *Physica* **27**, 490–496.

Piron C. (1964): Axiomatique quantique. *Helv. Phys. Acta* **37**, 439–468.

Pöschell J. (1989): *On Small Divisors with Spatial Structure*. Habilitationsschrift, Universität Bonn.

Poincaré H. (1892): *Les Methodes Nouvelles de la Mécanique Céleste, Vol. I* (Paris); reprinted in English by Dover, New York 1957.

Pollicott M. (1985): On the rate of mixing of axiom A flow. *Invent. Math.* **81**, 413–426; see also: Meromorphic extensions of generalized zeta functions. *Invent. Math.* **85**, 147–164 (1986).

Pollicott M. (1990): The differential zeta function for axiom A attractors. *Ann. Math.* **131**, 335–405.

Pour-el M. and Richards J. (1989): *Computability in Analysis and Physics* (Springer, Berlin).

Prigogine I. (1962): *Non-Equilibrium Statistical Mechanics* (Wiley, New York).

Prigogine I. (1980): *From Being to Becoming* (Freeman, San Francisco).

Prigogine I. (1992): Dissipative processes in quantum theory. *Phys. Rep.* **219**, 93–120.

Prigogine I. and Balescu R. (1959): Irreversible processes in gases. I. The diagram technique. *Physica* **25**, 281–301.

Prigogine I., George C., Henin F., and Rosenfeld L. (1973): A unified formulation of dynamics and thermodynamics. *Chem. Scr.* **4**, 5–32.

Prugovecki E. (1981): *Quantum Mechanics in Hilbert Space* (Academic Press, New York).

Ruelle D. (1986): Resonances of chaotic dynamical systems. *Phys. Rev. Lett.* **56**, 405–407; see also: Locating resonances for axiom A dynamical systems. *J. Stat. Phys.* **44**, 281–292 (1986).

Ruelle D. (1990): The thermodynamic formalism for expanding maps. *Commun. Math. Phys.* **125**, 239–262; see also: An extension of the theory of Fredholm determinants. *Publ. Math. IHES* **72**, 175–193 (1990).

Segal I. (1947): Postulates for general quantum mechanics. *Ann. Math.* **48**, 930–948.

Suchanecki Z., Antoniou I., Tasaki S., and Bandtlow O.F. (1996): Rigged Hilbert spaces for chaotic dynamical systems. *J. Math. Phys.*. In press.

Sz-Nagy B. and Foias C. (1970): *Harmonic Analysis of Operators in Hilbert Space* (North-Holland, Amsterdam).

Tangermann F. (1986): *Meromorphic Continuation of Ruelle Zeta Functions*. Thesis, Boston University.

Tasaki S., Suchanecki Z., and Antoniou I. (1993): Ergodic properties of piecewise linear maps on fractal repellers. *Phys. Lett. A* **179**, 103-110.

Tasaki S., Antoniou I., and Suchanecki Z. (1994): Spectral decomposition and fractal eigenvectors for a class of piecewise linear maps. *Chaos, Solitons and Fractals* **4**, 227-254.

Treves F. (1967): *Topological Vector Spaces, Distributions, and Kernels* (Academic Press, New York).

van Hove L. (1955): Energy corrections and persistent perturbation effects in continuous spectra. *Physica* **21**, 901-923.

van Hove L. (1956): Energy corrections and persistent perturbation effects in continuous spectra. II. The perturbed stationary states. *Physica* **22**, 343-354.

van Hove L. (1957): The approach to equilibrium in quantum statistics. *Physica* **23**, 441-480.

van Hove L. (1959): The ergodic behavior of quantum many-body systems. *Physica* **25**, 268-276.

Varadarajan V.S. (1993): Quantum theory and geometry: sixty years after von Neumann. *Int. J. Theor. Phys.* **32**, 1815-1834.

von Neumann J. (1955) *Mathematical Foundations of Quantum Mechanics* (Princeton University Press, Princeton).

Dynamical Entropy in Dynamical Systems

Harald Atmanspacher

Max-Planck-Institut für extraterrestrische Physik, D–85740 Garching, Germany

Abstract. The concept of a dynamical entropy h_T in the sense of Kolmogorov and Sinai (KS-entropy) is introduced (1) and discussed with respect to some of its most significant implications. These comprise (2) its role in intrinsically unstable processes like K-flows or deterministic chaos, (3) the distinction of ontic states (phase points) and epistemic states (phase volumes), (4) the issue of information flow in nonlinear dynamical systems, (5) the status of the KS-entropy in a non-commutative algebra of operators, and (6) its relevance for recent developments concerning the issue of temporal nonlocality. Finally (7), a few possible perspectives, particularly in view of interdisciplinary relationships, are briefly indicated.

1 KS-Entropy ...

In contrast to most familiar concepts of entropies like those according to Boltzmann, Gibbs, von Neumann, Shannon and others (Grad 1961, Wehrl 1978, Wehrl 1991), dynamical entropies are functions of the dynamics of a system, e.g., its history, rather than of its states. If a dynamical system is considered in a state space Ω with probability measure μ defined on a σ-algebra Σ of measurable subsets A in Ω, then the dynamics enters formally by introducing an automorphism $T : \Omega \rightarrow \Omega$. If the dynamics is reversible then $\mu(T^{-1}(A)) = \mu(T(A)) = \mu(A)$ for all $A \in \Sigma$. For a one-parameter group of such a μ-preserving invertible transformation, the evolution of a corresponding system (Ω, μ, T_t) is both forward and backward deterministic, if the parameter is chosen to be a (discrete or continuous) time $t \in \mathbb{R}$. In this case, there is no preferred direction of time. (This is, of course, different for any irreversible dynamics, where T_t becomes a one-parameter semigroup of non-invertible transformations.) The inclusion of T is the essential element distinguishing a *dynamical* system from a system in general. While non-dynamical entropies can be defined for systems (Ω, μ), dynamical entropies require that T is explicitly taken into account.

Dynamical entropies are entropies of a temporal evolution as such rather than entropies of states at a certain moment in such an evolution. Within the last years, a specific one among quite a number of dynamical entropies (Wehrl 1991) has received particular attention: the dynamical entropy according to Kolmogorov (1958) and Sinai (1959), briefly KS-entropy. The main reason for this popularity is that this kind of dynamical entropy has turned out to be an extremely useful tool in the characterization of systems showing chaotic behavior in the sense of *deterministic chaos*. This will be addressed in detail in the following section. The original proposals by Kolmogorov and Sinai did not explicitly mention this scope of interest. Instead, they were concerned with the way in which an entropy can be ascribed to the automorphism $T : \Omega \to \Omega$. This can be done by considering a partition P in Ω with disjoint measurable sets A_i $(i = 1, ..., m)$ and studying its temporal evolution $TP, T^2P, ...$ If the entropy $H(P)$ of P is given by

$$H(P) = -\sum \mu(A_i) \ln \mu(A_i), \tag{1}$$

then the dynamical KS-entropy h_T is defined as the supremum of $H(P,T)$ over all partitions P,

$$h_T = \sup_P H(P,T), \tag{2}$$

with

$$H(P,T) = \lim_{n\to\infty} \frac{1}{n} H(P \vee TP \vee ... \vee T^{n-1}P). \tag{3}$$

Three remarks: (1) The latter limit is well-defined because H is sub-additive, i.e., $H(P \vee P') \leq H(P) + H(P')$ for two partitions P, P'. (2) The partition providing the supremum of $H(P,T)$ is the so-called generating partition or, more specifically, the so-called K-partition (Cornfeld et al. 1982). The generating partition is constructively given by the dynamics of a system. (3) The KS-entropy is a relevant concept for commutative (Abelian) algebras of observables but cannot naively be taken over to non-commuting observables in the sense of conventional quantum theory. It can, however, acquire significant meaning for operator algebras in Koopman representations of classical systems (see Sects. 5 and 6). For non-commutative (non-Abelian) algebras of observables of conventional quantum systems, alternative concepts (mathematically generalizing the classical KS-entropy) have been introduced, e.g., by Connes, Narnhofer, and Thirring (1987), see also Hudetz (1988).

The KS-entropy can be considered as a paradigmatic case of the explicit consideration of genuinely *temporal* affairs in wide ranges of theoretical physics. By stressing *genuine* temporality, I mean to indicate that physics has more to say about time than is reflected by just using it as an arbitrary parameter. In the following I intend to address a number of consequences which the concept of a KS-entropy has for several currently discussed topics like chaos, information, nonlocality and others. I shall not entirely avoid formal details, but I shall try to make transparent how I understand the significance of the formalism. For a first quick reading, all "remarks" can be skipped.

2 ... and Intrinsic Instability

In the theory of dynamical systems, T_t is often called a flow $T_t(x) = T(x,t)$ on the state space (or phase space) Ω, where x is a phase point in Ω. This flow is said to be generated by a transformation F that can be discrete,

$$x(t+1) = F(x(t)), \tag{4}$$

or continuous in time t

$$\frac{dx(t)}{dt} = \dot{x}(t) = F(x(t)). \tag{5}$$

The continuous case represents a first-order, ordinary differential equation system as a very simple example which, however, is sufficient to illustrate the basic notions. The vector $x(t)$ characterizes the state of the system at time t; its components represent its continuous observables $(x_1, ..., x_d)$. F is a matrix containing the generally nonlinear coupling among the observables, whose number defines the dimension d of the state space Ω.

To characterize the flow T_t, i.e. the temporal evolution of $x(t)$ as the solution of (5), it is of crucial significance how T_t behaves under the influence of small perturbations δx. Such a characterization specifies the stability of the system and can be obtained in terms of a linear stability analysis. Avoiding too many details, a linear stability analysis yields local (in Ω) rates of amplification or damping of perturbations $\delta x(t)$ with respect to a reference state or a reference trajectory $x(t)$, respectively. From these local rates one can obtain a global dynamical invariant of T_t, essentially as a temporal average of the local rates. These global invariants are the so-called Lyapunov exponents:

$$\lambda_i = \lim_{t \to \infty} \frac{1}{t} \ln \left| \frac{\delta x_i(t)}{\delta x_i(0)} \right| \tag{6}$$

Under certain conditions the sum of all positive Lyapunov exponents can be identified as the KS-entropy h_T:

$$h_T = \sum_i \lambda_i^+ = \begin{cases} \sum \lambda_i & \text{if } \lambda_i > 0 \\ 0 & \text{otherwise} \end{cases} \tag{7}$$

Remark: Ledrappier and Young (1985) have proven that $\sum \lambda_i^+ D_i = h_T$ where D_i is the partial information dimension $0 \le D_i \le 1$ if T is a C^2-diffeomorphism and μ an associated T-invariant ergodic measure. Moreover: if T is hyperbolic and μ is absolutely continuous with respect to the Lebesgue measure along the unstable manifolds of T, then μ is called a Sinai-Ruelle-Bowen (SRB) measure and $D_i = 1$ such that $\sum \lambda_i^+ = h_T$. This is the Pesin identity (Pesin 1977). While the conditions for the result of Ledrappier and Young are fairly unrestrictive, the condition for Pesin's identity that the natural measure is a SRB measure is perhaps not always satisfied for practically relevant systems (cf. Tasaki et al. (1993) for a proposed extension of the SRB criterion). In any case we have the inequality $\sum \lambda_i^+ \ge h_T$.

The sum of all (d) Lyapunov exponents allows an elegant and fundamental classification of dynamical systems.

- $\sum \lambda_i > 0$ characterizes systems which are unstable in a global sense, for instance random systems. Their phase volume spreads over the entire phase space as $t \to \infty$.
- $\sum \lambda_i = 0$ characterizes conservative (e.g. Hamiltonian) systems. They are stable, but not asymptotically stable. Their phase volume remains constant in time (Liouville's theorem). Conservative systems with $h_T > 0$ are so-called K-flows.
- $\sum \lambda_i < 0$ characterizes dissipative systems. They have a shrinking phase volume and are asymptotically stable. It is intuitively suggestive (but not yet finally understood, see Ruelle 1981, Milnor 1985) that the flow T_t of a dissipative system is asymptotically restricted to a finite subspace of the entire phase space. This subspace is called an attractor for T_t. If $\lambda_i < 0 \,\forall i$ this attractor is a fixed point. If there are k vanishing Lyapunov exponents and the others are negative, then the attractor is a k-torus (limit cycle for $k = 1$). For systems with at least three degrees of freedom, $d \geq 3$, the condition of a negative sum of Lyapunov exponents can be satisfied by a combination of positive and negative ones. This situation defines a chaotic (strange) attractor in the sense of deterministic chaos.

From a historical point of view, it is interesting to note that chaotic behavior in the sense described above was for the first time explicitly mentioned in a paper by Koopman and von Neumann (1932): "... the states of motion corresponding to any set M in Ω become more and more spread out into an amorphous everywhere dense chaos. Periodic orbits, and such like, appear only as very special possibilities of negligible probability."

Both conservative K-flows and dissipative chaotic attractors provide what has now become well-known as sensitive dependence of the evolution of a system on small perturbations in the initial conditions. This dependence is due to an intrinsic instability that is formally reflected by the existence of positive Lyapunov exponents. The KS-entropy of a system is an operationally accessible quantity (Grassberger and Procaccia 1983). A positive (finite) KS-entropy is a necessary and sufficient condition for chaos in conservative as well as in dissipative systems (with a finite number of degrees of freedom). Chaos in this sense is not the same as randomness. It fills the gap between totally unpredictable random processes, such as white noise ($h_T \to \infty$) and regular (e.g., periodic, etc.) processes with $h_T = 0$.

Remark: The characterization of a dynamical system by its KS-entropy is not necessarily complete. For instance, systems with the same KS-entropy may approach equilibrium with different rates. Although their spectral and statistical properties are indistinguishable as far as expectation values (e.g., suitable limits) are concerned, they are not isomorphic concerning the way in which these expectation values (limits) are approached. See Antoniou and Qiao (1996) for a specific demonstration of this difference with respect to the spectral decomposition of the tent map. Another formal way to deal with problems like this is known as "large deviations statistics", a relatively new field of mathematical statistics which has recently been introduced into the context of dynamical systems (Oono 1989).

3   ... and the Ontic-Epistemic-Distinction

Deterministic chaos is deterministic, yet not determinable – at least not to arbitrary precision. What does this mean? In a certain sense, the distinction between determinism and determinablity refers to two different concepts of realism in physics. The issue of determinism relates to all sorts of inquiries into an independent ("when nobody looks") reality of the outside world. In contrast, to talk about determinability expresses a more moderate approach. The question "What is the outside world?" is replaced by "What can we know about the outside world?". Philosophically the distinction between these two questions is very much in the spirit of Kant's distinction of transcendental idealism and empirical realism. As an empirical science, physics addresses solely questions of the second kind. However, the mathematical formalism that constitutes the formal basis of physics often leads to a way of thinking very much in accordance with the first kind of question.

For a long time in the history of science, the two questions were not distinguished explicitly. Scientists and philosophers of science did not much worry about a possible difference between the world "as it really is", ontically, and the world "as it appears to us", epistemically. It is due to more recent interest in a systematic philosophy of quantum theory (unlike Bohr's) that the corresponding distinction is now beginning to be considered seriously and explicitly. The basic reference in this regard is Scheibe (1973). Later on, Primas (1990a) developed the ontic-epistemic-distinction into a powerful tool to understand consistently the relations between different interpretational schemes for quantum theory. A more general distinction entailing that of ontic and epistemic states is the distinction of internal (endo) and external (exo) perspectives (Primas 1994a). Avoiding technical details, I adopt the following compact characterizations (Primas 1990a, 1994a):

Ontic states in an ontic state space describe all properties of a physical system *completely* ("Completely" in this context means that an ontic state is "just the way it is", without any reference to epistemic knowledge or ignorance.) Ontic states are the referents of *individual* descriptions. Their properties are *empirically inaccessible* and formalized by *intrinsic observables* as elements of a C^*-algebra. They are subjects of a *universal endo-description* referring to an independent concept of reality. *Epistemic states* describe *our (usually incomplete) knowledge* of the properties of a physical system, i.e., based on a finite partition of the relevant state space. The referents of *statistical* descriptions are epistemic states. Their properties are *empirically accessible* and formalized by *contextual observables* as elements of a W^*-algebra.

They are subjects of a *contextual exo-description* refering to an empirical concept of reality depending on contexts.

> **Remark:** C^*-algebras and W^*-algebras are descriptive tools used in algebraic quantum theory; for more formal details and further references see Primas (1983, 1990a, 1994a). As compared to other approaches, this theory appears to provide a most consistent mathematical framework together with a most satisfying degree of insight into the structure of the material world. Major disadvantage: it is not always easy to handle for specific applications. Algebraic quantum theory covers both commutative ("classical" regime) and non-commutative ("quantum" regime) algebras of observables, and it entails both ontic and epistemic levels of description. With this scope, algebraic quantum theory is a promising candidate for solid and comprehensive approaches to a solution of the measurement problem (see, e.g., Hepp 1972, Lockhart and Misra 1986, Primas 1990b, Amann 1995).

> **Further remark:** The temporal evolution of ontic states is usually described by *universal, deterministic laws* given by an invariant Hamiltonian one-parameter group. Correspondingly, epistemic states are regarded to evolve according to *phenomenological, irreversible laws* which can be given by a dynamical one-parameter semigroup if the state space is properly chosen. However, I have elsewhere (Atmanspacher 1994a) argued that a straightforward assignment of the different kinds of dynamics to an ontic or epistemic level, respectively, might lead to inconsistencies. This is most obvious in the problem of formulating a dynamics of measurement. Since measurement is (by definition) the act leading from an ontic to an epistemic level, the dynamics of this act cannot be distinctly associated with one of these levels only. Note that this is not an objection against the ontic-epistemic-distinction as such but an indication that it is not sufficient (see also Sect. 6).

> **Yet another remark:** The distinction between ontic and epistemic states as used in quantum theory is *not* identical with that of micro- and macro-states as used in fields like statistical mechanics or thermodynamics. The main difference is that ontic quantum states are empirically inaccessible *in a fundamental, irreducible sense* whereas microstates in general can be considered epistemically, namely as macrostates with respect to an underlying micro-description. This leads to a hierarchy of levels, in which conceptual objects (e.g. molecules) emerge as macrostates by suitable limits (e.g. Born-Oppenheimer-approximation) with respect to the (e.g. atomic) microstate description (Primas 1994b). In a biologically oriented terminology, this resembles the emergence of phenotypes from genotypes.

Although the decisive momentum for an ontic-epistemic distinction in physics came from quantum theory, it is also important and useful in classical physics. Classical point mechanics provides an illustrative example for a "degeneracy" which confuses ontic and epistemic levels, whereas classical statistical mechanics is clearly epistemic. States in the sense of phase points

x in Ω and continuous trajectories $x(t)$, as they have been introduced in the preceding section, refer to an ontic description that can formally be expressed by an infinite refinement of Ω. Referring to empirically accessible states would require one to use of phase volumes associated with finite knowledge. Corresponding concepts like measurable subsets $A \in \Omega$ or partitions $P(\Omega)$ are relevant in epistemic descriptions. Insofar as our knowledge about a state of a system and its properties is incomplete in principle, epistemic states rather than ontic states have to be used for a description of the empirically accessible world. In this spirit, the notion of a perturbation δx together with an ontic reference state may be understood to constitute a measurable subset $A \in \Omega$, i.e. a phase volume $(\delta x)^d$. And, of course, such a volume can then be reasonably endowed with an interpretation in terms of a finite amount of information (see next section).

The temporal evolution of an ontic state remains unrecognizable as long as this ontic state belongs to the same epistemic state, i.e. as long as it stays in one and the same phase cell of a chosen or otherwise given partition. Refinements of partitions are possible, but they can never be infinite for all empirical purposes. If neighbouring phase points keep their initial distance from each other constant during the evolution of the system, they will (exceptions omitted) not change their status of indistinguishability with respect to a given partition. However, if this distance increases as a function of time, this is no longer so. Initially indistinguishable phase points may become distinguishable after a certain amount of time τ, since they may move into different phase cells, i.e. epistemic states. This does exactly apply to chaotic systems, for which τ is estimated by h_T^{-1}. Needless to say, this constitutes a measurement problem, though conceptually different from quantum theory (Crutchfield 1994, Atmanspacher et al. 1995). As we know very well today, classical point mechanics gets along with its ontic-epistemic-degeneracy only if chaotic processes are disregarded. Adopting (and misusing) a notion coined by Whittaker (1943), one might paraphrase the deterministic, yet not determinable behavior of such systems as "cryptodeterminism".

4   ... and Information Flow

In the terminology of the preceding section, information is a purely epistemic concept insofar as information is only finitely accessible and, thus, corresponds to incomplete knowledge. Dynamical systems can be interpreted as information-processing systems (Shaw 1981) with the KS-entropy h_T as the information flow rate (Goldstein 1981). This can be demonstrated replacing the notion of a perturbation δx in Sect. 2 by the notion of a corresponding uncertainty (incomplete knowledge). In this way the stability analysis of a system is changed into an informational analysis. At the same time, the discussion is shifted from a description using ontic (reference state as phase point) and epistemic (partition) elements to a purely epistemic level (uncertainty as phase volume). A good approximation of the resulting flow of Shannon information I is given by

$$I(t) = I(0) - h_T t. \tag{8}$$

It applies to conservative as well as dissipative systems. In information theoretical terms, the inverse of h_T estimates the time interval τ for which the behavior of the system can reasonably well be predicted from its deterministic equations.

> **Four remarks:** (1) Here and in the remainder of this article, the concept of information is restricted to Shannon information, i.e., it is solely used in a syntactic sense, without any reference to semantics or pragmatics. (2) The partition due to uncertainties is in general different from the partition P introduced in Section 1. For instance, the generating partition is generically given by the dynamics of a system, whereas the concept of an uncertainty refers to an experimental resolution or other external conditions. (3) The linearity of the information flow is "spurious" in the sense that it is merely a consequence of the linearity of the stability analysis on which its derivation is based. It is well-known that any linear analysis is only locally valid, hence the KS-entropy h_T, interpreted as an information flow rate, represents a (shifted) average of local information flow rates. (4) Strictly speaking, there is an additional contribution of the partial dimensions in the proportionality factor to t (Farmer 1982) (cf. the remark on Pesin's identity in Section 2). A more detailed discussion of the information flow in chaotic classical Hamiltonian systems has been given by Caves (1994).

The temporal decrease of $I(t)$ for $h_T > 0$ describes how an external observer loses information about the actual state of a system as time passes by. It is tempting to interpret this as an increasing amount of information in the system itself, generated by its intrinsic instability due to positive Lyapunov exponents enhancing initial uncertainties exponentially (Atmanspacher and Scheingraber 1987). Since such an internal view goes beyond the regime of a purely epistemic scenario, this temptation must be resisted if one wishes to stay within the scheme provided by a clean ontic-epistemic-distinction. The same argument holds if the notion of information is replaced by entropy (Elskens and Prigogine 1986). Weizsäcker's terminology uses *potential* information (Weizsäcker 1985), indicating exactly where the problem lies: the referent of this term becomes *actual* information if and only if it gets epistemic. The interplay between these concepts, including the transition from infinite to finite information and some of its consequences, has extensively been discussed in Atmanspacher (1989).

Another approach dealing with this problem has been proposed by Zurek (1989) (see also Caves 1993). He defines "physical entropy" as the sum of missing information plus known randomness according to $S = H + C$ where H is the conventional statistical entropy (outside view) and C is the algorithmic randomness (à la Kolmogorov and Chaitin, also called algorithmic information content or algorithmic complexity) of a data string produced by the system's evolution (inside view). The problem with the second term is that the corresponding states of the system must be "known" to some "information gathering and using system", short IGUS. Insofar as an IGUS is definitely epistemic if it is supposed to gather and use information (finitely!), it cannot be relevant at the ontic, internal level. However, Zurek's favorite IGUS, a universal Turing machine (UTM), has infinite capabilities of storing and processing information. This can justify an ontic interpretation of C but cuts the connection to empirical access. A UTM in this sense is nothing else than Laplace's, Maxwell's, or someone else's demon. In the framework of a strict distinction of ontic and epistemic levels of description, Zurek's approach thus appears conceptually problematic. (A similar problem seems to pervade his concept of decoherence (Zurek 1991) as far as it is claimed as a solution to the measurement problem.)

There is by definition no way of gathering (or using) information about a reality referred to by an ontic description since it is exactly the act of information gathering that leads to an epistemic concept of realism differing from its ontic counterpart. Of course, one may discuss how far knowledge about an ontic reality might be inferrable in an *indirect* manner. Rössler's conception of "endophysics" ("the study of demons"; Rössler 1987) seems to be inclined toward such a possibility. But eventually, endophysics according to Rössler is even more ambitious than addressing an ontic reality in the sense of quantum theory (for a corresponding discussion see Atmanspacher and Dalenoort 1994). Hence the question remains open whether the framework of an ontic-epistemic-distinction provides a suitable embedding for approaches like that by Rössler.

5   ... and Non-Commuting Operators

In 1931 Koopman suggested (Koopman 1931) how to describe classical systems and their evolution within a Hilbert space formalism equivalent to that used in quantum theory. The Koopman formalism is a special case (for classical systems) of the so-called standard representation of quantum mechanics (Haagerup 1975, see also Grelland 1993). The standard representation is isomorphic to phase space quantum mechanics à la Wigner-Moyal, as well as to the Liouville formalism of which Prigogine and coworkers have made extensive use in their approach toward a statistical theory of non-equilibrium systems (Prigogine 1962) and, more recently, in quantum theory (Prigogine 1980, 1992). The standard representation opens a formal way to treat "quantum" and "classical" contextual observables as elements of one and the same W^*-algebra, in one and the same Hilbert space representation.

Koopman's approach is applicable as a method providing an exact (not approximate!) linear representation of the evolution of nonlinear dynamical systems (see, e.g., Lasota and Mackey 1985, Kowalski 1992, Alanson 1992,

Peres and Terno 1996). The temporal evolution of a (epistemic) probability distribution $\rho(p, q)$ (or, more generally, $\rho(x_1, ..., x_d)$) in a canonical phase space Ω is then given by the Liouville equation

$$i\frac{\partial \rho}{\partial t} = \mathcal{L}\rho \tag{9}$$

where

$$\mathcal{L} = \sum \left(i\frac{\partial \mathcal{H}}{\partial q}\frac{\partial}{\partial p} - i\frac{\partial \mathcal{H}}{\partial p}\frac{\partial}{\partial q} \right) \tag{10}$$

is the (Hermitian) Liouville operator and the Hamilton function $\mathcal{H}$ is an energy observable. The primary meaning of $\mathcal{L}$ in this context is that of a temporal evolution operator according to

$$\rho(t) = \exp(-i\mathcal{L}t)\,\rho(0). \tag{11}$$

Since $\mathcal{L}\rho$ can be expressed by the classical Poisson bracket $\{\mathcal{H}, \rho\}$,

$$\mathcal{L}\rho = i\{\mathcal{H}, \rho\}, \tag{12}$$

it is tempting to interpret $\mathcal{L}$ also in terms of a difference of energies, or an energy bandwidth.

> **Remark:** The study of certain classes of non-integrable (quantum and classical) systems in the standard representation has recently made some progress. As an important result for unstable systems, one can find extensions of $\mathcal{L}$ with complex eigenvalues corresponding to resonances. The resulting spectral decomposition requires rigged Hilbert spaces instead of the conventional Hilbert space of square integrable functions (Antoniou and Tasaki 1993). Moreover, this framework yields a temporal symmetry breaking leading to two semigroups, one from past to future and the other from future to past (Antoniou and Prigogine 1993). Systems having been studied so far are so-called large Poincaré systems (quantum and classical); in particular the Renyi map, the baker transformation, and other chaotic systems. In general, these systems satisfy the algebraic Banach-*-property, but not the C^*-property (Antoniou and Suchanecki 1994).

Using the Liouville operator instead of the Hamilton operator as a generator of the dynamical evolution provides a promising shift in perspective on the notorious problem of a time operator (first stated by Pauli (1933), cf. the discussion in Atmanspacher (1994a)). This problem has its historical origin in the quantum mechanical uncertainty of energy and time, "heuristically" formulated by Heisenberg (1927). It has to do with the fact that the conventional Hamilton operator $\mathcal{H}$ has a twofold function of an energy operator *and* a generator of the unitary group providing the reversible Schrödinger dynamics. This conceptual degeneracy is lifted in the standard representation. The crucial point is that the spectrum of the Liouville operator is continuous and unbounded. This is essential with respect to Pauli's objection against the existence of a time operator. If T is required to be canonically conjugate to $\mathcal{L}$ rather than $\mathcal{H}$, then a time operator may indeed exist.

In the standard representation, such a time operator T can be defined as a shift operator according to

$$U_t^* T U_t = T + t \cdot \mathbb{1}, \qquad (13)$$

where U_t is the unitary evolution in Hilbert space that in Koopman's formalism replaces the evolution T_t in phase space, $U_t \rho(x) = \rho(T_t x)$. Precise conditions under which T exists have been first formulated by Misra (1978), shortly after two earlier pioneering papers (Tjøstheim 1976, Gustafson and Misra 1976). Basically, a positive KS-entropy, i.e., a chaotic flow, is a sufficient condition for the existence of a time operator T.

> **Remark:** $h_T > 0$ is always a sufficient condition for the existence of T, but it is not always necessary. There are systems for which $h_T > 0$ is sufficient and necessary (Goodrich et al. 1985, Suchanecki and Weron 1990), but the weakest necessary condition that has been derived is mixing (Misra 1978). A system is (strongly) mixing if its density ρ asymptotically reaches a well-defined expectation value, its correlations decay, and its transition probabilities become independent (Zaslavskii and Chirikov 1972). In contrast to ergodicity (phase averages equal temporal averages), these conditions exclude periodic or stationary systems. Mixing is therefore stronger than ergodicity, and chaos ($h_T > 0$) is stronger than mixing. (More exactly: the property $H(P,T) > 0 \, \forall P$ is stronger than mixing. This property implies, of course, $h_T > 0$ but not vice versa.) It is historically interesting that Krylov (1979) had done important work on this hierarchy of dynamical systems long before the papers of Kolmogorov and Sinai appeared.

One can now formally derive a relation

$$i[\mathcal{L}, T] = \mathbb{1}, \qquad (14)$$

expressing that $\mathcal{L}$ and T are non-commutative operators. However, the problem of implementing $\mathcal{L}$ and T in the sense of rigorous *observables* (instead of less specific formal operators) has not yet been solved in general. $\mathcal{L}$ is *not* an element of a W^*-algebra of observables as long as it is considered as nothing else than the generator of the dynamical evolution. Implementing it as an element of an algebra of observables would require a well-defined formulation of $\mathcal{L}$ in terms of a quantum mechanical energy observable, e.g., an energy bandwidth $\mathcal{L} = \mathcal{H} - \mathcal{H}'$. This is reasonably well motivated but not yet formally clarified in sufficient detail.

Another approach toward an understanding of the meaning of $\mathcal{L}$ has been proposed (Atmanspacher and Scheingraber 1987) by introducing an information operator $\mathcal{M}$ according to (cf. (8))

$$U_t^* \mathcal{M} U_t = \mathcal{M} - h_T t \cdot \mathbb{1}, \qquad (15)$$

which does not commute with $\mathcal{L}$ if and only if $h_T > 0$:

$$i[\mathcal{L}, \mathcal{M}] = h_T \cdot \mathbb{1}. \tag{16}$$

This means that $\mathcal{L}$ and $\mathcal{M}$ do not commute exactly when the sufficient condition for the existence of a time operator is satisfied. (Again, one has to be careful with proceeding to a commutation relation $i[\mathcal{L}, \mathcal{M}] = h_T\mathbb{1}$ in the rigorous sense of quantum mechanical *observables*.) This result offers some immediate hints toward an "explanatory surplus" as compared to the relationship between $\mathcal{L}$ and $\mathcal{T}$:

- the KS-entropy h_T as the commutator in $i[\mathcal{L}, \mathcal{M}] = h_T \cdot \mathbb{1}$ is an empirically accessible quantity instead of the unity operator in $i[\mathcal{L}, \mathcal{T}] = \mathbb{1}$.
- $\mathcal{M}$ is not an observable in the sense of conventional quantum mechanics: it might (tentatively) be characterized as a "meta"-observable insofar as it refers to the dynamics of information about conventional observables rather than to those observables themselves.
- The KS-entropy, specifying the linear change of this information, is a *statistical* quantity. It reflects an "average" *local* flow of information as a *global* invariant of a system, not its properties at any individual location in state space.
- Since h_T is explicitly system- and parameter-dependent (i.e., highly contextual), the "degree" of non-commutativity of $\mathcal{L}$ and $\mathcal{M}$ is not universally the same. This situation is at variance with conventional quantum mechanics with $\hbar$ as a universal commutator.
- As $\mathcal{M}$ has empirical meaning as a function of a parameter time t, the operator $\mathcal{T}$ suggests an "internal" time ("age") which can be related to the predictability interval h_T^{-1} if it is operationalized in terms of parameter time t.
- It may be speculated that the existence of a time operator $\mathcal{T}$ canonically conjugate to (not commuting with, respectively) an energy observable $\mathcal{L}$ serves as an indication for "nonlocality in time". This will be addressed in more detail in the subsequent section.

6   ... and Temporal Nonlocality

If an algebra of observables is non-commutative, nonlocal (Einstein-Podolsky-Rosen, EPR) correlations (Einstein et al. 1935) between these observables are generic and have been empirically confirmed beyond any reasonable doubt (Bell 1964, Aspect et al. 1982). Usually, the concept of nonlocal correlations is implicitly understood in a spatial sense, i.e., such that the extension of a system is not limited to a certain spatial region. This nonlocality, which gave rise to the concept of a holistic quantum reality, is in part abolished by measurement. Measuring an observable creates distinct empirical properties, separates objects from their environment, and provides justification for the concept of a local realism which was an unquestioned and unquestionable principle for centuries of physicists from Newton to Einstein. In the framework of the ontic-epistemic-distinction, holistic realism thus refers to an ontic level of description while locality is an epistemic concept depending on the cut between object and environment, the so-called Heisenberg cut.

> **Remark:** If EPR-correlations are present within a system S, then S is in a state ψ which is not factorizable according to $\psi = \psi_A \otimes \psi_B$, i.e., S is not separable into subsystems A and B. The formal reason is that ψ is a superposition, which involves a tensor product of the Hilbert spaces of A and B, not a Cartesian product as in classical mechanics. The physical reason consists in the existence of symmetry principles (e.g., conservation laws) within S and has nothing at all to do with "superluminal interactions". The superposition ψ can be decomposed into ψ_A and ψ_B by an act of measurement (of the first kind). Only after an act of measurement has been performed is it admissible to speak of subsystems A and B. The act of measurement creates objects by cutting (minimizing) EPR correlations within a system. The notion of a Heisenberg cut has become popular as a metaphor for this factual decomposition.

In view of the enormous consequences of a spatially dominated understanding of nonlocality for the current development of our world view, it is at least an interesting question to ask for its temporal counterpart. Nonetheless, temporal nonlocality and temporal EPR correlations do not seem to have been focussed on earlier than a decade ago. Inspired by a paper by Leggett and Garg (1985), Mahler has more recently contributed to the study of such correlations (Paz and Mahler 1993, Mahler 1994). His approach uses the formal framework of consistent (or inconsistent, respectively) histories, a viewpoint which clearly emphasizes dynamics over states. Although the dynamics are still dynamics of states, this approach leads to another logical level of description at which histories are the primary "objects" under study, not states as such. This reminds us of the tentative notion of meta-observables addressed in the preceding section. At the level of such meta-observables (as explicitly represented by $\mathcal{M}$), the distinction of ontic and epistemic states obviously loses its immediate relevance. However, the distinction of internal (endo) and external (exo) viewpoints remains attractive.

Some time before temporal EPR correlations began to attract interest, a concept of nonlocality in the context of a time operator was explicitly addressed by Misra and Prigogine (1983) and later generalized by Suchanecki et al. (1994). Here, nonlocality is a consequence of a nonunitary transformation Λ (see Misra et al. 1979 for an early reference), which implies that the concept of a uniquely defined trajectory is lost at the epistemic level of description. An intuitively intelligible discussion of this feature, omitting the Λ-formalism, has been given in Sects. 2 to 4. Λ-nonlocality is intimately related to the existence of a time operator $\mathcal{T}$, since both require $h_T > 0$, a positive KS-entropy. This is in turn equivalent with the non-commutativity of $\mathcal{L}$ and $\mathcal{M}$. Assuming a sensible implementation of $\mathcal{L}$ and $\mathcal{M}$ as elements of an overall algebra of observables, this would imply temporal nonlocality for any non-vanishing information flow.

Is there any solid relationship between temporal nonlocality à la EPR and temporal Λ-nonlocality? This question is still unanswered in detail, but some indications in the affirmative can be given. A crucial quantity in the Λ-approach is the inverse of the KS-entropy as a measure for the time interval $\tau \approx h_T^{-1}$, within which (on a temporal average) individual trajectories are inaccessible. Using another vocabulary, this has been denoted as a "finite duration of presence" (Atmanspacher and Scheingraber 1987) since empirically distinguishable facts can only be expected after a time interval $\Delta t > \tau$. Only if this is the case, then there is a history onto which a retrospective, external, observer-like perspective is possible. In this context, τ gives an estimate for a temporal cut establishing histories as objects in time, analogous to the Heisenberg cut establishing objects in space. As long as $\Delta t < \tau$, the perspective is necessarily internal, participatory. For processes with $h_T = 0$, e.g., a perfectly periodic clock, there is no history at all since $\tau \to \infty$. If a process is random in the sense of $h_T \to \infty$ then $\tau \to 0$. This may be considered as the limit of a presence with vanishing intrinsic duration – just the instant between past and future.

In the terminology of temporal Bell inequalities (Mahler 1994), another concept of τ exists which can be interpreted in a very similar manner. Here τ characterizes the time scale below which incoherent contributions to the equation of motion of the system can be neglected. For closed systems τ becomes infinitely large: there is no history at all since the cut that would be necessary to generate an external perspective in time is at $\Delta t \to \infty$. Accordingly, no event can "ever" be completed: "nothing happens". More generally, histories are not an element of empirical reality if $\Delta t < \tau$. This is the type of situations for which violations of temporal Bell inequalities can be expected (Mahler 1994). This means that the concept of histories, if it is nevertheless (unadmittedly) applied, can lead to inconsistencies like violations of probability sum rules. To my knowledge, corresponding experimental tests of this proposal have not yet led to results.

From a different viewpoint, Misra (1995) has recently shown that the Klein-Gordon evolution U_t of relativistic massless particles satisfies the conditions for the existence of a time operator for all real t. For massive particles, however, this condition can only be satisfied for a time-discrete evolution $U_{n\tau}$ where

$$\tau = \frac{h}{2mc^2}. \qquad (17)$$

This equation is more or less a reformulation of the uncertainty relation of energy and time. Misra concludes that there is no way of any direct observation of empirically relevant consequences from anything that might "happen" within τ (like the so-called "Zitterbewegung"). In fact, if "to happen" refers to "events" in the sense of a closed history, i.e., to an exo-perspective in the sense of a realism local in time, then simply nothing happens. This has also been discussed by Atmanspacher (1989), where the same value for τ is derived as the time interval (Δt in Eq.(37) of that reference) between successive realizations of elementary events, providing temporal nonlocality within τ. Whether another, internal viewpoint might be additionally relevant as an analog to a realism holistic in time remains beyond the scope of this argument. Some speculative indications in this direction can be found in Atmanspacher (1989).

> **Remark:** The three versions of τ due to Λ-nonlocality, due to incoherent dynamical terms, and due to an energy-time-uncertainty are not a priori identical. The first two concepts have so far been applied to examples in the sense of the lifetime of a state as related to its energy bandwidth. The third concept, however, addresses more or less the measurement process, i.e., the temporal actualization of an "object". Misra (1995) has demontrated that the time operator concept at the level at which Λ-nonlocality is usually discussed is not the same as would be required for a description of the process of measurement. Such a formulation is not yet available. All three concepts of τ share the significance of a distinction between endo- and exo-viewpoints. From within, τ is atemporal (no temporal order, no temporal direction, no events, no histories), whereas τ is accessible from outside, e.g., its size can be determined in terms of a parameter (clock) time t.

7 Interdisciplinary Perspectives

As indicated at the beginning of this paper, the concept of a dynamical entropy, e.g. the KS-entropy, represents an extension of earlier, "static" versions of entropy. The crucial point of this extension is an explicit focus on a dynamic evolution "as such" rather than a parametrized succession of states. Arguments have been presented leading to an analogous extension with respect to the concept of nonlocality. Traditionally, nonlocality is considered as a spatial concept expressing the insufficiency of a world view restricted to a spatially local realism. The introduction of a time operator and its ramifications, which are intimately related to the KS-entropy, imply a temporal kind of nonlocality which indicates a corresponding insufficiency of a world view restricted to a temporally local realism. (See Atmanspacher (1994b) for more details concerning this issue in historical and systematic perspective.)

This has a number of important physical consequences for the concept of time which are only incompletely understood as yet. Some of them are touched upon in this paper. However, since the subject of time and its riddles extends over many different fields of research and, thus, represents a paradigm for the need of interdisciplinary approaches, it may be appropriate to briefly mention two central areas out of the scope of physics in this last section. One of them is the concept of time in Whitehead's philosophy, and the other is the experience of time as addressed in the cognitive sciences.

20th-century philosophy has witnessed a steady but off-mainstream interest in concepts of time that can be covered by the notion of an "internal time". One of the early references is, of course, Bergson (1911); more recent authors are Gebser (1985) or Prigogine in his philosophically oriented publications (1979). The protagonist of internal time, however, is Whitehead (1979). He developed his philosophy of organism long before quantum theory provided its fascinating insights into the concept of a holistic realism as opposed to local realism together with its implications for the study of time as reported in this paper. For Whitehead the ultimate concrete entities in the universe are the "actual occasions". Actual occasions are not "objects" in the sense of building blocks of a local reality, but they are nonlocally and inseparately related to each other by so-called "prehensions". An actual occasion occupies a definite spatial region, and its temporal duration is finite. Both its temporal and spatial extension depend on contexts given by the prehensions. Analogies with modern physics have to some extent been discussed earlier (Shimony 1965, Stapp 1979, Malin 1988, Griffin 1986) and definitely deserve further studies (Klose, this volume, and Klose 1996).

Whitehead's philosophy of organism is tightly connected with his theory of perception. The finite duration of an actual occasion corresponds to the duration of the "specious present" in a perceptive mode denoted as "causal efficacy", complementing the spatially dominated mode of "presentational immediacy" (Klose 1996). Whitehead's concept of the specious present resembles very closely the idea of an extended "now" in the sense of current trends in cognitive science. At one level, such a "now" may correspond to the syntactic elementary integration units of approximately 30msec; at another level, an interesting candidate for comparative studies is the 3sec interval of semantically coherent perception (Ruhnau 1994). In my opinion, the distinction of an endo- and an exo-time according to a time operator T and its "empiricalized" counterpart M (including the KS-entropy) can provide interesting insights in these respects. In particular I think that a better understanding of the relationship between neurophysiological and neuropsychological results concerning the 30-msec window is achievable on such a basis.

Acknowledgments: I am grateful to Ioannis Antoniou, Jim Crutchfield, Thomas Hudetz, Günter Mahler, Hans Primas and Herbert Scheingraber for numerous discussions about the concepts presented in this article. In particular I thank Hans Primas for drawing my attention to the standard representation of quantum theory. This work has been supported by BMBF grant # 05 2ME62A(E) and by IGPP grant # 652110.

References

Alanson T. (1992): A "quantal" Hilbert space formulation for nonlinear dynamical systems in terms of probability amplitudes. *Phys. Lett. A* **163**, 41–45.

Amann A. (1995): Modeling the quantum mechanical measurement process. *Int. J. Theor. Phys.* **34**, 1187–1196.

Antoniou I.E. and Prigogine I. (1993): Intrinsic irreversibility and integrability of dynamics. *Physica A* **192**, 443–464.

Antoniou I. and Qiao B. (1996): Spectral decomposition of the tent maps and the isomorphism of dynamical systems. *Phys. Lett. A* **215**, 280–290.

Antoniou I. and Suchanecki Z. (1994): The logic of quantum systems with diagonal singularity. *Found. Phys.* **24**, 1439–1457.

Antoniou I.E. and Tasaki S. (1993): Generalized spectral decompositions of mixing dynamical systems. *Int. J. Quant. Chem.* **46**, 425–474.

Aspect A., Dalibard J., and Roger G. (1982): Experimental test of Bell's inequalities using time-varying analyzers. *Phys. Rev. Lett.* **49**, 1804–1807.

Atmanspacher H. (1989): The aspect of information production in the process of observation. *Found. Phys.* **19**, 553–577.

Atmanspacher H. (1994a): Is the ontic/epistemic distinction sufficient to represent quantum systems exhaustively? In *Symposium on the Foundations of Modern Physics 1994*, ed. by K.V. Laurikainen, C. Montonen, and K. Sunnarborg (Editions Frontières, Gif-sur-Yvette), 15–32.

Atmanspacher H. (1994b): Objectification as an endo-exo transition. In *Inside Versus Outside*, ed. by H. Atmanspacher and G.J. Dalenoort (Springer, Berlin), 15–32.

Atmanspacher H. and Dalenoort G.J. (1994): Introduction. In *Inside Versus Outside*, ed. by H. Atmanspacher and G.J. Dalenoort (Springer, Berlin), 1–12; in particular pp. 6–9.

Atmanspacher H. and Scheingraber H. (1987): A fundamental link between system theory and statistical mechanics. *Found. Phys.* **17**, 939–963.

Atmanspacher H., Wiedenmann G., and Amann A. (1995): Descartes revisited – the endo/exo-distinction and its relevance for the study of complex systems. *Complexity* **1**(3), pp. 15–21.

Bell J.S. (1964): On the Einstein Podolsky Rosen paradox. *Physics* **1**, 195–200.

Bergson H. (1911): *Creative Evolution* (Macmillan, London).

Caves C.M. (1993): Information and entropy. *Phys. Rev. A* **47**, 4010–4017.

Caves C. (1994): Information, entropy, and chaos. In *Physical Origins of Time Asymmetry*, ed. by J.J. Halliwell, J. Pérez-Mercader, and W.H. Zurek (Cambridge University Press, Cambridge), 47–89.

Cornfeld I.P., Fomin S.V., and Sinai Ya.G. (1982): *Ergodic Theory* (Springer, Berlin), 250–252, 280–284.

Connes A., Narnhofer H., and Thirring W. (1987): Dynamical entropy of C^*-algebras and von Neumann algebras. *Commun. Math. Phys.* **112**, 691–719.

Crutchfield J.P. (1994): Observing complexity and the complexity of observation. In *Inside Versus Outside*, ed. by H. Atmanspacher and G.J. Dalenoort (Springer, Berlin), 235–272; in particular sec. 2.1.

Einstein A., Podolsky B., and Rosen N. (1935): Can quantum-mechanical description of physical reality be considered complete? *Phys. Rev.* **47**, 777–780.

Elskens Y. and Prigogine I. (1986): From instability to irreversibility. *Proc. Natl. Acad. Sci. USA* **83**, 5756–5760.

Farmer D. (1982): Information dimension and the probabilistic structure of chaos. *Z. Naturforsch.* **37a**, 1304–1325.

Gebser J. (1985): *The Ever-Present Origin*, translated by N. Barstad (Ohio University Press, Athens). German original: *Ursprung und Gegenwart* (Deutsche Verlagsanstalt, Stuttgart, 1949/1953).

Goldstein S. (1981): Entropy increase in dynamical systems. *Israel J. Math.* **38**, 241–256.

Goodrich R.K., Gustafson K., and Misra B. (1985): On K-flows and irreversibility. *J. Stat. Phys.* **43**, 317–320.

Grad H. (1961): The many faces of entropy. *Commun. Pure Applied Math.* **14**, 323–354.

Grassberger P. and Procaccia I. (1983): Estimation of the Kolmogorov entropy from a chaotic signal. *Phys. Rev. A* **28**, 2591–2593.

Grelland H.H. (1993): Tomita representations of quantum and classical mechanics in a bra/ket formulation. *Int. J. Theor. Phys.* **32**, 905–925.

Griffin D.R., ed. (1986): *Physics and the Ultimate Significance of Time* (SUNY Press, Albany).

Gustafson K. and Misra B. (1976): Canonical commutation relations of quantum mechanics and stochastic regularity. *Lett. Math. Phys.* **1**, 275–280.

Haagerup U. (1975): The standard form of von Neumann algebras. *Math. Scand.* **37**, 271–283.

Heisenberg W. (1927): Über den anschaulichen Inhalt der quantentheoretischen Kinematik und Mechanik. *Zeitschr. f. Physik* **43**, 172–198. English translation in J.A. Wheeler and W.H. Zurek, eds., *Quantum Theory and Measurement* (Princeton Univesity Press, Princeton, 1983), 62–84.

Hepp K. (1972): Quantum theory of measurement and macroscopic observables. *Helv. Phys. Acta* **45**, 237–248.

Hudetz T. (1988): Spacetime dynamical entropy of quantum systems. *Lett. Math. Phys.* **16**, 151–161.

Klose J. (1996): *Die Struktur der Zeit in der Philosophie Alfred North Whiteheads* (PhD Thesis, University of Munich).

Kolmogorov A.N. (1958): A new metric invariant of transitive systems and automorphisms of Lebesgue spaces. *Dokl. Akad. Nauk SSSR* **119**, 861–864.

Koopman B. (1931): Hamiltonian systems and transformations in Hilbert space. *Proc. Natl. Acad. Sci. USA* **17**, 315–318.

Koopman B. and Neumann J. von (1932): Dynamical systems of continuous spectra. *Proc. Natl. Acad. Sci. USA* **18**, 255–263, here: p. 261.

Kowalski K. (1992): Linearization transformations for nonlinear dynamical systems: Hilbert space approach. *Physica A* **180**, 156–170.

Krylov N.S. (1979): *Works on the Foundations of Statistical Physics* (Princeton University Press, Princeton). See particularly the afterword by Sinai, pp. 239–281.

Lasota A. and Mackey M.C. (1985): *Probabilistic Properties of Deterministic Systems* (Cambridge University Press, Cambridge).

Ledrappier F. and Young L.-S. (1985): The metric entropy of diffeomorphisms. Part I: Characterization of measures satisfying Pesin's entropy formula. Part II: Relations between entropy, exponents, and dimension. *Ann. Math.* **122**, 509–574.

Leggett A.J. and Garg A. (1985): Quantum mechanics versus macroscopic realism: is the flux there when nobody looks? *Phys. Rev. Lett.* **54**, 857–860.

Lockhart C.M. and Misra B. (1986): Irreversibility and measurement in quantum mechanics. *Physica A* **136**, 47–76; cf. H. Primas, *Math. Rev.* **87k**, 81006 (1987).

Mahler G. (1994): Temporal Bell inequalities: a journey to the limits of "consistent histories". In *Inside Versus Outside*, ed. by H. Atmanspacher and G.J. Dalenoort (Springer, Berlin), 196–205.

Malin S. (1988): A Whiteheadian approach to Bell's correlations. *Found. Phys.* **18**, 1035–1044.

Milnor J. (1985): On the concept of attractor. *Commun. Math. Phys.* **99**, 177–195; correction and remark: *Commun. Math. Phys.* **102**, 517–519 (1985).

Misra B. (1978): Nonequilibrium entropy, Lyapounov variables, and ergodic properties of classical systems. *Proc. Ntl. Acad. Sci. USA* **75**, 1627–1631.

Misra B. (1995): From time operator to chronons. *Found. Phys.* **25**, 1087–1104.

Misra B. and Prigogine I. (1983): Irreversibility and nonlocality. *Lett. Math. Phys.* **7**, 421–429.

Misra B., Prigogine I., and Courbage M. (1979): From deterministic dynamics to probabilistic descriptions. *Physica A* **98**, 1–26; cf. A.S. Wightman; *Math. Rev.* **82e**, 58066 (1982).

Oono Y. (1989): Large deviations and statistical physics. *Prog. Theor. Phys. Suppl.* **99**, 165–205.

Pauli W. (1933): Die allgemeinen Prinzipien der Wellenmechanik. In *Handbuch der Physik, Vol. 24*, ed. by H. Geiger and K. Scheel (Springer, Berlin), 88–272; here: p. 140. Reprinted in S. Flügge, ed., *Encyclopedia of Physics, Vol. V, Part 1* (Springer, Berlin, 1958), 1–168; here: p. 60.

Paz J.P. and Mahler G. (1993): Proposed test for temporal Bell inequalities. *Phys. Rev. Lett.* **71**, 3235–3239.

Peres A. and Terno D. (1996): Evolution of the Liouville density of a chaotic system. *Phys. Rev. E* **53**, 284–290.

Pesin Ya.B. (1977): Characteristic Lyapunov exponents and smooth ergodic theory. *Russian Math. Surveys* **32**, 55–114. Russian original: *Uspekhi Mat. Nauk* **32**, 55–112.

Prigogine I. (1962): *Non-Equilibrium Statistical Mechanics* (Wiley, New York).

Prigogine I. (1980): *From Being to Becoming* (Freeman, San Francisco).

Prigogine I. (1992): Dissipative processes in quantum theory (including discussion session). *Phys. Rep.* **219**, 93–120.

Prigogine I. and Stengers I. (1979): *Order Out of Chaos* (Bantam, New York).

Primas H. (1983): *Chemistry, Quantum Mechanics, and Reductionism* (Springer, Berlin).

Primas H. (1990a): Mathematical and philosophical questions in the theory of open and macroscopic quantum systems. In *Sixty-Two Years of Uncertainty*, ed. by A.I. Miller (Plenum, New York), 233–257.

Primas H. (1990b): The measurement process in the individual interpretation of quantum mechanics. In *Quantum Theory Without Reduction*, ed. by M. Cini and J.M. Levy-Leblond (Adam Hilger, Bristol), 233–257.

Primas H. (1994a): Endo- and exotheories of matter. In *Inside Versus Outside*, ed. by H. Atmanspacher and G.J. Dalenoort (Springer, Berlin), 163–193.

Primas H. (1994b): Hierarchical quantum descriptions and their associated ontologies. In *Symposium on the Foundations of Modern Physics 1994*, ed. by

K.V. Laurikainen, C. Montonen, and K. Sunnarborg (Editions Frontières, Gif-sur-Yvette), 201–220.

Rössler O.E. (1987): Endophysics. In *Real Brains, Artificial Minds*, ed. by J.L. Casti and A. Karlqvist (North Holland, New York), 25–46.

Ruelle D. (1981): Small random perturbations of dynamical systems and the definition of attractors. *Commun. Math. Phys.* **82**, 137–151.

Ruhnau E. (1994): Time – a hidden window to dynamics. In *Inside Versus Outside*, ed. by H. Atmanspacher and G.J. Dalenoort (Springer, Berlin), 291–308.

Scheibe E. (1973): *The Logical Analysis of Quantum Mechanics* (Pergamon, Oxford), 82–88. German: *Die kontingenten Aussagen der Physik*. Athenäum, Frankfurt 1964.

Shaw R. (1981): Strange attractors, chaotic behavior, and information flow. *Z. Naturforsch.* **36a**, 80–112.

Shimony A. (1965): Quantum physics and the philosophy of Whitehead. In *Boston Studies in the Philosophy of Science, Vol. II*, ed. by R.S. Cohen and M.W. Wartovsky (Humanities Press, New York), 307–330.

Stapp H.P. (1979): Whiteheadian approach to quantum theory and the generalized Bell's theorem. *Found. Phys.* **9**, 1–25.

Sinai Ya.G. (1959): On the concept of entropy of a dynamical system. *Dokl. Akad. Nauk SSSR* **124**, 768–771.

Suchanecki Z., Antoniou I., and Tasaki S. (1994): Nonlocality of the Misra-Prigogine-Courbage semigroup. *J. Stat. Phys.* **75**, 919–928.

Suchanecki Z. and Weron A. (1990): Characterizations of intrinsically random dynamical systems. *Physica A* **166**, 220–228.

Tasaki S., Suchanecki Z., and Antoniou I. (1993): Ergodic properties of piecewise linear maps on fractal repellers. *Phys. Lett. A* **179**, 103–110.

Tjøstheim D. (1976): A commutation relation for wide sense stationary processes. *SIAM J. Appl. Math.* **30**, 115–122.

Wehrl A. (1978): General properties of entropy. *Rev. Mod. Phys.* **50**, 221–260.

Wehrl A. (1991): The many facets of entropy. *Rep. Math. Phys.* **30**, 119–129.

Weizsäcker C.F. von (1985): *Aufbau der Physik* (Hanser, München), Chap. 5

Whitehead A.N. (1979): *Process and Reality*, ed. by D.R. Griffin and D.W. Sherburne (Free Press, New York).

Whittaker E.T. (1943): Chance, freewill, and necessity in the scientific conception of the universe. *Proc. Phys. Soc. (London)* **55**, 459–471; here p. 461.

Zaslavskii G.M. and Chirikov B.V. (1972): Stochastic instability of non-linear oscillations. *Sov. Phys. Usp.* **14**, 549–568. Russian original: *Usp. Fiz. Nauk* **105**, 3–39 (1971).

Zurek W.H. (1989): Algorithmic randomness and physical entropy. *Phys. Rev. A* **40**, 4731–4751.

Zurek W.H. (1991): Decoherence and the transition from quantum to classical. *Physics Today*, October 1991, 36–44. See also comments on this article by J. Anderson, G.C. Ghirardi *et al.*, N. Gisin, D. Albert and G. Feinberg, P. Holland, V. Ambegaokar, and K.J. Epstein together with Zurek's reply in *Physics Today*, April 1993, 13–15, 81–90.

Nonlocality in Quantum Dynamics

Günter Mahler

Institut für Theoretische Physik, Universität Stuttgart, D–70550 Stuttgart,
Germany

Abstract. The "carrier" of nonlocality are quantized (matter) fields, the state
properties of which refer to single as well as to multiple subsystem indices: The
corresponding one- (local) and multi-node (nonlocal) properties can be incommen-
surable, like position and momentum in basic quantum mechanics. Nonlocality in
space (time) shows up as the index space is mapped onto real space locations (time
windows). Aspects of nonlocality may abound everywhere, but are, nevertheless,
hard to detect, unless one controls one's own little quantum universe down to each
individual subsystem.

1 Introduction

The dramatic effects of nonlocality, as demonstrated in the famous Aspect-
experiment on EPR-correlated photon-pairs (pseudo-spins) (Aspect et al.
1982), conceals the fact that nonlocality can and does emerge in many different ways (cf. Yurke and Stoler 1995). Though not always as striking, the
conclusions reached in each case tend to be a kind of "no-go"-statement. Typically, these statements can be formulated as a "test" such that, under given
constraints, a specific set of data can be shown to be incompatible with any
classical model. Here, as in any subsequent example, the meaning of "classical
model" has to be defined very carefully.[1] For the Aspect-experiment, e.g., one
usually refers to "local realism" (see Sect. 6), which can be tested by Bell's
inequality (Bell 1987) constraining pair-correlations in any 2-spin-state.

It is the purpose of this paper to outline the formal basis of quantum nonlocality as one of the main features of non-classicality and to give a number
of illustrations from experiments.

2 Index-Space as a Mathematical Description

The description of a quantum system requires the use of a certain "reference
frame" in terms of basis states and basis operators. We restrict ourselves to
a discrete system (finite number of states). In a system-theoretical approach,
one would like to leave most of the details open for later specification (and

---

[1] "Classical model" does certainly not refer to a simulation program run on a
classical computer. This possibility only indicates that quantum phenomena can
be computable.

experimental implementation). For the present purpose, though, it will be essential to focus on systems composed of subsystems ("nodes") right from the start (Mahler and Weberruss 1995).

Let the subsystem $\mu = 1, 2, ... N$ be defined in a Hilbert-space of dimension n, spanned by the orthonormal basis functions $|q(\mu)\rangle, q = 1, 2, ... n$. (A straightforward generalization would allow the subsystems to have different dimensions.) These basis functions are assumed to be the energy eigenfunctions of the subsystem if it were isolated. As new linear combinations of the n^2 projection operators

$$\hat{P}_{qr}(\mu) = |q(\mu)\rangle\langle r(\mu)|, \tag{1}$$

we can introduce the unit operator $\hat{1}(\mu)$ and $s = n^2 - 1$ traceless operators, the generators of $SU(n)$, $\hat{\lambda}_j(\mu)$. For $n = 2$, e.g., these are $\hat{\lambda}_1 = \hat{P}_{12} + \hat{P}_{21}$; $\hat{\lambda}_2 = \mathrm{i}(\hat{P}_{12} - \hat{P}_{21})$; $\lambda_3 = \hat{P}_{22} - \hat{P}_{11}$).

As the basis of the Hilbert-space for the total system, we can then use the n^N product states $|q(1)q(2)...q(N)\rangle$. Correspondingly, any operator can be expressed in terms of the n^{2N} products composed of $\hat{1}(\mu)$ and those generators $\hat{\lambda}_j(\mu)$ (Mahler and Weberruss 1995),

$$\hat{A} = \frac{1}{n^N} A_0 \hat{1} + \frac{1}{2n^{N-1}} \sum_{\mu=1}^{N} A_j^\mu \, \hat{\lambda}_j(\mu) + \frac{1}{4n^{N-2}} \sum_{\mu<\nu}^{N} A_{jk}^{\mu\nu} \, \hat{\lambda}_j(\mu) \, \hat{\lambda}_k(\nu)$$

$$+ \frac{1}{8n^{N-3}} \sum_{\mu<\nu<\sigma}^{N} A_{jkl}^{\mu\nu\sigma} \, \hat{\lambda}_j(\mu) \, \hat{\lambda}_k(\nu) \, \hat{\lambda}_l(\sigma) + ... \tag{2}$$

(summation over repeated indices) with

$$A_0 = \mathrm{Tr}\{\hat{A}\}$$
$$A_j^\mu = \mathrm{Tr}\{\hat{A} \, \hat{\lambda}_j(\mu)\}. \tag{3}$$

The multi-node parameters like $A_{jk}^{\mu\nu}$ are defined correspondingly. Any Hermitian operator is, thus, uniquely specified by a hierarchy of n^{2N} real numbers A. For $N = 3$ this hierarchy is sketched in Fig. 1.

Applied to the Hamilton operator $(\hat{A} \to \hat{H})$, one usually finds oneself restricted to one- and two-point parameters. A network of two-level systems, $n = 2$, driven by an optical field (frequency ω_{21}^μ, coupling constant g_{21}^μ), is specified in rotating wave approximation by

$$H_1^\mu = \mathrm{Tr}\{\hat{H}(\mu) \, \hat{\lambda}_1(\mu)\} = n^{N-1} \, \hbar \, g_{21}^\mu,$$
$$H_3^\mu = \mathrm{Tr}\{\hat{H}(\mu) \, \hat{\lambda}_3(\mu)\} = n^{N-1} \, \hbar \, \delta_{21}^\mu. \tag{4}$$

The inter-subsystem coupling, resulting, e.g., from dipole-dipole coupling, could be given by

$$H_{33}^{\mu\nu} = \mathrm{Tr}\{\hat{H}(\mu, \nu) \, \hat{\lambda}_3(\mu) \, \hat{\lambda}_3(\nu)\} = -2n^{N-2} \, \hbar \, C_R^{\mu\nu}. \tag{5}$$

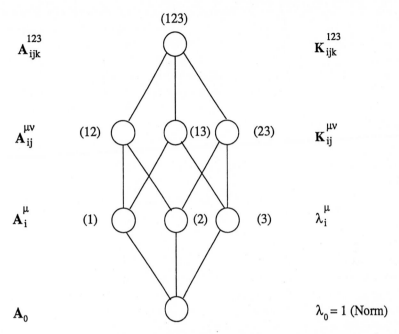

Fig. 1. Hierarchy of $SU(n)$-parameters ($N = 3$). Lhs: General operator $\hat{A}$, represented by its trace, A_0, the single node parameters, A_i^μ, the 2-node parameters $A_{ij}^{\mu\nu}$ and the 3-node parameters A_{ijk}^{123}. Rhs: Representation of density operator.

(All other parameters would be zero.)

The $SU(n)$-parameters for the density operator $\hat{\rho}$ consist of local and nonlocal expectation values up to Nth order: the trace (zero order),

$$\lambda_0 = \text{Tr}\{\hat{\rho}\} = 1, \tag{6}$$

the Bloch-vectors (first order),

$$\lambda_j^\mu = \text{Tr}\{\hat{\rho}\,\hat{\lambda}_j(\mu)\} = \langle\hat{\lambda}_j(\mu)\rangle, \tag{7}$$

and the multi-node expectation values (state tensors),

$$\begin{aligned} K_{jk}^{\mu\nu} &= \text{Tr}\{\hat{\rho}\,\hat{\lambda}_j(\mu)\,\hat{\lambda}_k(\nu)\} \\ K_{jkm}^{\mu\nu\sigma} &= \text{Tr}\{\hat{\rho}\,\hat{\lambda}_j(\mu)\,\hat{\lambda}_k(\nu)\,\hat{\lambda}_m(\sigma)\} \quad \text{etc.} \end{aligned} \tag{8}$$

Even though all these multi-node terms will, in general, be unequal to zero, they do not necessarily contain independent information. Defining the correlation proper

$$M_{ij}^{\mu\nu} = K_{ij}^{\mu\nu} - \lambda_i^\mu \lambda_j^\nu, \tag{9}$$

we see that $M_{ij}^{\mu\nu} = 0$ for product states. In this case, the two-node expectation values reduce to single-node ones. Higher order correlation tensors can easily be constructed along similar lines (Keller and Mahler 1994).

There are infinitely many different representations (different choices of subsystems, different choices of basis operators). So why should we prefer one over the other? There are no formal requirements to guide us. The physical situation itself has to impose constraints which transcend purely mathematical reasoning.

3   Index-Space as a Physical Reference Frame

We want the index-space to have more than just a formal meaning. Contrary to indistinguishable particles, for which an index is introduced only to conclude that this index cannot have a classical meaning, we do require addressability, i.e., for each subsystem μ there should exist at least one index-selective operator $\hat{A}$ with non-zero parameters A_j^μ for one specific μ only. The subsystem index thus becomes essentially a classical parameter. In many cases this parameter can be associated with a certain location in real space (confined photons, electrons, atoms etc) or with particular modes. Let us consider a few examples.

Example 1: Consider a network of two identical atoms (or molecules), the center-of-mass coordinates of which can be considered fixed (classical structure). Nevertheless, if their relative distance is small compared to the wavelength of a photon mode, to which each center might couple, this light field coupling is not index-selective (i.e., it does not allow the two centers to be distinguished, implying phenomena like super-radiance (Dicke 1954, Mahler and Weberruss 1995).

Example 2: A dimer of two distinct molecules, on the other hand, might be addressable in frequency space, even if their relative distance is still too small for selection in real space. This would constitute a valid ($N = 2$)-network (Keller and Mahler 1994).

Example 3: Cavity quantum electrodynamics (Berman 1994) often refers to a set of micro-resonators (for the confinement of photon modes) coupled via beams of (excited) Rydberg-atoms. Even though photons are indistinguishable, their localization in either cavity presents a decidable alternative. (The field outside the cavities is neglected.) The scenario thus establishes a valid network with the indexed cavities providing a static structure in real space and the indexed atoms (center-of-mass motion considered classical) providing transfer and coupling.

Example 4: Nonlinear crystals allow, via so-called down-conversion (Kwiat et al. 1995), an incoming photon of energy $h\nu$ to be decomposed into two photons, exiting at half energy in two different ray-directions. In this case, the photon modes are not localized, but can be distinguished by their respective direction.

4   A Reference Frame at Work

Addressability based on index-selective operators provides means to introduce local unitary transformations $\hat{U}(\mu)$ as well as local projections $\hat{P}_{rr}(\mu)$. While the former might result from the unitary time evolution generated by (time-dependent) local Hamiltonians, the latter can be implemented via a dissipative coupling to the environment. This coupling has specifically to be designed for the intended measurement.

Local damping channels (transition rates W_j^μ) can be characterized by operators $\hat{F}_j(\mu)$, which may be non-Hermitian. (A pertinent example would be $\hat{F}(\mu) = \hat{P}_{qr}(\mu)$.) The Liouville equation controlling the dynamics of closed systems then has to be replaced by a master equation. In the Markoff-approximation (i.e., without memory effects) one finds (Molmer and Castin 1996):

$$\dot{\hat{\rho}} = -\frac{i}{\hbar}(\hat{H}_{eff}\hat{\rho} - \hat{\rho}\hat{H}_{eff}^+) + \sum_{\mu=1}^{N}\sum_{j=1}^{s}\hat{L}_j(\mu)\hat{\rho}. \tag{10}$$

Here, the effective Hamilton operator, $\hat{H}_{eff} = \hat{H} + \sum_\mu \sum_j \Delta\hat{H}_j(\mu)$, includes the non-Hermitian supplements

$$\Delta\hat{H}_j(\mu) = \frac{i\hbar}{2}W_j^\mu \hat{F}_j^+(\mu)\hat{F}_j(\mu). \tag{11}$$

The second part in (10) is specified by

$$\hat{L}_j(\mu)\hat{\rho} = W_j^\mu \hat{F}_j(\mu)\hat{\rho}\hat{F}_j^+(\mu). \tag{12}$$

The master equation can be rewritten in terms of the $SU(n)$-parameters for the Hamiltonian, the density-operator, and the damping-operators $\hat{F}_j(\mu)$.

Quantum stochastics represents an extension of (ensemble) quantum dynamics. As such, it does not follow uniquely from the latter, just like quantum theory does not follow uniquely from classical mechanics. Experimental evidence supports the quantum jump model (continuous measurements), though more general versions might emerge in the future.

In this stochastic unravelling, the first term in (10) represents the deterministic and the second term the non-deterministic part. To see this, we consider the truncated dynamics

$$\hat{\rho}(t+dt) = \hat{\rho}(t) - \frac{i}{\hbar}(\hat{H}_{eff}\hat{\rho} - \rho\hat{H}_{eff}^+)dt, \tag{13}$$

which is connected with a reduction of the trace $\lambda_0(t+dt) = 1 - R$, where R is due to the combined effect of all damping channels, $R = \sum R_j^\mu$. We assume $R \ll 1$. Choosing a random number $z, 0 \leq z \leq 1$, we now replace $\hat{\rho}(t+dt)$ by its renormalized version,

$$\hat{\rho}'(t+dt) = \frac{\hat{\rho}(t+dt)}{\lambda_0(t+dt)}, \tag{14}$$

if $z \geq R$, or by one of the projected versions

$$\hat{\rho}'(t + dt) = \frac{\hat{L}_j(\mu)\,\hat{\rho}(t + dt)}{\mathrm{Tr}\{\hat{L}_j(\mu)\,\hat{\rho}(t + dt)\}}, \tag{15}$$

if $z < R$. The relative frequency for a specific jump j in subsystem μ is given by R_j^μ/R. The resulting quantum trajectories remain in the subspace of pure states. It should be noted that the local projections, in general, do modify local as well as nonlocal expectation values defined in (7,8).

The stochastic process is constructed in such a way that by averaging over Z quantum trajectories (for a given initial state) one recovers, in the limit $Z \to \infty$, the solution of the master equation. Of course, this must hold for any individual element of the (ensemble) density matrix and, thus, also for any of the local or nonlocal expectation values.

This averaging may be seen to be based on a one-time joint probability function f in the full space of all expectation values $\Gamma = \{\lambda_i^\mu, K_{ij}^{\mu\nu}, ...\}$. By construction we know that any linear moment like

$$\overline{\lambda_j^\mu(t)} = \int \lambda_j^\mu \, f(\Gamma; t)\, d\Gamma \tag{16}$$

has to coincide with the ensemble result of the master equation for the same initial conditions. This holds, correspondingly, for the correlation tensors. One should note, however, that this joint probability does not follow from the ensemble-density matrix alone, as would so-called quasi-probabilities like the Wigner-function.

Based on the function f and the coarse-graining

$$\Gamma = \sum_{m=1}^{M} \delta\Gamma_m, \tag{17}$$

$$\delta f_m = \int_{\delta\Gamma_m} f(\Gamma)d\Gamma, \tag{18}$$

we may define the "decomposition entropy" (Mahler et al. 1996)

$$\tilde{S}(\delta f_m) = -\sum_{m=1}^{M} \delta f_m \ln(\delta f_m). \tag{19}$$

As $M \gg n^N$, typically, this $\tilde{S} \geq 0$ can be much larger than the von Neumann entropy S, relating only to the the orthogonal eigenspace of the density operator. For a pure state, both entropies are zero. There are two ways to avoid the increase of $\tilde{S}$. The first is to suppress non-deterministic processes altogether; this would imply complete isolation of the system at the expense of being unable to get anything in or out. The alternative is to try to let the non-deterministic process never go unnoticed; this requires ideal and complete measurements.

5   Dissipation as a Source of Information

The stochastic rules allow the solution of the master equation to be unravelled in terms of quantum trajectories. As such, however, they give no hint as to how those jumps might be related with observable measurement events ("clicks"). Such information could become available only if the stochastic process within the open system were correlated with a process to be registered in the macroscopic environment. However, it is just this kind of correlations which has been suppressed in the bath approximation to derive the Markoff master equation.

These shortcomings can be circumvented if the pertinent state-space is extended, while the type of jumps is constrained. We recall that the coupling between system and environment (for one local channel $j = 1$ in subsystem μ) can be written in bilinear form as (Mahler et al. 1996)

$$\hat{V}(\mu) = \hat{F}_1(\mu)\,\hat{Q}_1 + \text{c.c.} \tag{20}$$

In the following, we shall restrict ourselves to an operator $\hat{F}_1(\mu)$ related to a relaxation, which will be connected with an environment operator $\hat{Q}_1$ that describes an excitation (generation of a photon). In an ideal photon counting measurement, it is assumed that the photon reference state is a pure state (the vacuum state) so that the occurrence of any photon, which is immediately registered and destroyed, can be interpreted to indicate a corresponding jump in the material system. The extended state space (including the zero-photon and one-photon subensemble) is indicated in Fig. 2. Their interaction can be treated in the weak-coupling limit, as the 1-photon substate is taken to be strongly damped. Its immediate reset to the zero-photon state is identified with the measurement "click". Transitions to few-photon subensembles with $N_{\text{photon}} > 1$ can safely be neglected. The information supplied by the measurement record thus consists of the time marks which, for a given initial state and together with the theoretical model at hand, then allow the complete network state to be inferred at any time.

Additional damping channels of that type could be connected with separate counting interfaces $\hat{Q}$ or – non-selectively – with one and the same register.

6   Classical Versus Quantum Correlations

Hierarchies of single- and multi-node correlations like in (7,8) also exist in classical statistics. One may wonder whether those parameters, meant to describe nonlocality, could in fact consistently be explained by classical statistics. Historically, local realism (Bell 1987) has been the first attempt of this kind. As will be argued below, it should now be seen in a broader context than previously.

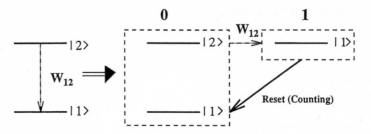

Fig. 2. Measurement interface. Incoherent transition in a two-level system, re-interpreted in product space including zero-photon and one-photon states. In the reset transition the 1-photon state is destructively measured.

Local realism accepts that the eigenvalues of non-commuting obervables may not be accessible to any single experiment, but should nevertheless "exist". Lack of control of these "hidden variables" would imply that we can talk only about statistical distributions (which would also account for the non-deterministic outcome of certain measurements). It is postulated that a joint distribution function g on local eigenvalues only should suffice to consistently describe any quantum state. Considering the hierarchy of parameters defined in (7,8), this is a very strong statement, indeed.

It has long since been known that the existence of such a g for nonclassical two-spin-states like the EPR-state,

$$|\psi\rangle = \frac{1}{\sqrt{2}}(|22\rangle - |11\rangle), \qquad (21)$$

is in conflict with Bell's inequalitiy. This can be verified by a *statistical test* based on homogeneous ensembles. So it came as a surprise that the $N = 3$ extension, the so-called GHZ-state (Mermin 1990),

$$|\psi\rangle = \frac{1}{\sqrt{2}}(|222\rangle - |111\rangle), \qquad (22)$$

would render the existence of g inconsistent even for any *individual* experiment. In terms of our multi-node expectation values, this state is characterized by (Mahler and Weberruss 1995)

$$K_{33}^{12} = K_{33}^{13} = K_{33}^{23} = 1, \qquad (23)$$
$$K_{122} = K_{212} = K_{221} = 1 = -K_{111}, \qquad (24)$$

which are all eigenvalues (i.e., there is no dispersion); the other terms (including the Bloch-vectors) are zero. In this case, the distribution

$$g = g(\lambda_j^1, \lambda_k^2, \lambda_l^3) \text{ with } \lambda_j^\mu = \mp 1 \qquad (25)$$

should give full weight to just one set of λ-values such that

$$K_{122} = \lambda_1^1 \lambda_2^2 \lambda_2^3 = 1 \quad \text{(and permutations of subindex 1).} \tag{26}$$

Now, using $(\lambda_2^\mu)^2 = 1$, one readily shows that

$$1 = K_{122} K_{212} K_{221} = \lambda_1^1 \lambda_1^2 \lambda_1^3. \tag{27}$$

The right-hand side is just the representation of K_{111} (in local realism), which, unequivocally, should be -1. Thus, no such g can consistently describe the GHZ-state!

Despite this failure, the idea of looking for classical contributions to the various expectation values remains a worthwhile attempt. The stochastic unravelling allows a constructive approach of this kind. Based on the joint probability function f we can decompose (9) as (Mahler et al. 1996)

$$M_{ij}^{\mu\nu} = \overline{M_{ij}^{\mu\nu}} + V_{ij}^{\mu\nu}, \tag{28}$$

where $V_{ij}^{\mu\nu} = \overline{\lambda_i^\mu \lambda_j^\nu} - \overline{\lambda_i^\mu} \; \overline{\lambda_j^\nu}$ describes a "classical part" due to statistical uncertainty in the space of local expectation values. This decomposition, which depends on the underlying stochastic process, illustrates the fact that $M_{ij}^{\mu\nu} \neq 0$ does not necessarily imply non-classical correlations (entanglement in the strict sense). Non-classicality tests would have to take these limits into consideration.

7   Changing Perspectives

The availability of addressible nodes not only motivates the use of our product-state language, it also allows to "change perspective" by referring to various combinations in index-space. It is this additional freedom of choice which, after all, makes nonlocality an operational concept. If we were forced to stick to just one perspective, the counterintuitive features to be discussed now would remain unobservable.

The following examples are all formulated in terms of the polarization degree of freedom (pseudo-spin) of free photons. The spatial degrees of freedom are considered only as a convenient vehicle for separating and combining the spin variables. In any of the Figs. 3-5, one representative spatial coordinate z, indicating the localization of the pair-source and two experimentalists A and B, is given together with the time-coordinate t pointing downward. We assume that we have access to a source of individual photon pairs each prepared in an EPR-state (21), characterized by $K_{ij} = -\delta_{ij}$. Furthermore, it is assumed that the individual photon can be stored, reflected, and its polarization state measured, if so required.

For the present purpose this implementation is most convenient, but by no means necessary. Localized $(N = 2)$-networks would also serve, but would lack the dramatic features connected with the potentially long distance travelled by the photons. Even a dissipative system with an appropriate coherent source could (accidentally) do the job if its damping channels constituted different perspectives.

Example 1: Conversion of nonlocal into local information (Ekert 1991).

While the two correlated photons are separate in real space, their individual polarization state remains undefined ($\lambda_j^\mu = 0$). When local measurements are performed in A and B, with respect to one and the same measurement angle, the results are undetermined but strictly anti-correlated: A and B are supplied with the same random sequence of data, which can be used as a key in cryptography. The additional interesting feature is that this key distribution can be made secure against eavesdropping. The reason is that anyone trying to extract information on one of the photons would destroy nonlocality. This, in turn, can statistically be tested by the legal users of the communication channel by reference to Bell's inequality (see Fig. 3).

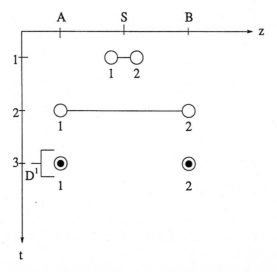

Fig. 3. Conversion of nonlocal into local information. D^1 = local detector; circles connected by a line indicate entanglement, filled circles indicate locally prepared states.

Example 2: Overdense coding (Bennett and Wiesner 1992).

In this case, A does not measure his photon, but rather applies one of four local unitary transformations (unit operation, giving state $|2\rangle$ a minus sign; exchanging the two basis states and then performing one of the two options as before). Locally, the last two operations are indistinguishable from the first two. However, the pair is thus prepared in one of four different *orthogonal* states. After A has sent his photon back to B, B can detect these four alter-

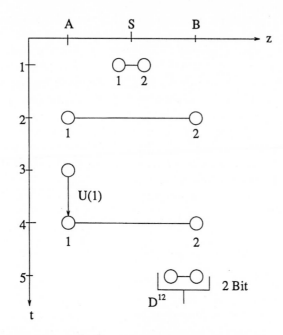

Fig. 4. Overdense coding. $\hat{U}_i(1)$ = local unitary transformation; D^{12} = bilocal detector (Bell measurement).

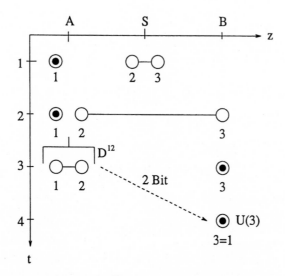

Fig. 5. Quantum teleportation. D^{12} = bilocal detector; broken line: transmission of classical information; $\hat{U}_i(3)$ = local unitary transformation.

natives (see Fig. 4). In this way, one photon (in polarization space) is able to carry 2 bits of information!

Example 3: Quantum teleportation (Bennett et al. 1993).

In this case, A already has photon 1 in a local polarization state, which may be unknown to him. After having received photon 2 as part of the EPR-state, he performs a nonlocal measurement, producing one of four possible outcomes. B, after learning the result of this measurement, can perform a local transformation on his photon 3 such that this photon is now in the same state as photon 1 was before (see Fig. 5). The polarization state of photon 1 at A is thereby teleported to B (note that in a conventional scheme this would require to transmit many bits of information, as the angle of direction is continuous). In this way, the original perspective of (1) versus (23) has been changed to (12) versus (3).

8 Virtual Snap-Shots: Nonlocality in Time

If one considers both time and space as parameters one may wonder to what extent the concept of nonlocality could be carried over to the time domain. It turns out that this possibility rests upon the concept of pulses (time slots) as counterparts to spatial filters (i.e., node-selective couplings) (Wawer et al. 1996).

Up to now, the parameters describing the Hamiltonian have been considered constants. We are working in the Schrödinger picture, so that dynamical equations exist for the density operator only. However, this does not exclude manipulations of the Hamilton parameters via a separate and autonomous dynamics, i.e., switching a laser, changing its frequency, changing the distance between individual nodes to modify their interaction, etc.

We consider a $(N = 3)$-network composed of an object system $\mu = 2$ and two memories, $\mu = 1, 3$. The object is continuously driven by an external laser field g_{21}^2 with zero detuning. The resulting Rabi-oscillation (cf. Fig. 6a) is

$$\lambda_j^2 = \{0, \sin g_{21}^2(t - t_0), \cos g_{21}^2(t - t_0)\}, \tag{29}$$

with the two-time correlation function

$$K_{33}^{22}(t_1, t_2) = \cos g_{21}^2(t_1 - t_2). \tag{30}$$

Now, during a π-pulse on memory 1, we switch on the interaction between the object and memory 1, as specified by (5). This type of interaction can be used so that the state of the object sets a condition, whether or not the memory is in resonance with its driving field. The length of the pulse is short compared to the time for one Rabi-cycle, so that this pulse can be considered instantaneous and specified by a time t_1. We repeat this process with memory 3 at time t_2. One can show that under these conditions

$$K_{33}^{22}(t_1, t_2) = K_{33}^{13}, \tag{31}$$

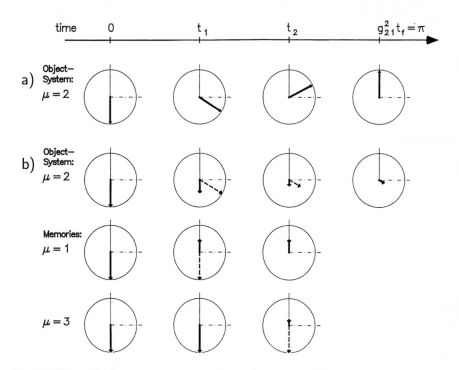

Fig. 6. "Reversible" projections in an $(N = 3)$-network. (a) Driven object system (π-pulse). (b) Driven object system undergoing two reversible projections at time t_1, t_2, respectively. At time t_f the object has now finite probability to be measured in state $|1\rangle$ (down). All three subsystems are entangled (not shown).

i.e., the two-time correlation of the object is mapped onto the two-node correlation of the memories. The resulting dynamics in terms of the Bloch-vectors ("quantum clocks") is shown in Fig. 6b. We see that all the Bloch-vector lengths are reduced, indicating nonlocal correlation. We also see that the interaction with the memory implies a "projection" of the object Bloch vector on the 3-axis, just as if it were "measured", but without reaching a decision yet between up or down. Note that up to the time t_f the dynamics is coherent and reversible.

In Fig. 7, the object system is measured first (projected on "down" position), then memory 3, followed by memory 1 (not shown explicitly). Recalling how the network has been prepared, the measurement sequence refers to a travel backwards in time! This imposes no logical problems, though, as the decisions in the past will always be compatible with what is already established later on (the "clocks" are updated as information is retrieved, exploiting the nonlocal correlations, see Fig. 7).

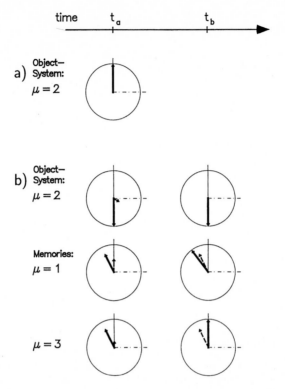

Fig. 7. "Time travel". Entangled 3-level system as of Fig. 6 (at time $t_f > t_2 > t_1$). The state of memory system $\mu = 1$ refers to the object state of time t_1, the state of memory system $\mu = 3$ to that of time $t_2 > t_1$. t_a: Measurement of object system $\mu = 2$ at time t_f ("now"), affecting both memories. t_b: Measurement of memory $\mu = 3$ (affecting memory $\mu = 1$). t_c: Measurement of memory $\mu = 1$ (not shown). In this way decisions are reached backward in time.

9   Superposition of Paths

Standard devices (beam-splitters, mirrors), allow particle wave-packets to be sent in a certain superposition of paths. When finally recombined, the resulting interference pattern depends on whether or not the "which way"-information is in principle available.

In Fig. 8, we see a Mach-Zehnder-type scenario with two 50 : 50 beam-splitters and two mirrors. Individual photons (or other elementary particles) enter from the left and experience phase shifts depending on whether the particle is reflected (phase shift $\Delta\phi = \pi/2$) or transmitted ($\Delta\phi = 0$). While the geometric phase shifts are the same along any path, the accumulated additional phase shifts ($e^{i\Delta\phi} = i, 1$) due to the optical devices are different. In this example, there is constructive interference for the paths leading to

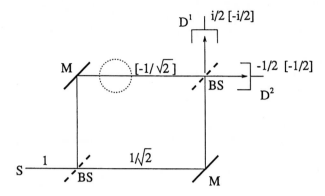

Fig. 8. Superposition of paths. M = mirror, BS = beam-splitter, D_i = detectors. We assume a phase shift of $\pi/2$ for each reflection. Given are the amplitudes for unit input. The amplitudes in brackets are missing, if the upper path is blocked.

detector 2, and destructive interference for detector 1. This result can be understood only in the wave picture. In the particle picture, both detectors have the same chance to be hit by a photon.

When the upper path is interrupted by an object (which may be a photon detector with 100% efficiency), only the lower path leads to the two original detectors. Thus, in 50% of the cases the photon will neither be seen by detector 1 nor 2, while in 25% of the cases, detector 1 will now see a photon. This will prove the presence of the object, even though the photon did not "interact" with the object – indicated by the fact that the ideal photodetector did not register (Vaidman 1994)!

This detection strategy rests upon prior knowledge about the original scenario, the nonlocality of which is broken by the presence of the object. This counterintuitive result is less striking when we realize that we do not measure the *presence* of the object but rather the *absence* of nonlocality. This scenario may, thus, be considered as another variant of the transfer between local and nonlocal information.

10   Summary

Any investigation of quantum systems is plagued by the fact that we need not only a description of the system, its state, and its evolution, but also means to verify all that with respect to a given reference frame. This reference frame is not only a mathematical device, but requires physical implementation (embedding), which will react back on what we want to study.

Here we have been concerned with quantum systems that can be decomposed into selectively addressable subunits. We wanted to account for scenarios in which we can meaningfully say "something happens here, something else happens there." Only under these conditions might we gain access to nonlocal coherence (entanglement) in addition to local coherence.

For individual quantum objects, one has to distinguish between dissipation as a source of suppressing control, as a source of decoherence, and as a source of information.

The loss of control is due to the fact that the unravelling of the master equation leads to a process consisting of a deterministic and a nondeterministic part. For known initial conditions, the predictability of the state will deteriorate, implying an increase of entropy with time. The nondeterministic part implies random projections, which tend to destroy coherence. At the same time the projections might be connected with measurement events, the registration of which would reduce our lack of information about the present state (and thus quench the increase of entropy).

Nonlocality in space should not be confused with classical correlations. Tests to distinguish between the two require a clear classical reference. Nonlocality in time should not be confused with memory effects. The former refers to a situation in which decisions on what has "happened" at two instants of time, say, have not yet been reached, while the correlations between those events are already specified. The latter means that the evolution of a system depends not only on its present state, but rather on a whole sequence of past states.

Acknowledgment: I thank the Deutsche Forschungsgemeinschaft for financial support and M. Keller, J. Schlienz, and R. Wawer for valuable discussions.

References

Aspect A., Grangier P., and Roger G. (1982): Experimental realization of Einstein-Podolsky-Rosen-Bohm- Gedankenexperiment: a new violation of Bell's inequalities. *Phys. Rev. Lett.* **49**, 91–94.

Bell J.S. (1987): *Speakable and Unspeakable in Quantum Mechanics* (Cambridge University Press, Cambridge).

Bennett C.H. and Wiesner S.J. (1992): Communication via one- and two-particle operators on Einstein-Podolsky-Rosen states. *Phys. Rev. Lett.* **69**, 2881-2884.

Bennett C.H., Brassard G., Crépeau C., Jozsa R., Peres A., and Wooters W.K. (1993): Teleporting an unknown quantum state via dual classical and Einstein-Podolsky-Rosen channels. *Phys. Rev. Lett.* **70**, 1895–1899.

Berman P.R., ed. (1994): *Cavity Quantum Electrodynamics* (Academic, San Diego).

Dicke R.H. (1954): Coherence in spontaneous radiation processes. *Phys. Rev.* **93**, 99–110.

Ekert A.K. (1991): Quantum cryptography based on Bell's theorem. *Phys. Rev. Lett.* **67**, 661–663.

Keller M. and Mahler G. (1994): Nanostructures, entanglement, and the physics of quantum control. *J. Mod. Optics* **41**, 2537-2555.

Kwiat P.G., Mattle K., Weinfurter H., Zeilinger A., Sergienko A.V., and Shih Y. (1995): New high-intensity source of polarization-entangled photon pairs. *Phys. Rev. Lett.* **75**, 4337–4341.

Mahler G., Keller M., and Wawer R. (1996): SU(n)-networks. I. Master equation and local measurements. *Phys. Rev. A*, submitted.

Mahler G. and Weberruss V.A. (1995): *Quantum Networks – Dynamics of Open Nanostructures* (Springer, Berlin).

Mermin N.D. (1990): What's wrong with these elements of reality? *Phys. Today,* June 1990, p. 9.

Molmer K. and Castin Y. (1996): Monte Carlo wave functions in quantum optics. *Quantum & Semicl. Optics* **8**, 49–72.

Wawer R., Keller M., and Mahler G. (1996): SU(n)-networks. II. Quantum Zenon effects. *Phys. Rev. A*, submitted.

Vaidman L. (1994): On the realization of interaction-free measurements. *Quantum Optics* **6**, 115–124.

Yurke B. and Stoler D. (1995): Bell's inequality experiment employing four harmonic oscillators. *Phys. Rev. A* **51**, 3437–3444.

Process and Time

Basil J. Hiley and Marco Fernandes

Physics Department, Birkbeck College, University of London,
Malet Street, London WC1E 7HX, England

Abstract. In this paper we outline an attempt to provide a mathematical description of process from which both space and time can be abstracted. We indicate how spatial properties can be abstracted from the orthogonal Clifford algebra and the symplectic Clifford algebra, algebras that are at the heart of quantum mechanics. The approach is generalised by constructing a bi-algebra from which we are able to abstract a notion of time. The relation of this approach to the work of Prigogine and Umezawa is discussed.

1 Introduction

In a recent paper, Haag (1990) has pointed out that even after a collective and sustained agonising over quantum field theory, there still remain a number of dark spots that veil some fundamental internal incompatibility within the conceptual framework of the theory. This unease becomes particularly significant when attempts are made to quantize the gravitational field (see Isham 1987). Some of these concerns have influenced our thinking over recent years and it is for this reason that we have been exploring somewhat different approaches to these questions. One approach has been extremely conservative since it tries to explore the precise differences between quantum and classical phenomena by rewriting the Schrödinger equation in a form that is closer to the Hamilton-Jacobi equation of classical mechanics (Bohm and Hiley 1993), while others have been of a more radical nature (Bohm and Hiley 1984; Frescura and Hiley 1984; Hiley and Monk 1993).

The more radical approaches have some features in common with Haag's own proposals, particularly in sharing his general view that we should regard the space-time continuum not as a priori given, but to be a structure derived from something more primitive which contains within it features that are more appropriate to the quantum domain. Our interest in this field was considerably influenced by the Penrose twistor programme (1972) although we were first alerted to more radical possibilities after reading the following three sentences in Eddington's classic work "The Mathematical Theory of Relativity" (Eddington 1937):

> "The hiatus (in trying to unite general relativity with electromagnetism) probably indicates something more than a temporary weakness of the rigorous deduction. It means that space and time are only approximate conceptions, which must ultimately give way to a more

general conception of ordering of events in nature not expressible in terms of a fourfold co-ordinate system. It is in this direction that some physicists hope to find a solution of the contradictions of the quantum theory."

Clearly, in this view, it is not sufficient merely to strive for frame independent expressions since they assume that the relevant structure of the underlying physical processes can be captured within the very order of the continuum that supports a co-ordinate frame in the first place. As Isham (1987) has argued, it is possible that the order contained in the structure of the differential manifold itself is not appropriate for exploring the structure of space-time near the Planck length and that any alteration to the underlying structure may have profound implications at much larger scales. The proposal is that we should make a thorough re-examination of the underlying assumptions of our theories which should include a re-examination of the topology in the small, with a view to developing more appropriate structures such as "quantum topology" or more generally what Bohm (1965) has called a "quantum topo-chronology". The choice of this word is intended to signify that not only must we emphasise spatial topology i.e., the study of the order of placing one thing in relationship to another, but the study of how one event or moment acts physically in another. In other words topo-chronology is not limited to the concepts of neighbourhood, incidence, boundary and closure, but will include novel notions of relationship, structure and order involving process in general.

This was one of the motivations behind Bohm's (1980) introduction of his novel orders, namely, the implicate and explicate orders, notions that have been criticised as "vague", a criticism implying that these orders can have no value in physics. The intention behind the introduction of these new orders was simply to provide a framework within which to develop new physical theories together with the appropriate mathematical structures that will lead to new insights into the behaviour of matter and ultimately to new experimental tests. This paper is an attempt to develop these arguments further and to suggest appropriate ways of proceeding.

As we have already remarked, the idea of topo-chronology means that we must study order, not only in the static sense, but in an active sense. Thus nature is to be regarded as a structure or order in an evolving process. This process is not to be regarded as a process evolving in space-time, but space-time is to be regarded as a higher order abstraction arising from this process involving events and abstracted notions of "space (or space-like) points". These points are active in the sense that each point is a process that preserves its identity and its incidence relations with neighbouring points. In Hiley and Monk (1993) we described how this could be achieved in a very simple algebraic structure, namely, the discrete Weyl algebra. Thus like Haag (1990), our basic structure assumes an element of process that can be described by an element of some suitable algebra. For example, it could be a Clifford algebra

(Frescura and Hiley 1980a), an Einstein algebra (Geroch 1972), or a Hopf algebra (Zhong 1995).

The basic idea of assuming elements of an algebra can represent structure process goes back to the end of the previous century when Grassmann (1894), Hamilton (1967), and Clifford (1882) were laying the foundations of what were to become the Grassmann, the quaternion and the Clifford algebras. We feel that it is no coincidence that these algebras play a fundamental role in our present physical theories and are structures that are valid in both the classical and quantum domains. These early motivations stemmed from an assumption that metaphysics was a legitimate tool for exploring new ideas both in mathematics and physics; hence the emphasis that these pioneers placed on process and activity. For example, Hankins (1976) recalls that at the 1835 meeting of the British Association in Dublin, Hamilton read a paper entitled "Theory of Conjugate functions, or Algebraic Couples; with a Preliminary and Elementary Essay on the Algebra as a Science of Pure Time" in which he freely uses metaphysical arguments to develop his ideas. Again in one of his notebooks he writes:

"In all Mathematical Science we consider and compare relations. In algebra the relations which we first consider and compare, are relations between successive states of some changing thing or thought. And numbers are the names or nouns of algebra; marks or signs, by which one of these successive states may be remembered and distinguished from another. .... Relations between successive thoughts thus viewed as successive states of one more general and changing thought, are the primary relations of algebra. .... For with Time and Space we connect all continuous change, and by symbols of Time and Space we reason on and realise progression. Our marks of temporal and local site, our then and there, are once signs and instruments by which thoughts become things ...."

Unfortunately the concept of process is difficult to work with and already by the 1880s the active role of the elements of the algebra had been lost. The elements of the Grassmann algebra had become the static forms of the vector, the bivector, etc., together the exterior product that we now associate with a Grassmann algebra. It was for this reason that Clifford (1882) found it necessary to emphasise once again that the quaternion product was based on movement or activity rather than a static structure as the Grassmann algebra had become. Indeed in developing his algebra, Clifford constantly emphasises the active role of the elements of the algebra, giving them names such as "rotors" and "motors".

We wish to revive the spirit of those early explorations by taking process as fundamental and to construct space-time, fields and matter from this basic process. To do this we will follow Hamilton and assume that process is describable by elements of an algebra, while the relevant structure process is defined by the algebra itself. The relevance of these ideas to quantum phenomena is suggested by the realisation that the quantum formalism is, in essence, an algebraic theory, and the insistence that it is a description that requires Hilbert space is forced on it by the interpretation that the wave function is related to probability. In our view Hilbert space has become overemphasised in physics and this in spite of the fact that in order to accommodate wave functions for position and momentum, we are forced to a larger structure, the rigged Hilbert space. In a much neglected paper, Dirac (1965) argues that in quantum field theory

> "...the Heisenberg picture is a good picture, the Schrödinger picture
> is a bad picture, and the two pictures are not equivalent (in field
> theory), as physicists usually suppose."

By staying in the Heisenberg picture, he then goes on to show that it is possible to obtain solutions of physical problems in this picture when no solution is possible in the Schrödinger picture. In this way Dirac concluded that the Heisenberg picture is more general and since it is an algebraic approach to quantum phenomena, it suggests that the algebra may be of more significance. Unfortunately the difficulty that Dirac saw is that we have no interpretation of a purely algebraic approach. The approach we are suggesting here is an attempt to provide such an interpretation and to do this we must develop a structure in which process is taken as basic, opening up the possibility of interpreting the elements of the algebra in terms of this underlying process. We are far from realising such an approach but at least it is possible to make a start as we will now show.

Let us begin with the assumption that underlying quantum phenomena there is a structure that requires us to take the notion of process as the basic form. Rather than getting into a philosophical discussion as to the meaning of the term "process", let us see if a careful examination of the algebraic structure that lies behind the quantum theory can reveal insights into the nature of this underlying process. After all it was quantum mechanics that led to a revival of interest in the Clifford algebra with the success of the Dirac theory. Here the dynamical variables require us to use elements of the Clifford algebra and the wave functions themselves can be expressed in terms of spinors. As has been pointed out by Frescura and Hiley (1980a) and by Benn and Tucker (1983), these spinors arise in the algebra itself as elements of a minimal left ideal, so that it is not necessary to use the Hilbert space formalism in a fundamental way. Likewise the Penrose twistor programme, although it has not reached the same status as the Dirac theory, exploits the spinors of the conformal Clifford algebra which, in turn, contains the Dirac Clifford algebra as a substructure. In fact Bohm and Hiley (1984) have already shown how the algebraic structure contained in the twistor formalism can lead to a novel approach to the notion of pre-space.

What we regard as an additional significant point is that any Clifford algebra can be constructed from a pair of dual Grassmann algebras. The elements of these Grassmann algebras are analogous to the annihilation and creation operators used in the Dirac approach to fermion fields, thus giving us the possibility of a new significance to these operators in terms of the underlying process, rather than merely as the creation and annihilation of excitations of the quantum fields.

Recently we have been exploring an analogous structure, namely, the symplectic Clifford algebra (Fernandes and Hiley 1996) which can be constructed from boson annihilation and creation operators. This algebra contains the Heisenberg algebra, again suggesting that it will strongly feature in a process orientated approach to quantum theory. Indeed it was these possibilities that lead Hiley and Monk (1993) to explore a simpler finite structure, the discrete Weyl algebra that we have already referred to above. What we have recently been made aware of is the possibility of using a pair of dual boson operators to introduce the notion of time, a feature that was lacking in our own approach (Umezawa 1993). We will discuss these ideas toward the end of this paper.

2   The Algebra of Process

Let us start by illustrating the type of approach that we have been following. We have already intimated above that by starting with Grassmann's original motivations, it is possible to generate a Clifford algebra of movements (for more details see Hiley 1995). Let us regard process as a continuous, undivided flux or flow. To be able to discuss such a flow, it is necessary to distinguish certain features within the flux. The key question to ask is: How is one distinctive feature related to another? In a succession of features, is each feature independent of the others or is there some essential dependence of one on the other? The answer to the first part of the question is clearly "no" because each feature is reflected in the others in an inseparable way. Thus one feature can contain the potentialities for a succeeding feature and each feature will contain a trace of its predecessor.

Let us label two distinguished features by P_1 and P_2, and regard them as the opposite poles of an indivisible process. To emphasise the indivisibility of this process, we will follow Grassmann and write the mathematical expression for this process between a pair of braces as $[P_1 P_2]$. Here the braces emphasise that P_1 and P_2 cannot be separated. Note that the order of the elements is significant, implying the order of succession, i.e., $[P_1 P_2] = -[P_2 P_1]$. This could be construed as implying that a primitive notion of time is being introduced. However no notion of a universal time order is implied. When applied to space, these braces were called extensives by Grassmann.

It should be noted that any attempt to discuss and describe order without introducing time is very difficult. For example, it is clear that the order in any map is timeless, nevertheless it is easier to describe its order in terms of a sequence of directions such as "Turn right when you come to the next crossroads" and so on. Similarly when we come to discuss the order of physical process, we trace out the order in time for convenience, but this does not imply that the process itself evolves in that time order. Thus in discussing the order of distinguished points in our structure process, it is convenient to discuss the order in terms of "succession" even though no time may be involved in the order we are discussing. Our distinguished points are not then a sequence of independent points, but each successive point is the opposite pole of its immediate predecessors so that points become essentially related in an active way.

For more complex structures, we can generalise these basic processes to $[P_1P_2]$, $[P_1P_2P_3]$, $[P_1P_2P_3P_4]$ etc. In this way we have a field of extensives from which we can construct a multiplex of relations within the overall process or activity. The sum total of all such relations constitutes what Bohm has termed the holomovement. Thus, in this view, space cannot be a static receptacle for matter; it is a dynamic, active structure.

We now argue that process forms an algebra over the real field in the following sense:

1. Multiplication by a real scalar denotes the strength of the process.
2. The addition of two processes produces a new process. A mechanical analogy of this is the motion that arises when two harmonic oscillations at right angles are combined. It is well-known that these produce an elliptical motion when the phases are adjusted appropriately. This addition of processes can be regarded as an expression of the order of co-existence.
3. To complete the algebra there is a inner multiplication of processes defined by $[P_1P_2][P_2P_3] = [P_1P_3]$ which will be called the order of "succession" using this word in the sense explained above.

3   The Clifford Algebra of Process

We now illustrate briefly how this algebra can carry the directional properties of space within it without the need to introduce a co-ordinate system. Let us start by assuming there are three basic movements, which corresponds to the fact that space has three dimensions. We leave open the question as to why only three space dimensions are needed. The scheme can be generalised easily to higher dimensional spaces with different metrics.

Let these three movements be $[P_0P_1]$, $[P_0P_2]$, $[P_0P_3]$. To describe movements that take us from $[P_0P_1]$ to $[P_0P_2]$, from $[P_0P_1]$ to $[P_0P_3]$ and from $[P_0P_2]$ to $[P_0P_3]$, we need a set of movements $[P_0P_1P_0P_2]$, $[P_0P_1P_0P_3]$ and $[P_0P_2P_0P_3]$. At this stage the notation is looking a bit clumsy so it will be simplified by writing the six basic movements as $[a]$, $[b]$, $[c]$, $[ab]$, $[ac]$, $[bc]$.

We now use the order of succession to establish

$$[ab][bc] = [ac],$$
$$[ac][cb] = [ab], \tag{1}$$

$$[ba][ac] = [bc],$$

where the rule for the product (contracting) is self evident. There exist in the algebra three two-sided units, $[aa]$, $[bb]$, and $[cc]$. For simplicity we will replace these elements by the unit element 1. This can be justified since $[aa][ab] = [ab]$ and $[ba][aa] = [ba]$; $[bb][ba] = [ba]$, etc., so that

$$
\begin{aligned}
[ab][ab] &= -[ab][ba] = -[aa] = -1, \\
[ac][ac] &= -[ac][ca] = -[cc] = -1, \\
[bc][bc] &= -[bc][cb] = -[bb] = -1.
\end{aligned}
\tag{2}
$$

There is the possibility of forming $[abc]$. This gives

$$
\begin{aligned}
[abc][abc] &= -[abc][acb] = [abc][cab] = [ab][ab] = -1, \\
[abc][cb] &= [ab][b] = [a];
\end{aligned}
\tag{3}
$$

etc. Thus the algebra closes on itself, and it is a straightforward exercise to show that it is isomorphic to the Clifford algebra $C(2)$ which was called the Pauli-Clifford algebra in Frescura and Hiley (1980a).

The significance of this algebra is that it carries the rotational symmetries and for this reason it is called the directional calculus. The movements $[ab]$, $[ac]$, and $[bc]$ generate the Clifford group which is homomorphic to the Lie group $SO(3)$, the group of ordinary rotations. For good measure this algebra also contains the spinors as minimal left ideals which form a linear sub-space in the algebra. This relationship shows that the spinors arise naturally in an algebra of movements. An extension of this algebra to the Dirac algebra then enables us to discuss the light cone structure so that the light ray itself can be given an algebraic meaning. The background to all of this has been discussed in Frescura and Hiley (1984).

4   The Algebra of Points

The calculus of directions is based on the existence of a single point around which the directions emanate. In the Dirac algebra this would be equivalent to considering a light cone at a single point. In order to relate light cones at different points, we have to extend these ideas. We then have two possibilities. We can enlarge the algebra to the conformal Clifford algebra and use the corresponding spinors (i.e., the twistors) to relate light cones at different points in space. This is the basic idea that lies at the heart of the twistor programme. An alternative way involves generalising the algebraic structure to include the kinematics. To do this we introduced a generalised phase space algebra containing translations within it. These translations can also be used to relate light cones at different points (Bohm and Hiley 1981). The algebra that contains these translations is the symplectic Clifford algebra (see

Crumeyrolle 1990) or at least an extension of it which has been developed by Frescura and Hiley (1984, 1980b) and Fernandes and Hiley (1996).

We do not want to discuss the details of this structure here as they can be found in the above papers, but it should be noted that this algebra contains the Heisenberg algebra as an automorphism algebra. This clearly shows that our overall approach uses another important aspect of quantum theory that is missing from the twistor programme. Furthermore the symplectic Clifford algebra can be generated by bosonic creation and annihilation operators in a manner that is very similar to the way one generates the orthogonal Clifford algebra by a pair of dual Grassmann algebras as mentioned above. However it should be emphasised once again that in our approach we are not talking about the creation and annihilation of particles, but the generation and annihilation of elements in our underlying process. In this sense the underlying structure process can be thought of as being quantum in essence and therefore can be regarded as a kind of quantum geometry from which the classical space-time will emerge by some suitable averaging.

As we have already mentioned these ideas can be illustrated using a "toy" algebra, namely, the discrete Weyl algebra (see Hiley 1991, Hiley and Monk 1993). Although this algebra is very limited, it nevertheless gives an indication of how the idea works. Here one can construct the points of space directly in terms of the minimum left ideals of the algebra so that the spinors of the Weyl algebra represent the "points" of our generalised phase space.

In order to get some feeling of how these general concepts can be given meaning, let us consider some details. The discrete Weyl algebra is a finite polynomial algebra generated by a set of elements $1, e_1^0, e_0^1$ subject to the relations

$$e_0^1 e_1^0 = \omega e_1^0 e_0^1, \qquad (e_0^1)^n = 1, \qquad (e_1^0)^n = 1, \qquad (4)$$

where $\omega = \exp(2i\pi/n)$.

One of the reasons for choosing this algebra is that in the limit when $n \rightarrow \infty$, we obtain the symplectic Clifford algebra referred to above. The fact that the algebra is finite has the added advantage that it makes the mathematics very easy to manipulate. It also has the advantage that it is much easier to see exactly how a discrete space emerges from the algebra. It is then straight forward to see how the continuum emerges in the limit.

To show how it is possible to abstract a phase space from this algebra, we must first explain how it is possible to abstract a 'generalised point' from the algebra. An important clue as to how to proceed has been given by Eddington (1958), who argued that, within a purely algebraic approach to physical phenomena, there are elements of existence defined, not in terms of some metaphysical concept of existence, but in the sense that existence is represented by an algebraic element that contains only two possibilities: existence or non-existence. He assumed that existence could be represented by an idempotent element in an appropriate algebra. In our view it is not so much that points exist, but that they *persist*. They persist not in the sense that they are static, but rather that each point continually transforms into itself. Now the elements that transform into themselves are the idempotents, e, which are defined by the relation $e \cdot e = e$. (Recall that in our approach a product represents succession.) Thus, in the algebra, generalised points correspond to a set of idempotents in the algebra.

Again since the Weyl algebra is finite, it is possible to find a complete set of pairwise orthogonal primitive idempotents e_i. One such set is

$$e_i = \frac{1}{n} \sum_k \omega^{-ik} e_0^k. \tag{5}$$

The e_i will satisfy the relation $\sum_i e_i = 1$ and $e_i^2 = e_i$. The set e_i will constitute a set of generalised points in our phase space. To show that this set can be used to represent a finite set of points in 'space', let us introduce a 'position' operator X defined by

$$X = \frac{1}{n} \sum_{jk} j\, \omega^{-jk}\, e_1^0 = \sum_j j\, e^j, \tag{6}$$

so that

$$X\, e^j = j\, e^j. \tag{7}$$

Thus each primitive idempotent is labelled by the eigenvalue of the "position" operator. If we choose e_0 as the point of the origin and e_1 as its neighbour then it can be shown that there exists a translation operator, T, such that

$$e_1 = T\, e_0\, T^{-1}. \tag{8}$$

In this way we can generate all the points by successive application of T. Thus we generate all the points in the space from the algebra itself. Of course in this very simple model the space is only one dimensional, but the approach can be generalised to any dimension by using the full symplectic Clifford algebra.

It is perhaps worth mentioning here that this set of idempotents is not unique. It is always possible to find another set under a suitable inner auto-morphism of the algebra. Indeed there are pairs of spaces within the algebra that are related in such a way that one set can be chosen to be the position space while the dual pair will be the momentum space. Notice that what is called complementarity in standard quantum mechanics takes on an on-tological significance in our approach in the sense that it is not possible to 'display' the position space at the same time as 'displaying' the momentum space. This is very reminiscent of David Bohm's notion of the explicate or-der (Bohm 1980). Recall that not all explicate orders can be made manifest together. So in our view the algebra is a description of the implicate order, with one explicate order representing the position space, while the momentum space is represented by another complementary explicate order. Thus within the algebra there is the possibility of finding many different generalised phase spaces.

In the finite algebra, all these spaces are equivalent, but when we go to an algebra with an infinite number of degrees of freedom, as is the case for quantum field theory, there emerges the possibility of many inequivalent spaces (see Umezawa 1993). Within these we would expect the ultimate emergence of a preferred space-time arising as a result of some as yet unknown symmetry breaking process.

The continuous phase space algebra can be generated by $[1, a, a^\dagger, \{P\}]$, where a and $a^\dagger$ satisfy the commutation relation $[a, a^\dagger] = 1$. $\{P\}$ is an idempotent that satisfies the relations

$$\{P\}^2 = P; \qquad a\{P\} = 0; \qquad \{P\}a^\dagger = 0. \tag{9}$$

Using the translation operator T which can be written in the form

$$T(\alpha) = \exp\left[\alpha^* a + \alpha a^\dagger\right], \tag{10}$$

we can generate a continuum of points by the inner automorphism

$$\{P_n\} = T(\alpha)^n \{P\} T(\alpha)^{-n}. \tag{11}$$

By increasing the number generators $a_j, a_j^\dagger$ we can construct higher dimensional spaces within which an algebraic geometry can be constructed from the elements of the algebra as shown in Fernandes (1996) and Fernandes and Hiley (1996). This structure has many similarities with the Einstein algebra introduced by Geroch (1972).

5   Process and Time

In the above discussions, we have been exploring algebras that are actually used in quantum mechanics. When it comes to time, we cannot use these methods because nowhere does time appear as an element of an operator algebra in quantum mechanics. Thus it is not possible simply to exploit the mathematics as we have done in the previous section. The problem of finding a suitable time operator is by now well documented and various possibilities of finding a suitable operator within more general schemes have been proposed (Unruh and Wald 1989; Isham 1993). Of the various attempts that have been made, the best developed are the proposals of Prigogine and his group (see, e.g., Petrosky and Prigogine 1990, 1994). Our approach shares some of the techniques used by this group although their work is far more extensive.

There is another interesting approach emerging from the work of Umezawa (1993) who has been exploring an algebraic approach based on thermal field theory. Within this approach a specific time operator has been proposed by Ban (1991) and it is this work that is most closely related to the ideas that we will now discuss. Ban's approach enables us to develop further the ideas on time proposed by Bohm (1986). It is this extension that we now want to address.

One of the key new ideas described in the previous section was the way it is possible to abstract the notion of a "point" from the algebra itself. This offered a way to build a space from the algebra itself. When considering the question of time, it is tempting to try to exploit a similar idea, namely, to attempt to construct the notion of an instant of time from some suitable algebra, but we have two problems. Firstly, as we have already remarked, we have no suitable algebra containing an element which can be identified with time. Secondly the notion of an "instant" of time cannot be regarded as a "point" that persists. Time is about becoming, not persisting. In our approach we are attempting to find time in nature so that it emerges from process. We are not trying to explain the development of nature in time. In the latter, time appears as a "metaphysical enigma" without any verifiable content. As Whitehead (1930) puts it:

"There is time because there are happenings, and apart from these happenings there is nothing."

Here a "happening" is not sharply defined. It is more like a moment which has an inner structure of its own. Each moment contains both the memories of the past with the potentialities for future development.

We can obtain a better feeling for the notion of a moment by examining our own perception of time. Any process of which we become aware has already past, even if it is only for a fraction of a second. What we are conscious of is a memory of the past tinged with an expectation of the future. In the words of Bohm (1986):

"Although the present is, it cannot be specified in words or thoughts without slipping into the past."

Further recall that in order to perceive objects in motion, we need to integrate the information that reaches the retina over about a fifth of a second. For example, slow down the speed at which a cinefilm is run and the image of the motion becomes a series of discontinuous jumps. Speed the film up and we perceive a smooth unfolding motion as normally experienced. The brain has merged the images to produce not only the visual experience of motion, but it can also produce the physical experience of motion if the conditions are right. Thus time is experienced not as an instant, but as an extended moment, a notion that is of necessity ambiguous.

This ambiguity does not only arise from philosophical and psychological considerations. It also arises directly from quantum physics itself. Firstly the ambiguous nature of a moment receives support through the energy-time uncertainty relationship. Recall Einstein's gedanken experiment of determining the time at which a photon leaves a box. The time this occurs is determined by observing when the box changes its weight. To determine this weight change, the box is suspended by a spring fixed to a rigid external frame. However because of the position-momentum uncertainty, there is an ambiguity in the position of the box relative to the fixed frame and thus the moment the photon leaves the box is ambiguous. This means that the timing of any change that occurs in the box is ambiguous relative to the timing of changes occurring in the fixed frame.

There is a different but related ambiguity that has been pointed out by Peres (1980) who discusses the problems involved in constructing a quantum clock. One of the significant conclusions in his paper is that nonlocality in time is an essential feature of all quantum clocks. He goes on to conclude:

"It seems that the Schrödinger wave function $\psi(t)$ with its continuous time evolution given by $i\hbar\dot{\psi} = H\psi$, is an idealisation rooted in classical theory. It is operationally ill defined (except in the limiting case of stationary states) and should probably give way to a more complicated dynamical formalism, perhaps one nonlocal in time. Thus in retrospect, the Hamiltonian approach to quantum physics carries the seeds of its own demise."

These considerations suggest that we must give up the idea of trying to describe motion as a precise point-to-point development in time. We propose that an appropriate description will entail integrating over what would, in the usual approach, be regarded as a finite period of external time.

Bohm and Hiley (1981) have already provided some support for these proposals. We started by examining the differences between the way classical and quantum phase space motion is generated. In contrast to the classical picture where the motion involves a point-to-point substitution through the group of canonical transformations, quantum processes require that the contribution to a point on the quantum motion involves an integration over an extended range of points in a generalised phase space.

To arrive at this conclusion it is necessary to assume that it is the density matrix rather than the wave function that provides the most complete description of an evolving quantum process. A similar assumption is introduced by Prigogine (1993) in his work on irreversible quantum processes. To generalise the dynamics further, we replace the density matrix $\rho(x, x')$ by a non-Hermitean matrix which we called the characteristic matrix $\xi(x, x')$. The characteristic matrix is then used to provide a description of the evolution of the generalised quantum state. We began by assuming that the characteristic matrix satisfies the Liouville equation

$$i\frac{\partial \xi}{\partial t} = H\xi - \xi H = [H(x, p) - H^*(x', p')]\xi = \hat{L}\xi, \tag{12}$$

where $H(x, p)$ is the Hamiltonian of the system and $\hat{L}$ is the Liouville superoperator given by

$$\hat{L} = H \otimes \mathbb{1} - \mathbb{1} \otimes H. \tag{13}$$

The last step in (12) is achieved by writing the characteristic square matrix as a column matrix. This means that we are working with a superalgebra where $\hat{L}$ takes the form given by (13). In order to proceed it is necessary to change co-ordinates to $X = (x + x')/2$, and $\eta = x - x'$, and then introduce the Wigner-Moyal transformation

$$F(X, P) = \int U(P, \eta)\, \xi(X, \eta)\, d\eta, \qquad (14)$$

where

$$U(P, \eta) = \frac{1}{2\pi} \exp\left[iP\eta\right], \qquad (15)$$

with $P = (p + p')/2$. The Liouville equation then becomes

$$\frac{\partial F(X, P)}{\partial t} + \frac{P}{m}\frac{\partial F(X, P)}{\partial X} + \int L(P, P')\, F(X, P')\, dP' = 0 \qquad (16)$$

where

$$L(P, P') = iV_{2(P-P')} \exp\left[2(P - P')X\right] - V_{2(P'-P)} \exp\left[2(P' - P)X\right], \qquad (17)$$

$V_{2(P-P')}$ being the Fourier component of the potential. $F(X, P)$ is a complex function which is analogous to the classical quasi-probability density used in the Wigner-Moyal interpretation of phase space, but we do not use it as a probability density.

In Bohm and Hiley (1981) a positive definite probability is obtained by constructing a statistical matrix $w(x, x')$ from the characteristic matrix

$$\xi(x, x') = \sum_i \sqrt{p_i}\, e^{i\phi}\psi_i^*(x')\psi_i(x), \qquad (18)$$

so that

$$w(x, x') = \int \tilde{\xi}(x', x'')\, \xi(x'', x)dx''. \qquad (19)$$

Here $\tilde{\xi} = (\xi^*)^T$ with T being the transpose. Thus the characteristic matrix is a kind of "square-root" of the density matrix. The details of this approach to the classical limit will not concern us further in this paper.

What we wish to point out here is that the generalised motion requires *two* points in configuration space rather than one as used in classical physics. In other words we need both phase space points (x, p) and (x', p'). Then in discussing a quantum process, (x, p) and (x', p') become operators so that in algebraic terms this means we have a pair of symplectic algebras, one attached to each point of phase space. We then assume that this pair of algebras form a superalgebra with

$$\begin{aligned} X &= (x \otimes \mathbb{1} + \mathbb{1} \otimes x')/2; & \eta &= x \otimes \mathbb{1} - \mathbb{1} \otimes x'; \\ P &= (p \otimes \mathbb{1} + \mathbb{1} \otimes p')/2; & \pi &= p \otimes \mathbb{1} - \mathbb{1} \otimes p'; \end{aligned} \qquad (20)$$

which give the commutation relations

$$[X, P] = [\eta, \pi] = [X, \eta] = [P, \pi] = 0; \qquad [X, \pi] = [\eta, P] = \mathrm{i}. \qquad (21)$$

Furthermore if we follow through the ideas outlined in the previous section, we can now abstract a phase space from the underlying process where neighbouring points are related, not by externally imposed neighbourhood relations, but by relationships provided by the generalised dynamics itself. Thus the structure of the abstracted phase space is generated by the underlying quantum processes themselves.

Recalling that a symplectic algebra can be generated by the Fock boson algebra, our structure thus contains two Fock algebras. The elements (x, p) and (x', p') are then replaced by pairs of elements $(a, a^\dagger)$ and $(\tilde{a}, \tilde{a}^\dagger)$, with

$$a = x + \mathrm{i}p, \quad a^\dagger = x - \mathrm{i}p; \qquad \tilde{a} = x' + \mathrm{i}p', \quad \tilde{a}^\dagger = x' - \mathrm{i}p'. \qquad (22)$$

The algebraic structure that we have introduced is now identical to the structure introduced by Takahashi and Umezawa (1975) and forms the basis of what they have called thermo field dynamics (see Umezawa 1993). In their original work, because the emphasis was placed on the field theoretic meaning of a and $a^\dagger$ (i.e., that they are particle annihilation and creation operators), Takahashi and Umezawa were forced to regard the additional pair of operators as arising from a "fictitious system", a notion that does not encourage confidence. However from the point of view that we are adopting here, it appears as a natural consequence of the way we must describe the evolution of quantum systems in a phase space. Furthermore it provides the means of investigating the possible connections between thermal physics and quantum mechanics at a deep level. Before proceeding along these lines we need to generalise the notion of time.

To begin the discussion on time, it should first be noted that in the evolution described by the Schrödinger equation, and, in consequence, in the Liouville equation, nothing ever actually happens. By this we mean that the unitary evolution is simply a re-description of the process. This redescription is parameterised by an external variable, t, which we can correlate to our classical clocks and therefore we call the parameter "time". Now to make something happen within quantum theory, we need a measurement in order to produce an actual result. In an ontological description this means we need some non-unitary transformation to actualise the process. A description of this feature is something that the standard interpretation does not provide.

Let us proceed naively and argue that we need to consider a two-time characteristic matrix $\xi(x, x'; t, t')$, the two times enabling us to incorporate the notion of a moment which, in the first approximation, is characterised by a mean time $T = (t + t')/2$ and a time difference $\tau = t - t'$. Since there is an ambiguity in the relationship between time and energy, we must necessarily have an ambiguity in the energy of the system. We can characterise this ambiguity by a difference in energy $\varepsilon = E' - E$, about a mean energy

$\bar{E} = (E' + E)/2$. In order to extend these ideas to an algebraic theory, we use the analogy with the commutation relations (21) derived above and we define a set of commutation relations for $T, \tau, \bar{E}$ and ε as follows:

$$[T, \bar{E}] = [\tau, \varepsilon] = [T, \tau] = [\bar{E}, \varepsilon] = 0; \qquad [T, \varepsilon] = [\tau, \bar{E}] = i. \qquad (23)$$

Thus if we treat the operator T as a "time", then the commutation relation $[T, \varepsilon] = i$ is identical to the time operator introduced by Ban (1991). This gives an ambiguity in specifying the mean time of a process passing between two energies. Unfortunately the second commutator in (23) can only be regarded as an approximation because $\bar{E}$ is bounded from below. Nevertheless this commutation relation can be regarded as indicating that there is some ambiguity in how long a system can remain in a state with a given spread of its mean energy and this is clearly related to the life-time of the process.

In requiring a description which needs a pair of algebras to capture the structure of quantum processes we are essentially constructing a bi-algebra, the superalgebra, and our proposal is that this structure is the appropriate mathematical descriptive form that we need to describe a process based physics. This means that we can write $\varepsilon = H \otimes \mathbb{1} - \mathbb{1} \otimes H$ which we immediately recognise as the Liouville operator introduced above. Thus we can also write the commutator in the form $[T, \hat{L}] = i$, which is essentially the same relation as introduced by Prigogine (1980). In Prigogine's approach T was regarded as the "age" of the process. We prefer to call it internal time.

If we regard L as generating the evolution of the process, then the Heisenberg equation of motion for the operator T is

$$\frac{dT}{dt} = i[\hat{L}, T] = \mathbb{1}, \qquad (24)$$

which has the solution $T = t + t_0$. This means that the internal time increases linearly with external time. We now have the possibility of developing an algebraic theory of time but this will require us to consider more carefully the structure of the bi-algebra and to develop a connection between time and entropy.

To begin a discussion of this relationship let us first see in what sense a moment can be described by a two-time density matrix. Consider two wave functions $\Psi_1(x, t)$ and $\Psi_2(x', t')$. After expanding these wave functions in terms of a complete set of energy eigenfunctions, we can form the density matrix

$$\rho_{EE'}(x, x', t, t') = \Psi_{E'}^*(x') \Psi_E(x) \exp\left[-i(Et - E't')\right]/\hbar. \qquad (25)$$

If we now change the time co-ordinates to T and τ, we find

$$\rho_{EE'}(x, x', T, \tau) = \Psi_{E'}^*(x') \Psi_E(x) \exp\left[-i(\varepsilon T + \bar{E}\tau)\right]/\hbar. \qquad (26)$$

Thus the density matrix is characterised by a mean time T and a time difference τ which, as we have already suggested, should be regarded as the time spent passing between the two energy states.

Now let us turn to thermal field theory. Here the inner automorphism

$$U(\beta) = \exp\left[\beta(a\tilde{a} - \tilde{a}^\dagger a^\dagger)\right] \tag{27}$$

produces a representation of the algebra in which the new vacuum state can be characterised by a temperature $\theta = 1/k\beta$. This vacuum state is represented by a superwave function from which we can project an ordinary wave function which can be written in the form:

$$\Psi(x,t) = \sum_E \exp\left[-\beta E/2\right] \Psi_E(x) \exp\left[-iEt/\hbar\right]. \tag{28}$$

We can interpret this wave function by assuming that there is an underlying process which generates a movement that is equivalent to a thermally induced diffusion, so that we can charaterise this movement with an equivalent temperature θ. The corresponding density matrix constructed from a pair of wave functions of the type (28) is

$$\rho_{EE'}(x,x',T,\beta) = \sum_{EE'} \exp\left[-\beta E/2\right] \Psi_{E'}^*(x') \Psi_E(x) \exp\left[-i\varepsilon T/\hbar\right]. \tag{29}$$

By comparing the two expressions (26) and (29) for $\rho_{EE'}$, we see that τ can be identified with $\hbar\beta/2$. In other words the time between states can be characterised by the temperature θ . Now let us consider the correlation function $< t+\tau, t >$ defined by

$$< t+\tau, t > = \int \Psi^*(x, t+\tau)\Psi(x,t)dx$$
$$= \int \sum_{EE'} \exp\left[-\frac{\beta}{2}(E+E')\right] \Psi_{E'}^*(x') \Psi_E(x) \exp\left[i\left(\frac{E(t+\tau)}{2\hbar} - \frac{Et}{\hbar}\right)\right] dx. \tag{30}$$

Using the orthogonality of the set $\Psi_E(x)$, this expression reduces to

$$< t+\tau, t >= \sum_E \exp\left[-\beta E\right] \exp\left[-iE\tau/\hbar\right]. \tag{31}$$

Therefore the correlation $< t+\tau, t >$ goes to zero in a time characterised by $\tau > \hbar\beta$. This implies that, as the diffusive process proceeds, successive wave functions become orthogonal. The superwave functions from which the wave function (28) is derived can be characterised by its entropy, thus wave functions which differ significantly in entropy will be orthogonal. Therefore if each moment of time is characterised by one of these wave functions we can begin to understand how successive moments and the increase of entropy become related.

For a small enough entropy change there is a direct connection from one moment to the next. However when the entropy difference becomes large enough, then the moments are independent of each other. It was this orthogonality that allowed Bohm (1986) to argue that in the implicate order all moments of time were present together in what can be regarded as a timeless order. In spite of this, the order of time could unfold through a series of explicate orders. Thus what seemed like a vague general notion can now be given a more explicit mathematical form. In our view, this form contains sufficient structure to enable us to provide an algebraic description of process from which space-time can be abstracted. It is hoped that this new approach will provide some insight into how gravity can be encompassed as a quantum phenomenon.

Acknowledgements: M. Fernandes wishes to thank the Brazilian National Research Council (CAPES) for their support while this work was being completed. We should also like to thank C. Papatheodorou and O. Cohen for extensive discussions on key aspects of this work.

References

Ban M. (1991): Relative number state representation and phase operator for physical systems. *J. Math. Phys.* **32**, 3077–3087.

Benn I.M. and Tucker R.W. (1983): Fermions without spinors. *Commun. Math. Phys.* **89**, 341–362.

Bohm D. (1965): A proposed topological formulation of the quantum theory. In *The Scientist Speculates*, ed. by I.J. Good (Putnam, New York), 302–314.

Bohm D. (1980): *Wholeness and the Implicate Order* (Routledge, London).

Bohm D. (1986): Time, the implicate order, and pre-space. In *Physics and the Ultimate Significance of Time*, ed. by D.R. Griffin (SUNY Press, New York), 177–208.

Bohm D. and Hiley B.J. (1981): On a quantum algebraic approach to a generalized phase space. *Found. Phys.* **11**, 179–203.

Bohm D. and Hiley B.J. (1984): Generalization of the twistor to Clifford algebras as a basis for geometry. *Revista Brasileira de Fisica*, Volume Especial (Julho 1984), Os 70 anos de Mário Schönberg, 1–26.

Bohm D. and Hiley B.J. (1993): *The Undivided Universe: an Ontological Interpretation of Quantum Theory* (Routledge, London).

Clifford W.K. (1882): *Collected Mathematical Works*, ed. by R. Tucker (Macmillan, London), 265–275.

Crumeyrolle A. (1990): *Orthogonal and Symplectic Clifford Algebras: Spinor Structures* (Kluwer, Dordrecht).

Dirac P.A.M. (1965): Quantum electrodynamics without dead wood. *Phys. Rev. B* **139**, 684–690.

Eddington, A.S. (1937): *The Mathematical Theory of Relativity* (Cambridge University Press, Cambridge), p. 225.

Eddington, A.S. (1958): *The Philosophy of Physical Science* (University of Michigan Press, Michigan), p. 162.

Fernandes M. (1996): *Geometric Algebras and the Foundations of Quantum Theory* (PhD Thesis, London University).

Fernandes M. and Hiley B.J. (1996): The metaplectic group, the symplectic spinor, and the Goy phase. To be published.

Frescura F.A.M. and Hiley B.J. (1980a): The implicate order, algebras, and the spinor. *Found. Phys.* **10**, 7–31.

Frescura F.A.M. and Hiley B.J. (1980b): The algebraization of quantum mechanics and the implicate order. *Found. Phys.* **10**, 705–722.

Frescura F.A.M. and Hiley B.J. (1981): Geometric interpretation of the Pauli spinor. *Am. J. Phys.* **49**, 152-157.

Frescura F.A.M. and Hiley B.J. (1984): Algebras, quantum theory, and pre-space. *Revista Brasileira de Fisica*, Volume Especial (Julho 1984), Os 70 anos de Mário Schönberg, 49–86.

Geroch R. (1972): Einstein algebras. *Commun. Math. Phys.* **26**, 271–275.

Grassmann H.G. (1894): *Gesammelte Math. und Physik. Werke* (Leipzig).

Haag R. (1991): Thoughts on the synthesis of quantum physics and general relativity and the role of space-time. *Nuclear Phys. B* (Proc. Suppl.) **18**, 135–140.

Hamilton W.R. (1967): *Mathematical Papers, Vol. 3: Algebra*, ed. by H. Halberstam and R.E. Ingram (Cambridge University Press, Cambridge).

Hankins T.L. (1976): Algebra as pure time: William Rowan Hamilton and the foundations of algebra. In *Motion and Time, Space and Matter*, ed. by P.K. Machamer and R.G. Turnbull (Ohio State University Press, Ohio), 327–357.

Hiley, B.J. (1991): Vacuum or holomovement. In *The Philosophy of Vacuum*, ed. by S. Saunders and H.R. Brown (Clarendon Press, Oxford), 217–249.

Hiley, B.J. (1995): The algebra of process. In *Consciousness at the Crossroads of Cognitive Science and Philosophy*, ed. by B. Borstner and J. Shawe-Taylor (Imprint Academic, Thorverton), 52–67.

Hiley B.J. and Monk N. (1993): Quantum phase space and the discrete Weyl algebra. *Mod. Phys. Lett. A* **8**, 3625–3633.

Isham C.J. (1987): Quantum gravity – GRG11 review talk. In *Quantum Gravity, General Relativity and Gravitation*, Proc. 11th Int. Conf. on General Relativity and Gravitation (GR11), Stockholm, 1986, ed. by M. MacCallum (Cambridge University Press, Cambridge).

Isham C.J. (1993): Canonical quantum gravity and the problem of time. In *Integrable Systems, Quantum Groups, and Quantum Field Theories*, ed. by L.A. Ibert and M.A. Rodriguez (Kluwer, Amsterdam), 157–288.

Penrose R. and MacCallum M.A.H. (1972): Twistor theory: an approach to the quantisation of fields and space-time. *Phys. Reports* **6**, 241–315.

Peres A. (1980): Measurement of time by quantum clocks. *Am. J. Phys.* **48**, 552–557.

Petrosky T.Y. and Prigogine I. (1990): Laws and events – the dynamical basis of self-organization. *Can. J. Phys.* **68**, 670–682.

Petrosky T.Y. and Prigogine I. (1994): Quantum chaos, complex spectral representations, and time symmetry breaking. *Chaos, Solitons and Fractals* **4**, 311–359.

Prigogine I. (1980): *From Being to Becoming* (Freeman, San Francisco).

Prigogine I. (1993): Time, dynamics, and chaos. In *Nobel Conference XXVI – Chaos: The New Science*, ed. by J. Holte (University Press of America), 55–84.

Takahashi Y. and Umezawa H. (1975): Thermo Field Dynamics. *Collect. Phenom.* **2**, 55–80.

Umezawa H. (1993): *Advanced Field Theory, Macro, Micro, and Thermal Physics* (American Institute of Physics, New York).

Unruh W.G. and Wald R.M. (1989): Time and the interpretation of quantum gravity. *Phys. Rev. D* **40**, 2598–2614.

Whitehead A.N. (1930): *The Concept of Nature* (Cambridge University Press, Cambridge), p. 65.

Zhong Z.-Z. (1995): Complete algebraization of quantum mechanics in terms of quantum groups and noncommutative geometry, and q-implicate order. *Phys. Rev. A* **52**, 2564–2568.

Index

Springer
and the
environment

At Springer we firmly believe that an international science publisher has a special obligation to the environment, and our corporate policies consistently reflect this conviction.

We also expect our business partners – paper mills, printers, packaging manufacturers, etc. – to commit themselves to using materials and production processes that do not harm the environment. The paper in this book is made from low- or no-chlorine pulp and is acid free, in conformance with international standards for paper permanency.

Springer

Printing: Saladruck, Berlin
Binding: Buchbinderei Lüderitz & Bauer, Berlin